T0189596

Problem Books in Mathematics

Edited by K. Bencsáth
P.R. Halmos

Springer
New York
Berlin
Heidelberg
Barcelona
Hong Kong
London
Milan
Paris
Singapore
Tokyo

Problem Books in Mathematics

Series Editors: K. Bencsáth and P.R. Halmos

Polynomials
by *Edward J. Barbeau*

Problems in Geometry
by *Marcel Berger, Pierre Pansu, Jean-Pic Berry, and Xavier Saint-Raymond*

Problem Book for First Year Calculus
by *George W. Bluman*

Exercises in Probability
by *T. Cacoullos*

Probability Through Problems
by *Marek Capi• ski and Tomasz Zastawniak*

An Introduction to Hilbert Space and Quantum Logic
by *David W. Cohen*

Unsolved Problems in Geometry
by *Hallard T. Croft, Kenneth J. Falconer, and Richard K. Guy*

Berkeley Problems in Mathematics (2nd ed.)
by *Paulo Ney de Souza and Jorge-Nuno Silva*

Problem-Solving Strategies
by *Arthur Engel*

Problems in Analysis
by *Bernard R. Gelbaum*

Problems in Real and Complex Analysis
by *Bernard R. Gelbaum*

Theorems and Counterexamples in Mathematics
by *Bernard R. Gelbaum and John M.H. Olmsted*

Exercises in Integration
by *Claude George*

Algebraic Logic
by *S.G. Gindikin*

(continued after index)

Loren C. Larson

Problem-Solving
Through Problems

With 104 Illustrations

Springer

Loren C. Larson
Department of Mathematics
St. Olaf College
Northfield, MN 55027
USA

Editors
Paul R. Halmos
Department of Mathematics
Santa Clara University
Santa Clara, CA 95053
USA

Katalin Bencsáth
Department of Mathematics
Manhattan College
Riverdale, NY 10471
USA

Mathematics Subject Classification (2000): 00A07

Library of Congress Cataloging-in-Publication Data
Larson, Loren C., 1937–
    Problem-solving through problems.
    (Problem books in mathematics)
    1. Mathematics—Problems, exercises, etc.   2. Problem solving.
I. Title. II. Series.
QA43.L37   1983   510   82-19493

Printed on acid-free paper.

© 1983 by Springer-Verlag New York Inc.
All rights reserved. This work may not be translated or copied in whole or in part without the written permission of the publisher (Springer-Verlag, 175 Fifth Avenue, New York, NY 10010, USA), except for brief excerpts in connection with reviews or scholarly analysis. Use in connection with any form of information storage and retrieval, electronic adaptation, computer software, or by similar or dissimilar methodology now known or hereafter developed is forbidden.
The use of general descriptive names, trade names, trademarks, etc., in this publication, even if the former are not especially identified, is not to be taken as a sign that such names, as understood by the Trade Marks and Merchandise Marks Act, may accordingly be used freely by anyone.

Typeset by Computype, St. Paul, Minnesota.
Printed and bound by Edwards Brothers, Inc., Ann Arbor, Michigan.
Printed in the United States of America.

9 8

ISBN 0-387-96171-2
ISBN 3-540-96171-2   SPIN 10866432

Springer-Verlag   New York Berlin Heidelberg
A member of BertelsmannSpringer Science+Business Media GmbH

To Elizabeth

Preface

The purpose of this book is to isolate and draw attention to the most important problem-solving techniques typically encountered in undergraduate mathematics and to illustrate their use by interesting examples and problems not easily found in other sources. Each section features a single idea, the power and versatility of which is demonstrated in the examples and reinforced in the problems. The book serves as an introduction and guide to the problems literature (e.g., as found in the problems sections of undergraduate mathematics journals) and as an easily accessed reference of essential knowledge for students and teachers of mathematics.

The book is both an anthology of problems and a manual of instruction. It contains over 700 problems, over one-third of which are worked in detail. Each problem is chosen for its natural appeal and beauty, but primarily to provide the context for illustrating a given problem-solving method. The aim throughout is to show how a basic set of simple techniques can be applied in diverse ways to solve an enormous variety of problems. Whenever possible, problems within sections are chosen to cut across expected course boundaries and to thereby strengthen the evidence that a single intuition is capable of broad application. Each section concludes with "Additional Examples" that point to other contexts where the technique is appropriate.

The book is written at the upper undergraduate level. It assumes a rudimentary knowledge of combinatorics, number theory, algebra, analysis, and geometry. Much of the content is accessible to students with only a year of calculus, and a sizable proportion does not even require this. However, most of the problems are at a level slightly beyond the usual contents of textbooks. Thus, the material is especially appropriate for students preparing for mathematical competitions.

The methods and problems featured in this book are drawn from my experience of solving problems at this level. Each new issue of *The American Mathematical Monthly* (and other undergraduate journals) contains material that would be just right for inclusion. Because these ideas continue to find new expression, the reader should regard this collection as a starter set and should be encouraged to create a personal file of problems and solutions to extend this beginning in both breadth and depth. Obviously, we can never hope to develop a "system" for problem-solving; however, the acquiring of ideas is a valuable experience at all stages of development.

Many of the problems in this book are old and proper referencing is very difficult. I have given sources for those problems that have appeared more recently in the literature, citing contests whenever possible. I would appreciate receiving exact references for those I have not mentioned.

I wish to take this opportunity to express my thanks to colleagues and students who have shared many hours of enjoyment working on these problems. In this regard I am particularly grateful to O. E. Stanaitis, Professor Emeritus of St. Olaf College. Thanks to St. Olaf College and the Mellon Foundation for providing two summer grants to help support the writing of this manuscript. Finally, thanks to all individuals who contributed by posing problems and sharing solutions. Special acknowledgement goes to Murray S. Klamkin who for over a quarter of a century has stood as a giant in the area of problem-solving and from whose problems and solutions I have learned a great deal.

March 21, 1983   LOREN C. LARSON

Contents

Contents

Chapter 1. Heuristics

Strategy or tactics in problem-solving is called *heuristics*. In this chapter we will be concerned with the heuristics of solving mathematical problems. Those who have thought about heuristics have described a number of basic ideas that are typically useful. The five classics on problem-solving by George Polya are masterpieces devoted entirely to the practical study of heuristics in mathematics. Among the ideas developed in these books, we shall focus on the following:

(1) Search for a pattern.
(2) Draw a figure.
(3) Formulate an equivalent problem.
(4) Modify the problem.
(5) Choose effective notation.
(6) Exploit symmetry.
(7) Divide into cases.
(8) Work backward.
(9) Argue by contradiction.
(10) Pursue parity.
(11) Consider extreme cases.
(12) Generalize.

Our interest in this list of problem-solving ideas is not in their description but in their implementation. By looking at examples of how others have used these simple but powerful ideas, we can expect to improve our problem-solving skills.

Before beginning, a word of advice about the problems at the end of the sections: Do not be overly concerned about using the heuristic treated in that section. Although the problems are chosen to give practice in the use of the heuristic, a narrow focus may be psychologically debilitating. A single problem usually admits several solutions, often employing quite different heuristics. Therefore, it is best to approach each problem with an open mind rather than with a preconceived notion about how a particular heuristic should be applied. In working on a problem, solving it is what matters. It is the accumulated experience of all the ideas working together that will result in a heightened awareness of the possibilities in a problem.

1.1. Search for a Pattern

Virtually all problem solvers begin their analysis by getting a feel for the problem, by convincing themselves of the plausibility of the result. This is best done by examining the most immediate special cases; when this exploration is undertaken in a systematic way, patterns may emerge that will suggest ideas for proceeding with the problem.

1.1.1. Prove that a set of n (different) elements has exactly 2^n (different) subsets.

When the problem is set in this imperative form, a beginner may panic and not know how to proceed. Suppose, however, that the problem were cast as a query, such as

(i) How many subsets can be formed from a set of n objects?
(ii) Prove or disprove: A set with n elements has 2^n subsets.

In either of these forms there is already the implicit suggestion that one should begin by checking out a few special cases. This is how each problem should be approached: remain skeptical of the result until convinced.

Solution 1. We begin by examining what happens when the set contains $0, 1, 2, 3$ elements; the results are shown in the following table:

n	Elements of S	Subsets of S	Number of subsets of S
0	none	$\emptyset$	1
1	x_1	$\emptyset, \{x_1\}$	2
2	x_1, x_2	$\emptyset, \{x_1\}, \{x_2\}, \{x_1, x_2\}$	4
3	x_1, x_2, x_3	$\emptyset, \{x_1\}, \{x_2\}, \{x_1, x_2\}$ $\{x_3\}, \{x_1, x_3\}, \{x_2, x_3\}, \{x_1, x_2, x_3\}$	8

Our purpose in constructing this table is not only to verify the result, but also to look for patterns that might suggest how to proceed in the general case. Thus, we aim to be as systematic as possible. In this case, notice when $n = 3$, we have listed first the subsets of $\{x_1, x_2\}$ and then, in the second line, each of these subsets augmented by the element x_3. This is the key idea that allows us to proceed to higher values of n. For example, when $n = 4$, the subsets of $S = \{x_1, x_2, x_3, x_4\}$ are the eight subsets of $\{x_1, x_2, x_3\}$ (shown in the table) together with the eight formed by adjoining x_4 to each of these. These sixteen subsets constitute the entire collection of possibilities; thus, a set with 4 elements has 2^4 ($= 16$) subsets.

A proof based on this idea is an easy application of mathematical induction (see Section 2.1).

Solution 2. Another way to present the idea of the last solution is to argue as follows. For each n, let A_n denote the number of (different) subsets of a set with n (different) elements. Let S be a set with $n + 1$ elements, and designate one of its elements by x. There is a one-to-one correspondence between those subsets of S which do not contain x and those subsets that do contain x (namely, a subset T of the former type corresponds to $T \cup \{x\}$). The former types are all subsets of $S - \{x\}$, a set with n elements, and therefore, it must be the case that

$$A_{n+1} = 2A_n.$$

This recurrence relation, true for $n = 0, 1, 2, 3, \ldots$, combined with the fact that $A_0 = 1$, implies that $A_n = 2^n$. ($A_n = 2A_{n-1} = 2^2 A_{n-2} = \cdots = 2^n A_0 = 2^n$.)

Solution 3. Another systematic enumeration of subsets can be carried out by constructing a "tree". For the case $n = 3$ and $S = \{a, b, c\}$, the tree is as shown below:

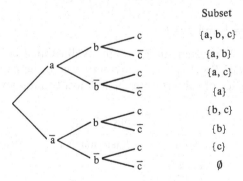

	Subset
	$\{a, b, c\}$
	$\{a, b\}$
	$\{a, c\}$
	$\{a\}$
	$\{b, c\}$
	$\{b\}$
	$\{c\}$
	$\emptyset$

Each branch of the tree corresponds to a distinct subset of S (the bar over the name of the element means that it is not included in the set corresponding to that branch). The tree is constructed in three stages, corresponding to the three elements of S. Each element of S leads to two possibilities: either it is in the subset or it is not, and these choices are represented by two branches. As each element is considered, the number of branches doubles.

Thus, for a three-element set, the number of branches is $2 \times 2 \times 2 = 8$. For an n-element set the number of branches is

$$\underbrace{2 \times 2 \times \cdots \times 2}_{n} = 2^n;$$

thus, a set with n elements has 2^n subsets.

Solution 4. Suppose we enumerate subsets according to their size. For example, when $S = \{a, b, c, d\}$, the subsets are

Number of elements		Number of subsets
0	$\varnothing$	1
1	$\{a\}, \{b\}, \{c\}, \{d\}$	4
2	$\{a,b\}, \{a,c\}, \{a,d\}, \{b,c\}, \{b,d\}, \{c,d\}$	6
3	$\{a,b,c\}, \{a,b,d\}, \{a,c,d\}, \{b,c,d\}$	4
4	$\{a,b,c,d\}$	1

This beginning could prompt the following argument. Let S be a set with n elements. Then

$$\text{No. of subsets of } S = \sum_{k=0}^{n} (\text{No. of subsets of } S \text{ with } k \text{ elements})$$

$$= \sum_{k=0}^{n} \binom{n}{k} = 2^n.$$

The final step in this chain of equalities follows from the binomial theorem,

$$(x + y)^n = \sum_{k=0}^{n} \binom{n}{k} x^k y^{n-k},$$

upon setting $x = 1$ and $y = 1$.

Solution 5. Another systematic beginning is illustrated in Table 1.1, which lists the subsets of $S = \{x_1, x_2, x_3\}$. To understand the pattern here, notice the correspondence of subscripts in the leftmost column and the occurrence

Table 1.1

Subset	Triple	Binary number	Decimal number
$\varnothing$	$(0,0,0)$	0	0
$\{x_3\}$	$(0,0,1)$	1	1
$\{x_2\}$	$(0,1,0)$	10	2
$\{x_2, x_3\}$	$(0,1,1)$	11	3
$\{x_1\}$	$(1,0,0)$	100	4
$\{x_1, x_3\}$	$(1,0,1)$	101	5
$\{x_1, x_2\}$	$(1,1,0)$	110	6
$\{x_1, x_2, x_3\}$	$(1,1,1)$	111	7

of 1's in the second column of triples. Specifically, if A is a subset of $S = \{x_1, x_2, \ldots, x_n\}$, define a_i, for $i = 1, 2, \ldots, n$, by

$$a_i = \begin{cases} 1 & \text{if } a_i \in A, \\ 0 & \text{if } a_i \notin A. \end{cases}$$

It is clear that we can now identify a subset A of S with $(a_1, a_2, \ldots, a_n)$, an n-tuple of 0's and 1's. Conversely, each such n-tuple will correspond to a unique subset of S. Thus, the number of subsets of S is equal to the number of n-tuples of 0's and 1's. This latter set is obviously in one-to-one correspondence with the set of nonnegative binary numbers less than 2^n. Thus, each nonnegative integer less than 2^n corresponds to exactly one subset of S, and conversely. Therefore, it must be the case that S has 2^n subsets.

Normally, we will give only one solution to each example—a solution which serves to illustrate the heuristic under consideration. In this first example, however, we simply wanted to reiterate the earlier claim that a single problem can usually be worked in a variety of ways. The lesson to be learned is that one should remain flexible in the beginning stages of problem exploration. If an approach doesn't seem to lead anywhere, don't despair, but search for a new idea. Don't get fixated on a single idea until you've had a chance to think broadly about a variety of alternative approaches.

1.1.2. Let $S_{n,0}$, $S_{n,1}$, and $S_{n,2}$ denote the sum of every third element in the nth row of Pascal's Triangle, beginning on the left with the first element, the second element, and the third element respectively. Make a conjecture concerning the value of $S_{100,1}$.

Solution. We begin by examining low-order cases with the hope of finding patterns that might generalize. In Table 1.2, the nonunderlined terms are those which make up the summands of $S_{n,0}$; the singly underlined and

Table 1.2

Pascal's triangle	n	$S_{n,0}$	$S_{n,1}$	$S_{n,2}$
1	0	1^+	0	0
1 1	1	1	1	0^-
1 2 1	2	1	2^+	1
1 3 3 1	3	2^-	3	3
1 4 6 4 1	4	5	5	6^+
1 5 10 10 5 1	5	11	10^-	11
1 6 15 20 15 6 1	6	22^+	21	21
1 7 21 35 35 21 7 1	7	43	43	42^-

doubly underlined terms are those of $S_{n,1}$ and $S_{n,2}$, respectively. The three columns on the right show that, in each case, two of the sums are equal, whereas the third is either one larger (indicated by a superscript $+$) or one smaller (indicated by a superscript $-$). It also appears that the unequal term in this sequence changes within a cycle of six. Thus, from the pattern established in the first rows, we expect the anomaly for $n = 8$ to occur in the middle column and it will be one less than the other two.

We know that $S_{n,0} + S_{n,1} + S_{n,2} = 2^n$ (see 1.1.1). Since $100 = 6 \times 16 + 4$, we expect the unequal term to occur in the third column ($S_{100,2}$) and to be one more than the other two. Thus $S_{100,0} = S_{100,1} = S_{100,2} - 1$, and $S_{100,1} + S_{100,1} + S_{100,1} + 1 = 2^{100}$. From these equations we are led to conjecture that

$$S_{100,1} = \frac{2^{100} - 1}{3}.$$

A formal proof of this conjecture is a straightforward application of mathematical induction (see Chapter 2).

1.1.3. Let $x_1, x_2, x_3, \ldots$ be a sequence of nonzero real numbers satisfying

$$x_n = \frac{x_{n-2} x_{n-1}}{2x_{n-2} - x_{n-1}}, \qquad n = 3, 4, 5, \ldots .$$

Establish necessary and sufficient conditions on x_1 and x_2 for x_n to be an integer for infinitely many values of n.

Solution. To get a feel for the sequence, we will compute the first few terms, expressing them in terms of x_1 and x_2. We have (omitting the algebra)

$$x_3 = \frac{x_1 x_2}{2x_1 - x_2},$$

$$x_4 = \frac{x_1 x_2}{3x_1 - 2x_2},$$

$$x_5 = \frac{x_1 x_2}{4x_1 - 3x_2}.$$

We are fortunate in this particular instance that the computations are manageable and a pattern emerges. An easy induction argument establishes that

$$x_n = \frac{x_1 x_2}{(n-1)x_1 - (n-2)x_2},$$

which, on isolating the coefficient of n, takes the form

$$x_n = \frac{x_1 x_2}{(x_1 - x_2)n + (2x_2 - x_1)}.$$

In this form, we see that if $x_1 \neq x_2$, the denominator will eventually exceed the numerator in magnitude, so x_n then will not be an integer. However, if $x_1 = x_2$, all the terms of the sequence are equal. Thus, x_n is an integer for infinitely many values of n if and only if $x_1 = x_2$.

1.1.4. Find positive numbers n and $a_1, a_2, \ldots, a_n$ such that $a_1 + \cdots + a_n = 1000$ and the product $a_1 a_2 \cdots a_n$ is as large as possible.

Solution. When a problem involves a parameter which makes the analysis complicated, it is often helpful in the discovery stage to replace it temporarily with something more manageable. In this problem, we might begin by examining a sequence of special cases obtained by replacing 1000 in turn with $2, 3, 4, 5, 6, 7, 8, 9, \ldots$. In this way we are led to discover that in a maximum product

(i) no a_i will be greater than 4,
(ii) no a_i will equal 1,
(iii) all a_i's can be taken to be 2 or 3 (because $4 = 2 \times 2$ and $4 = 2 + 2$),
(iv) at most two a_i's will equal 2 (because $2 \times 2 \times 2 < 3 \times 3$ and $2 + 2 + 2 = 3 + 3$).

Each of these is easy to establish. Thus, when the parameter is 1000 as in the problem at hand, the maximum product must be $3^{332} \times 2^2$.

1.1.5. Let S be a set and $*$ be binary operation on S satisfying the two laws

$$x * x = x \qquad \text{for all } x \text{ in } S,$$

$$(x * y) * z = (y * z) * x \qquad \text{for all } x, y, z \text{ in } S.$$

Show that $x * y = y * x$ for all x, y in S.

Solution. The solution, which appears so neatly below, is actually the end result of considerable scratch work; the procedure can only be described as a search for pattern (the principle pattern is the cyclic nature of the factors in the second condition). We have, for all x, y in S, $x * y = (x * y) * (x * y)$
$= [y * (x * y)] * x = [(x * y) * x] * y = [(y * x) * x] * y = [(x * x) * y] * y$
$= [(y * y)] * (x * x) = y * x$.

Problems

Develop a feel for the following problems by searching for patterns. Make appropriate conjectures, and think about how the proofs might be carried out.

1.1.6. Beginning with 2 and 7, the sequence $2, 7, 1, 4, 7, 4, 2, 8, \ldots$ is constructed by multiplying successive pairs of its members and adjoining the result as the next one or two members of the sequence, depending on whether the product is a one- or a two-digit number. Prove that the digit 6 appears an infinite number of times in the sequence.

1.1.7. Let S_1 denote the sequence of positive integers $1, 2, 3, 4, 5, 6, \ldots$, and define the sequence S_{n+1} in terms of S_n by adding 1 to those integers in S_n which are divisible by n. Thus, for example, S_2 is $2, 3, 4, 5, 6, 7, \ldots$, S_3 is $3, 3, 5, 5, 7, 7, \ldots$. Determine those integers n with the property that the first $n - 1$ integers in S_n are n.

1.1.8. Prove that a list can be made of all the subsets of a finite set in such a way that

(i) the empty set is first in the list,
(ii) each subset occurs exactly once, and
(iii) each subset in the list is obtained either by adding one element to the preceding subset or by deleting one element of the preceding subset.

1.1.9. Determine the number of odd binomial coefficients in the expansion of $(x + y)^{1000}$. (See 4.3.5.)

1.1.10. A well-known theorem asserts that a prime $p > 2$ can be written as a sum of two perfect squares ($p = m^2 + n^2$, with m and n integers) if and only if p is one more than a multiple of 4. Make a conjecture concerning which primes $p > 2$ can be written in each of the following forms, using (not necessarily positive) integers x and y: (a) $x^2 + 16y^2$, (b) $4x^2 + 4xy + 5y^2$. (See 1.5.10.)

1.1.11. If $\langle a_n \rangle$ is a sequence such that for $n \geqslant 1$, $(2 - a_n)a_{n+1} = 1$, what happens to a_n as n tends toward infinity? (See 7.6.4.)

1.1.12. Let S be a set, and let $*$ be a binary operation on S satisfying the laws

$$x * (x * y) = y \qquad \text{for all } x, y \text{ in } S,$$

$$(y * x) * x = y \qquad \text{for all } x, y \text{ in } S.$$

Show that $x * y = y * x$ for all x, y in S.

Additional Examples

Most induction problems are based on the discovery of a pattern. Thus, the problems in Sections 2.1, 2.2, 2.3, 2.4 offer additional practice in this heuristic. Also see 1.7.2, 1.7.7, 1.7.8, 2.5.6, 3.1.1, 3.4.6, 4.3.1, 4.4.1, 4.4.3, 4.4.15, 4.4.16, 4.4.17.

1.2. Draw a Figure

Whenever possible it is helpful to describe a problem pictorially, by means of a figure, a diagram, or a graph. A diagrammatic representation usually makes it easier to assimilate the relevant data and to notice relationships and dependences.

1.2.1. A chord of constant length slides around in a semicircle. The midpoint of the chord and the projections of its ends upon the base form the vertices of a triangle. Prove that the triangle is isosceles and never changes its shape.

Solution. Let AB denote the base of the semicircle, let XY be the chord, M the midpoint of XY, C and D the projections of X and Y on AB (Figure 1.1). Let the projection of M onto AB be denoted by N. Then N is the midpoint of CD and it follows that $\triangle CMD$ is isosceles.

To show that the shape of the triangle is independent of the position of the chord, it suffices to show that $\angle MCD$ remains unchanged, or equivalently, that $\angle XCM$ is constant, for all positions of XY. To see that this is the case, extend XC to cut the completed circle at Z (Figure 1.2). Then CM is parallel to ZY (C and M are the midpoints of XZ and XY, respectively),

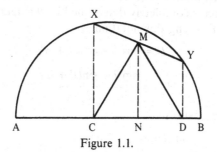

Figure 1.1.

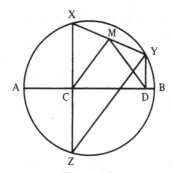

Figure 1.2.

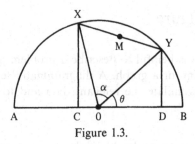

Figure 1.3.

and consequently $\angle XCM = \angle XZY$. But $\angle XZY$ equals one-half the arc XY, and this arc depends only on the length of the chord XY. This completes the proof.

One might ask: How in the world did anyone ever think to extend XC in this way? This is precisely the step that makes the argument so pretty, and it is indeed a very difficult step to motivate. About all that can be said is that the use of auxiliary lines and arcs (often found by reflection, extension, or rotation) is a common practice in geometry. Just the awareness of this fact will add to the possible approaches in a given problem.

Another interesting approach to this problem is to coordinatize the points and to proceed analytically. To show that the shape of the triangle is independent of the position of the chord, it suffices to show that the height-to-base ratio, MN/CD, is constant.

Let O denote the midpoint of AB, and let $\theta = \angle YOB$. It is clear that the entire configuration is completely determined by θ (Figure 1.3).

Let $\alpha = \angle XOY$. Using this notation,

$$CD = \cos\theta - \cos(\theta + \alpha),$$

$$MN = \frac{\sin\theta + \sin(\theta + \alpha)}{2},$$

and the height-base ratio is

$$F(\theta) = \frac{\sin\theta + \sin(\theta + \alpha)}{2(\cos\theta - \cos(\theta + \alpha))}, \qquad 0 \leqslant \theta \leqslant \pi - \alpha.$$

It is not immediately clear that this quantity is independent of θ; this is the content of 1.8.1 and 6.6.7.

1.2.2. A particle moving on a straight line starts from rest and attains a velocity v_0 after traversing a distance s_0. If the motion is such that the acceleration was never increasing, find the maximum time for the transverse.

Solution. Focus attention on the graph of the velocity $v = v(t)$ (Figure 1.4). We are given that $v(0) = 0$, and the graph of v is never concave upward (because the acceleration, dv/dt, is never increasing). The area under the

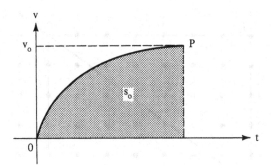

Figure 1.4.

curve is equal to s_0 (distance traversed $= \int_0^t v(t)\,dt$). From this representation, it is clear that we will maximize the time of traverse when the curve $v(t)$ from 0 to P is a straight line (Figure 1.5). At the maximum time t_0, $\frac{1}{2} t_0 v_0 = s_0$, or equivalently, $t_0 = 2s_0/v_0$.

1.2.3. If a and b are positive integers with no common factor, show that

$$\left[\!\left[\frac{a}{b}\right]\!\right] + \left[\!\left[\frac{2a}{b}\right]\!\right] + \left[\!\left[\frac{3a}{b}\right]\!\right] + \cdots + \left[\!\left[\frac{(b-1)a}{b}\right]\!\right] = \frac{(a-1)(b-1)}{2}.$$

Solution. When $b = 1$, we will understand that the sum on the left is 0 so the result holds.

It is not clear how a figure could be useful in establishing this purely arithmetic identity. Yet, the statement involves two independent variables, a and b, and a/b, $2a/b$, $3a/b$, ... are the values of the function $f(x) = ax/b$ when $x = 1, 2, 3, \ldots$, respectively. Is it possible to interpret $[\![a/b]\!]$, $[\![2a/b]\!]$, ... geometrically?

To make things concrete, consider the case $a = 5$ and $b = 7$. The points $P_k = (k, 5k/7)$, $k = 1, 2, \ldots, 6$, each lie on the line $y = 5x/7$, and $[\![5k/7]\!]$ equals the number of lattice points on the vertical line through P_k which lie above the x-axis and below P_k. Thus, $\sum_{k=1}^6 [\![5k/7]\!]$ equals the

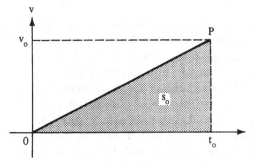

Figure 1.5.

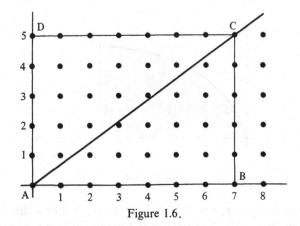

Figure 1.6.

number of lattice points interior to $\triangle ABC$ (see Figure 1.6). By symmetry, this number is one-half the number of lattice points in the interior of rectangle $ABCD$. There are $4 \times 6 = 24$ lattice points in $ABCD$, which means that triangle ABC contains 12 interior lattice points.

The same argument goes through in the general case. The condition that a and b have no common factor assures us that none of the lattice points in the interior of $ABCD$ will fall on the line $y = ax/b$. Thus,

$$\sum_{k=1}^{b-1} \left[\!\left[\frac{ka}{b} \right]\!\right] = \tfrac{1}{2}(\text{No. of lattice points in the interior of } ABCD)$$
$$= \frac{(a-1)(b-1)}{2}.$$

1.2.4 (The handshake problem). Mr. and Mrs. Adams recently attended a party at which there were three other couples. Various handshakes took place. No one shook hands with his/her own spouse, no one shook hands with the same person twice, and of course, no one shook his/her own hand.

After all the handshaking was finished, Mr. Adams asked each person, including his wife, how many hands he or she had shaken. To his surprise, each gave a different answer. How many hands did Mrs. Adams shake?

Solution. Although a diagram is not essential to the solution, it is helpful to view the data graphically in the following fashion. Represent the eight individuals by the eight dots as shown in Figure 1.7.

Now the answers to Mr. Adams' query must have been the numbers $0, 1, 2, 3, 4, 5, 6$. Therefore, one of the individuals, say A, has shaken hands with six others, say B, C, D, E, F, G. Indicate this on the graph by drawing line segments from A to these points, as in Figure 1.8.

From this diagram, we see that H must be that person who has shaken no one's hand. Furthermore, A and H must be spouses, because A has shaken hands with six others, not counting his/her own spouse.

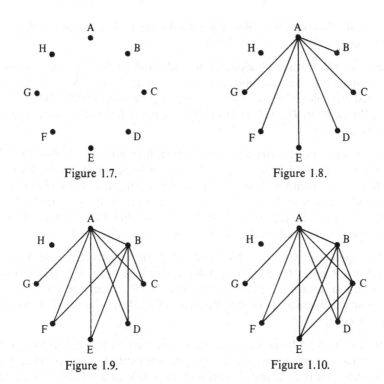

Figure 1.7.

Figure 1.8.

Figure 1.9.

Figure 1.10.

By supposition, one of B, C, D, E, F, G, has shaken five hands. By relabeling if necessary we may assume this person is B. Also, we may assume without loss of generality that the five with whom B has shaken hands are labeled A, C, D, E, F. This is shown in Figure 1.9. From this sketch we easily see that G is the only person who could have answered "one", and B and G must be spouses.

Again, as before, by relabeling the points C, D, E if necessary, we may assume that C shook four hands and that they belonged to A, B, D, E. The corresponding diagram is given in Figure 1.10. Using the same reasoning as above, F and C are spouses, and consequently, D and E are spouses.

Each of D and E has shaken hands with three others. Since Mr. Adams did not receive two "three" answers, D and E must correspond to Mr. and Mrs. Adams; that is to say, Mrs. Adams shook hands with three others.

Problems

1.2.5. Two poles, with heights a and b, are a distance d apart (along level ground). A guy wire stretches from the top of each of them to some point P on the ground between them. Where should P be located to minimize the total length of the wire? (Hint: Let the poles be erected at points C and D, and their tops be labeled A and B, respectively. We wish to minimize $AP + PB$. Augment this diagram by reflecting it in the base line CD.

Suppose B reflects to B' ($PB = PB'$). Now the problem is: Where should P be located to minimize $AP + PB'$?)

1.2.6. Let ABC be an acute-angled triangle, and let D be on the interior of the segment AB. Locate points E on AC and F on CB such that the inscribed triangle DEF will have minimum perimeter. (Hint: Reflect D in line AC to a point D'; reflect D in CB to a point D'' and consider the line segment $D'D''$.)

1.2.7. A rectangular room measures 30 feet in length and 12 feet in height, and the ends are 12 feet in width. A fly, with a broken wing, rests at a point one foot down from the ceiling at the middle of one end. A smudge of food is located one foot up from the floor at the middle of the other end. The fly has just enough energy to *walk* 40 feet. Show that there is a path along which the fly can walk that will enable it to get to the food.

1.2.8. Equilateral triangles ABP and ACQ are constructed externally on the sides AB and AC of triangle ABC. Prove that $CP = BQ$. (Hint: For a nice solution, rotate the plane of the triangle 60° about the point A, in a direction which takes B in the direction of C. What happens to the line segment CP?)

1.2.9. Let a and b be given positive real numbers with $a < b$. If two points are selected at random from a straight line segment of length b, what is the probability that the distance between them is at least a? (Hint: Let x and y denote the randomly chosen numbers from the interval $[0, b]$, and consider these independent random variables on two separate axes. What area corresponds to $|x - y| \geq a$?)

1.2.10. Give a geometric interpretation to the following problem. Let f be differentiable with f' continuous on $[a, b]$. Show that if there is a number c in $(a, b]$ such that $f'(c) = 0$, then we can find a number d in (a, b) such that

$$f'(d) = \frac{f(d) - f(a)}{b - a}.$$

1.2.11. Let a and b be real numbers, $a < b$. Indicate geometrically the precise location of each of the following numbers: $(a + b)/2$ ($= \frac{1}{2}a + \frac{1}{2}b$); $\frac{2}{3}a + \frac{1}{3}b$; $\frac{1}{3}a + \frac{2}{3}b$; $[m/(m + n)]a + [n/(m + n)]b$, where $m > 0$ and $n > 0$. (The latter number corresponds to the center of gravity of a system of two masses—one, of mass m, located at a, and the other, of mass n, located at b.)

1.2.12. Use the graph of $y = \sin x$ to show the following. Given triangle ABC,

(a) $\dfrac{\sin B + \sin C}{2} \leqslant \sin \dfrac{B + C}{2}$,

(b) $\dfrac{m}{m + n} \sin B + \dfrac{n}{m + n} \sin C \leqslant \sin\left(\dfrac{m}{m + n} B + \dfrac{n}{m + n} C\right)$, $m > 0$, $n > 0$.

1.2.13. Use a diagram (a rectangular array $(a_i a_j)$) to show that

(a)
$$\sum_{i=0}^{n} \sum_{j=0}^{n} a_i a_j = \sum_{j=0}^{n} \sum_{i=0}^{n} a_i a_j ,$$

(b)
$$\sum_{j=0}^{n} \sum_{i=j}^{n} a_i a_j = \sum_{i=0}^{n} \sum_{j=i}^{n} a_i a_j ,$$

(c)
$$\left(\sum_{i=0}^{n} a_i \right)^2 = \sum_{i=0}^{n} \sum_{j=0}^{n} a_i a_j = 2 \sum_{i=0}^{n} \sum_{j=0}^{i} a_i a_j - \sum_{i=0}^{n} a_i^2 .$$

Additional Examples

Most of the problems in Chaper 8 (Geometry); also 1.3.11, 1.9.2, 1.9.4, 1.11.3, 2.1.3, 2.5.5, 2.6.11, 5.1.2, 6.2.2, 6.4.1, 6.6.3, 6.8.1, 7.1.14, 7.4.19, 7.6.1, 8.1.1.

1.3. Formulate an Equivalent Problem

The message of the preceding section is that the first step in problem solving is to gather data, to explore, to understand, to relate, to conjecture, to analyze. But what happens when it is not possible to do this in a meaningful way, either because the computations become too complicated or because the problem simply admits no special cases that shed any insight? In this section we will consider some problems of this type. The recommendation of this section is to try to reformulate the problem into an equivalent but simpler form. The appeal is to one's imagination and creativity. Some standard reformulation techniques involve algebraic or trigonometric manipulation, substitution or change of variable, use of one-to-one correspondence, and reinterpretation in the language of another subject (algebra, geometry, analysis, combinatorics, etc.).

1.3.1. Find a general formula for the nth derivative of $f(x) = 1/(1 - x^2)$.

Solution. A common simplifying step when working with rational functions is to write the function as a sum of partial fractions. In this case,

$$f(x) = \frac{1}{2} \left[\frac{1}{1 - x} + \frac{1}{1 + x} \right],$$

and in this form it is easy to show that

$$f^{(n)}(x) = \frac{n!}{2}\left[\frac{1}{(1-x)^{n+1}} + \frac{(-1)^n}{(1+x)^{n+1}}\right].$$

1.3.2. Find all solutions of $x^4 + x^3 + x^2 + x + 1 = 0$.

Solution. This equation can be solved by dividing by x^2, then substituting $y = x + 1/x$, and then applying the quadratic formula. Thus, we have

$$x^2 + \frac{1}{x^2} + x + \frac{1}{x} + 1 = 0,$$

$$\left(x^2 + 2 + \frac{1}{x^2}\right) + \left(x + \frac{1}{x}\right) + (1-2) = 0,$$

$$\left(x + \frac{1}{x}\right)^2 + \left(x + \frac{1}{x}\right) - 1 = 0,$$

$$y^2 + y - 1 = 0.$$

The roots of this equation are

$$y_1 = \frac{-1+\sqrt{5}}{2}, \qquad y_2 = \frac{-1-\sqrt{5}}{2}.$$

It remains to determine x by solving the two equations

$$x + \frac{1}{x} = y_1 \quad \text{and} \quad x + \frac{1}{x} = y_2,$$

which are equivalent to

$$x^2 - y_1 x + 1 = 0 \quad \text{and} \quad x^2 - y_2 x + 1 = 0.$$

The four roots found by solving these are

$$x_1 = \frac{-1+\sqrt{5}}{4} + i\frac{\sqrt{10+2\sqrt{5}}}{4},$$

$$x_2 = \frac{-1+\sqrt{5}}{4} - i\frac{\sqrt{10+2\sqrt{5}}}{4},$$

$$x_3 = \frac{-1-\sqrt{5}}{4} + i\frac{\sqrt{10-2\sqrt{5}}}{4},$$

$$x_4 = \frac{-1-\sqrt{5}}{4} - i\frac{\sqrt{10-2\sqrt{5}}}{4}.$$

Another approach to this problem is to multiply each side of the original equation by $x - 1$. Since $(x-1)(x^4+x^3+x^2+x+1) = x^5 - 1$, an equivalent problem is to find all x (other than $x = 1$) which satisfy $x^5 = 1$. These are the five fifth roots of unity, given by

$$x_1 = \cos\tfrac{2}{5}\pi + i\sin\tfrac{2}{5}\pi,$$

$$x_2 = \cos\tfrac{4}{5}\pi + i\sin\tfrac{4}{5}\pi,$$

$$x_3 = \cos\tfrac{6}{5}\pi + i\sin\tfrac{6}{5}\pi,$$

$$x_4 = \cos\tfrac{8}{5}\pi + i\sin\tfrac{8}{5}\pi,$$

$$x_5 = 1.$$

As a by-product of having worked this problem two different ways, we see that

$$\cos\tfrac{2}{5}\pi + i\sin\tfrac{2}{5}\pi = \frac{-1+\sqrt5}{4} + i\frac{\sqrt{10+2\sqrt5}}{4}.$$

Equating real and imaginary parts yields

$$\cos 72° = \frac{-1+\sqrt5}{4}, \qquad \sin 72° = \frac{\sqrt{10+2\sqrt5}}{4}.$$

(Similar formulas can be found for x_2, x_3, and x_4.)

1.3.3. P is a point inside a given triangle ABC; D, E, F are the feet of the perpendiculars from P to the lines BC, CA, AB, respectively. Find all P for which

$$\frac{BC}{PD} + \frac{CA}{PE} + \frac{AB}{PF}$$

is minimal.

Solution. Denote the lengths of BC, AC, AB by a, b, c, respectively, and PD, PE, PF by p, q, r, respectively (see Figure 1.11). We wish to minimize $a/p + b/q + c/r$.

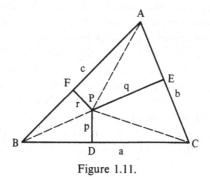

Figure 1.11.

Notice that

$$\text{Area } \triangle ABC = \text{Area } \triangle BCP + \text{Area } \triangle CAP + \text{Area } \triangle ABP$$

$$= \tfrac{1}{2}ap + \tfrac{1}{2}bq + \tfrac{1}{2}cr$$

$$= \frac{ap + bq + cr}{2}.$$

Thus, $ap + bq + cr$ is a constant, independent of the placement of P. Therefore, instead of minimizing $a/p + b/q + c/r$, we will minimize $(ap + bq + cr)(a/p + b/q + c/r)$. (This step will appear more natural after a study of inequalities with constraints taken up in Section 7.3.) We have

$$(ap + bq + cr)\left(\frac{a}{p} + \frac{b}{q} + \frac{c}{r}\right)$$

$$= a^2 + b^2 + c^2 + ab\left(\frac{p}{q} + \frac{q}{p}\right) + bc\left(\frac{q}{r} + \frac{r}{q}\right) + ac\left(\frac{p}{r} + \frac{r}{p}\right)$$

$$\geq a^2 + b^2 + c^2 + 2ab + 2bc + 2ac$$

$$= (a + b + c)^2.$$

The inequality in the second step follows from the fact that for any two positive numbers x and y we have $x/y + y/x \geq 2$, with equality if and only if $x = y$. As a result of this fact, $(ap + bq + cr)(a/p + b/q + c/r)$ will attain its minimum value $(a + b + c)^2$ when, and only when, $p = q = r$. Equivalently, $a/p + b/q + c/r$ attains a minimum value when P is located at the incenter of the triangle.

1.3.4. Prove that if m and n are positive integers and $1 \leq k \leq n$, then

$$\sum_{i=0}^{k} \binom{n}{i}\binom{m}{k-i} = \binom{m+n}{k}.$$

Solution. The statement of the problem constitutes one of the fundamental identities involving binomial coefficients. On the left side is a sum of products of binomial coefficients. Obviously, a direct substitution of factorials for binomial coefficients provides no insight.

Quite often, finite series (especially those which involve binomial coefficients) can be summed combinatorially. To understand what is meant here, transform the series problem into a counting problem in the following manner. Let $S = A \cup B$, where A is a set with n elements and B is a set, disjoint from A, with m elements. We will count, in two different ways, the number of (distinct) k-subsets of S. On the one hand, this number is $\binom{m+n}{k}$. On the other hand, the number of k-subsets of S with exactly i elements from A (and $k - i$ elements from B) is $\binom{n}{i}\binom{m}{k-i}$. It follows that

$$\binom{m+n}{k} = \text{No. of } k\text{-subsets of } S$$

$$= \sum_{i=0}^{k} (\text{No. of } k\text{-subsets of } S \text{ with } i \text{ elements from } A)$$

$$= \sum_{i=0}^{k} \binom{n}{i}\binom{m}{k-i}.$$

(Another solution to this problem, based on the properties of polynomials, is given in 4.3.2.)

Counting problems can often be simplified by "identifying" (by means of a one-to-one correspondence) the elements of one set with those of another set whose elements can more easily be counted. The next three examples illustrate the idea.

1.3.5. On a circle n points are selected and the chords joining them in pairs are drawn. Assuming that no three of these chords are concurrent (except at the endpoints), how many points of intersection are there?

Solution. The cases for $n = 4, 5, 6$ are shown in Figure 1.12. Notice that each (interior) intersection point determines, and is determined by, four of the given n points along the circle (these four points will uniquely produce two chords which intersect in the interior of the circle). Thus, the number of intersection points is $\binom{n}{4}$.

1.3.6. Given a positive integer n, find the number of quadruples of integers (a, b, c, d) such that $0 \leqslant a \leqslant b \leqslant c \leqslant d \leqslant n$.

Solution. The key idea which makes the problem transparent is to notice that there is a one-to-one correspondence between the quadruples of our set

1

5

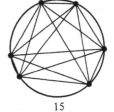

15

Figure 1.12.

and the subsets of four objects taken from $\{0, 1, \ldots, n + 3\}$. Specifically, let (a, b, c, d), $0 \leqslant a \leqslant b \leqslant c \leqslant d \leqslant n$, be identified with the subset $\{a, b + 1, c + 2, d + 3\}$. It is easy to see that this correspondence is one-to-one—each element of our set corresponds to exactly one subset of four from $\{0, 1, \ldots, n + 3\}$, and vice versa. Thus, the desired number is $\binom{n+4}{4}$.

1.3.7. The number 5 can be expressed as a sum of 3 natural numbers, taking order into account, in 6 ways, namely, as $5 = 1 + 1 + 3 = 1 + 3 + 1 = 3 + 1 + 1 = 1 + 2 + 2 = 2 + 1 + 2 = 2 + 2 + 1$. Let m and n be natural numbers such that $m \leqslant n$. In how many ways can n be written as a sum of m natural numbers, taking order into account?

Solution. Write n as a sum of n ones:

$$n = \underbrace{1 + 1 + \cdots + 1}_{n}.$$

The number we seek is the number of ways of choosing $m - 1$ plus signs from the $n - 1$; that is, $\binom{n-1}{m-1}$.

Problems

1.3.8. Show that $x^7 - 2x^5 + 10x^2 - 1$ has no root greater than 1. (Hint: Since it is generally easier to show that an equation has no *positive* root, we are prompted to consider the equivalent problem obtained by making the algebraic substitution $x = y + 1$.)

1.3.9. The number 3 can be expressed as a sum of one or more positive integers, taking order into account, in four ways, namely, as 3, $1 + 2$, $2 + 1$, and $1 + 1 + 1$. Show that any positive integer n can be so expressed in 2^{n-1} ways.

1.3.10. In how many ways can 10 be expressed as a sum of 5 *nonnegative* integers, when order is taken into account? (Hint: Find an equivalent problem in which the phrase "5 nonnegative integers" is replaced by "5 positive integers".)

1.3.11. For what values of a does the system of equations

$$x^2 = y^2,$$
$$(x - a)^2 + y^2 = 1$$

have exactly zero, one, two, three, four solutions, respectively? (Hint: Translate the problem into an equivalent geometry problem.)

1.3.12. Given n objects arranged in a row. A subset of these objects is called *unfriendly* if no two of its elements are consecutive. Show that the number of unfriendly subsets each having k elements is $\binom{n-k+1}{k}$. (Hint: Adopt an idea similar to that used in 1.3.6.)

1.3.13. Let $a(n)$ be the number of representations of the positive integer n as a sum of 1's and 2's taking order into account. Let $b(n)$ be the number of representations of n as a sum of integers greater than 1, again taking order into account and counting the summand n. The table below shows that $a(4) = 5$ and $b(6) = 5$:

a-sums	b-sums
$1 + 1 + 2$	$4 + 2$
$1 + 2 + 1$	$3 + 3$
$2 + 1 + 1$	$2 + 4$
$2 + 2$	$2 + 2 + 2$
$1 + 1 + 1 + 1$	6

(a) Show that $a(n) = b(n + 2)$ for each n, by describing a one-to-one correspondence between the a-sums and b-sums.
(b) Show that $a(1) = 1$, $a(2) = 2$, and for $n > 2$, $a(n) = a(n - 1) + a(n - 2)$.

1.3.14. By finding the area of a triangle in two different ways, prove that if p_1, p_2, p_3 are the altitudes of a triangle and r is the radius of its inscribed circle, then $1/p_1 + 1/p_2 + 1/p_3 = 1/r$.

1.3.15. Use a counting argument to prove that for integers r, n, $0 < r \leqslant n$,

$$\binom{r}{r} + \binom{r+1}{r} + \binom{r+2}{r} + \cdots + \binom{n}{r} = \binom{n+1}{r+1}.$$

Additional Examples

1.2.3, 5.1.5, 5.1.14, 7.4.6, 8.2.6. There are so many examples of this heuristic that it is difficult to single out those that are most typical. Noteworthy are the indirect proofs in Section 1.9, 1.10, 1.11, the congruence problems in Section 3.2, the limit problems in Section 6.8. Other examples of partial fractions (see 1.3.1) are 4.3.23, 5.3.1, 5.3.2, 5.3.3, 5.3.6, 5.3.12, 5.4.9, 5.4.13, 5.4.20, 5.4.24, 5.4.25. Examples based on the identity $x = \exp(\log x)$ are 5.3.7(c), 6.3.3, 6.7.1, 6.7.4, 6.7.5, 6.7.7, 6.9.5, 7.4.1, 7.4.2, 7.4.9, 7.4.20.

1.4. Modify the Problem

In the course of work on problem A we may be led to consider problem B. Characteristically, this change in problems is announced by such phrases as "it suffices to show that . . ." or "we may assume that . . ." or "without loss of generality . . .". In the last section we looked at examples in which A and B were equivalent problems, that is, the solution of either one of them implied the solution of the other. In this section we look at cases where the solution of the modified (or auxiliary) problem, problem B, implies the solution of A, but not necessarily vice versa.

1.4.1. Given positive numbers a, b, c, d, prove that

$$\frac{a^3 + b^3 + c^3}{a + b + c} + \frac{b^3 + c^3 + d^3}{b + c + d} + \frac{c^3 + d^3 + a^3}{c + d + a} + \frac{d^3 + a^3 + b^3}{d + a + b}$$
$$\geqslant a^2 + b^2 + c^2 + d^2.$$

Solution. Because of the symmetry in the problem, *it is sufficient to prove* that for all positive numbers x, y, and z

$$\frac{x^3 + y^3 + z^3}{x + y + z} \geqslant \frac{x^2 + y^2 + z^2}{3}.$$

For if this were the case, the left side of the original inequality is at least

$$\frac{a^2 + b^2 + c^2}{3} + \frac{b^2 + c^2 + d^2}{3} + \frac{c^2 + d^2 + a^2}{3} + \frac{d^2 + a^2 + b^2}{3}$$
$$= a^2 + b^2 + c^2 + d^2.$$

Now, to prove this latter inequality, *there is no loss of generality* in supposing that $x + y + z = 1$. For if not, simply divide each side of the inequality by $(x + y + z)^2$, and let $X = x/(x + y + z)$, $Y = y/(x + y + z)$, and $Z = z/(x + y + z)$.

Thus, the original problem reduces to the following *modified* problem: Given positive numbers X, Y, Z such that $X + Y + Z = 1$, prove that

$$X^3 + Y^3 + Z^3 \geqslant \frac{X^2 + Y^2 + Z^2}{3}.$$

(For a proof of this inequality, see 7.3.5.)

1.4.2. Let C be any point on the line segment AB between A and B, and let semicircles be drawn on the same side of AB with AB, AC, and CB as diameters (Figure 1.13). Also let D be a point on the semicircle having diameter AB such that CD is perpendicular to AB, and let E and F be points on the semicircles having diameters AC and CB, respectively, such

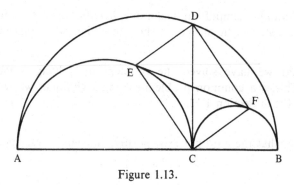

Figure 1.13.

that *EF* is a segment of their common tangent. Show that *ECFD* is a rectangle.

Solution. Note that *it is sufficient to show* that *A*, *E*, and *D* are collinear (*the same argument* would show that *B*, *F*, and *D* are collinear). For if this were the case, $\angle AEC = 90°$ (*E* is on circle *AEC*), $\angle ADB = 90°$, $\angle CFB = 90°$, and the result holds. It turns out, however, that without some insight, there are many ways of going wrong with this approach; it's difficult to avoid assuming the conclusion.

One way of gaining insight into the relationships among the parameters in a problem is to notice the effect when one of them is allowed to vary (problem modification). In this problem, let *D* vary along the circumference. Let *G* and *H* (Figure 1.14) denote the intersections of the segments *AD* and *BD* with the circles with diameters *AC* and *CB* (and centers *O* and *O′*) respectively. Then $\angle AGC = \angle ADB = \angle CHB = 90°$, so that *GDHC* is a rectangle. Furthermore, $\angle OGC = \angle OCG$ ($\triangle OGC$ is isosceles), and $\angle CGH = \angle GCD$ because *GH* and *CD* are diagonals of a rectangle. Therefore, $\angle OGH = \angle OCD$. Now, as *D* moves to make *CD* perpendicular to *AB*, $\angle OGH$ will also move to 90°, so that *GH* is tangent to circle *O*, and *G* coincides with *E*. A similar argument shows *GH* is tanget to circle *O′*, so *H* = *F*. This completes the proof. (Note the phrase "a similar

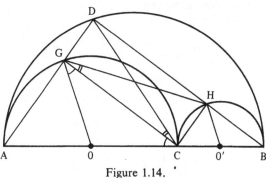

Figure 1.14.

argument," another simplifying technique, has the same effect when placed after an argument as "it suffices to show that" has when placed before the argument.)

Note that we have solved the problem by solving a more general problem. This is a common problem-solving technique; we will see more examples of it in Section 1.12.

1.4.3. Prove that there do not exist positive integers x, y, z such that

$$x^2 + y^2 + z^2 = 2xyz.$$

Solution. Suppose x, y, and z are positive integers such that $x^2 + y^2 + z^2 = 2xyz$. Since $x^2 + y^2 + z^2$ is even ($= 2xyz$), either two of x, y, and z are odd and the other even, or all three are even. Suppose x, y, z are even. Then there are positive integers x_1, y_1, z_1 such that $x = 2x_1$, $y = 2y_1$, $z = 2z_1$. From the fact that $(2x_1)^2 + (2y_1)^2 + (2z_1)^2 = 2(2x_1)(2y_1)(2z_1)$ it follows that x_1, y_1, z_1 satisfy $x_1^2 + y_1^2 + z_1^2 = 2^2 x_1 y_1 z_1$. Again, from this equation, if x_1, y_1, z_1 are even, *a similar argument* shows there will be positive integers x_2, y_2, z_2 such that $x_2^2 + y_2^2 + z_2^2 = 2^3 x_2 y_2 z_2$.

Continue in this way. Eventually we must arrive at an equation of the form $a^2 + b^2 + c^2 = 2^n abc$ where not all of a, b, c are even (and hence two of a, b, c are even and one is odd).

Thus, we are led to consider the following modified problem: Prove there do not exist positive integers x, y, z and n, with x, y odd, such that

$$x^2 + y^2 + z^2 = 2^n xyz.$$

(This is Problem 1.9.3.)

1.4.4. Evaluate $\int_0^\infty e^{-x^2}\, dx$.

Solution. The usual integration techniques studied in first-year calculus will not work on this integral. To evaluate the integral we will transform the single integral into a double integral.

Let $I = \int_0^\infty e^{-x^2}\, dx$. Then

$$I^2 = \left[\int_0^\infty e^{-x^2}\, dx \right]\left[\int_0^\infty e^{-y^2}\, dy \right]$$

$$= \int_0^\infty \left[\int_0^\infty e^{-x^2}\, dx \right] e^{-y^2}\, dy$$

$$= \int_0^\infty \int_0^\infty e^{-x^2} e^{-y^2}\, dx\, dy$$

$$= \int_0^\infty \int_0^\infty e^{-(x^2+y^2)}\, dx\, dy.$$

Now change to an equivalent integral by switching to polar coordinates. We then have

$$I^2 = \int_0^{\pi/2} \int_0^{\infty} e^{-r^2} r \, dr \, d\theta$$

$$= \int_0^{\pi/2} -\frac{1}{2} e^{-r^2} \Big]_0^{\infty} d\theta$$

$$= \frac{1}{2} \int_0^{\pi/2} d\theta$$

$$= \frac{1}{4} \pi.$$

It follows that $I = \sqrt{\pi}/2$.

A modified (auxiliary) problem can arise in many ways. It may come about with a change in notation (as in 1.4.4; see Section 1.5) or because of symmetry (as in 1.4.1; see Section 1.6). Often it is the result of "working backward" (see Section 1.8) or arguing by contradiction (as in 1.4.3; see Section 1.9). It is not uncommon to consider a more general problem at the outset (as in 1.4.2; see Section 1.12). Thus we see that problem modification is a very general heuristic. Because of this, we will defer adding more examples and problems, putting them more appropriately in the more specialized sections which follow.

1.5. Choose Effective Notation

One of the first steps in working a mathematics problem is to translate the problem into symbolic terms. At the outset, all key concepts should be identified and labeled; redundancies in notation can be eliminated as relationships are discovered.

1.5.1. One morning it started snowing at a heavy and constant rate. A snowplow started out at 8:00 A.M. At 9:00 A.M. it had gone 2 miles. By 10:00 A.M. it had gone 3 miles. Assuming that the snowplow removes a constant volume of snow per hour, determine the time at which it started snowing.

Solution. It is difficult to imagine there is enough information in the problem to answer the question. However, if there is a way, we must proceed systematically by first identifying those quantities that are unknown. We introduce the following notation: Let t denote the time that has elapsed since it started snowing, and let T be the time at which the plow goes out (measured from $t = 0$). Let $x(t)$ be the distance the plow has gone at time t (we are only interested in $x(t)$ for $t \geqslant T$). Finally, let $h(t)$ denote the depth of the snow at time t.

We are now ready to translate the problem into symbolic terms. The fact that the snow is falling at a constant rate means that the depth is increasing at a constant rate; that is,

$$\frac{dh}{dt} = c, \qquad c \text{ constant.}$$

Integrating each side yields

$$h(t) = ct + d, \qquad c, d \text{ constants.}$$

Since $h(0) = 0$, we get $d = 0$. Thus $h(t) = ct$.

The fact that the plow removes snow at a constant rate means that the speed of the plow is inversely proportional to the depth at any time t (for example, twice the depth corresponds to half the speed). Symbolically, for $t \geqslant T$,

$$\frac{dx}{dt} = \frac{k}{h(t)}, \qquad k \text{ constant}$$

$$= \frac{k}{ct} = \frac{K}{t}, \qquad K = \frac{k}{c} \text{ constant.}$$

Integrating each side yields

$$x(t) = K \log t + C, \qquad C \text{ constant.}$$

We are given three conditions: $x = 0$ when $t = T$, $x = 2$ when $t = T + 1$, and $x = 3$ when $t = T + 2$. With two of these conditions we can evaluate the constants K and C, and with the third, we can solve for T. It turns out (the details are not of interest here) that

$$T = \frac{\sqrt{5} - 1}{2} \approx 0.618 \text{ hours} \approx 37 \text{ minutes, 5 seconds.}$$

Thus, it started snowing at 7:22:55 A.M.

1.5.2.

(a) If n is a positive integer such that $2n + 1$ is a perfect square, show that $n + 1$ is the sum of two successive perfect squares.

(b) If $3n + 1$ is a perfect square, show that $n + 1$ is the sum of three perfect squares.

Solution. By introducing proper notation, this reduces to a simple algebra problem. For part (a), suppose that $2n + 1 = s^2$, s an integer. Since s^2 is an odd number, so also is s. Let t be an integer such that $s = 2t + 1$. Then $2n + 1 = (2t + 1)^2$, and solving for n we find

$$n = \frac{(2t + 1)^2 - 1}{2} = \frac{4t^2 + 4t}{2} = 2t^2 + 2t.$$

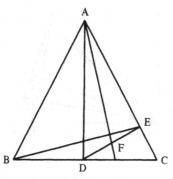

Figure 1.15.

Consequently,

$$n + 1 = 2t^2 + 2t + 1 = t^2 + (t + 1)^2.$$

(b) Suppose $3n + 1 = s^2$, s an integer. Evidently, s is not a multiple of 3, so $s = 3t \pm 1$ for some integer t. Then $3n + 1 = (3t \pm 1)^2$, and therefore

$$n = \frac{(3t \pm 1)^2 - 1}{3} = \frac{9t^2 \pm 6t}{3} = 3t^2 \pm 2t.$$

Hence,

$$n + 1 = 3t^2 \pm 2t + 1 = 2t^2 + (t \pm 1)^2 = t^2 + t^2 + (t \pm 1)^2.$$

1.5.3. In triangle ABC, $AB = AC$, D is the midpoint of BC, E is the foot of the perpendicular drawn D to AC, and F is the midpoint of DE (Figure 1.15). Prove that AF is perpendicular to BE.

Solution. We can transform the problem into algebraic terms by coordinatizing the relevant points and by showing that the slopes m_{BE} and m_{AF} are negative reciprocals.

One way to proceed is to take the triangle as it appears in Figure. 1.15: take D as the origin $(0,0)$, $A = (0, a)$, $B = (-b, 0)$, and $C = (b, 0)$. This is a natural labeling of the figure because it takes advantage of the bilateral symmetry of the isosceles triangle (see the examples in Section 1.6). However, in this particular instance, this notation leads to some minor complications when we look for the coordinates of E and F.

A better coordinatization is to take $A = (0, 0)$, $B = (4a, 4b)$, $C = (4c, 0)$, as in Figure 1.16. Then $a^2 + b^2 = c^2$, $D = (2a + 2c, 2b)$, $E = (2a + 2c, 0)$, and $F = (2a + 2c, b)$. (Almost no computation here; all relevant points are coordinatized.) It follows that

$$m_{AF}m_{BE} = \left(\frac{b}{2(a + c)} \right)\left(\frac{4b}{4a - (2a + 2c)} \right) = \frac{b^2}{a^2 - c^2} = -1,$$

and the proof is complete.

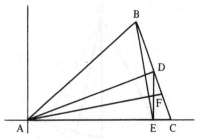

Figure 1.16.

1.5.4. Let $-1 < a_0 < 1$ and define recursively

$$a_n = \left(\frac{1 + a_{n-1}}{2} \right)^{1/2}, \qquad n > 0.$$

Let $A_n = 4^n(1 - a_n)$. What happens to A_n as n tends to infinity?

Solution. Direct attempts to express a_n in terms of a_0 lead to hopelessly complicated expressions containing nested sequences of radicals, and there is no way to condense them into a closed form.

The key insight needed is to observe that there is a unique angle θ, $0 < \theta < \pi$, such that $a_0 = \cos\theta$. For this θ,

$$a_1 = \left(\frac{1 + \cos\theta}{2} \right)^{1/2} = \cos\left(\frac{\theta}{2} \right).$$

Similarly,

$$a_2 = \left(\frac{1 + \cos(\theta/2)}{2} \right)^{1/2} = \cos\left(\frac{\theta}{4} \right), \quad \ldots, \quad a_n = \cos\left(\frac{\theta}{2^n} \right).$$

We can now compute

$$A_n = 4^n(1 - \cos(\theta/2^n))$$

$$= \frac{4^n(1 - \cos(\theta/2^n))(1 + \cos(\theta/2^n))}{1 + \cos(\theta/2^n)}$$

$$= \frac{4^n \sin^2(\theta/2^n)}{1 + \cos(\theta/2^n)}$$

$$= \left(\frac{\theta^2}{1 + \cos(\theta/2^n)} \right) \left(\frac{\sin(\theta/2^n)}{\theta/2^n} \right)^2.$$

As n becomes large, $\theta^2/(1 + \cos(\theta/2^n))$ tends to $\theta^2/2$, and $(\sin(\theta/2^n))/(\theta/2^n)$ approaches 1 (recall that $(\sin x)/x \to 1$ as $x \to 0$), and therefore, A_n converges to $\theta^2/2$ as n tends to infinity.

Problems

1.5.5. Write an equation to represent the following statements:

(a) At Mindy's restaurant, for every four people who ordered cheesecake, there were five who ordered strudel.
(b) There are six times as many students as professors at this college.

1.5.6. Guy wires are strung from the top of each of two poles to the base of the other. What is the height from the ground where the two wires cross?

1.5.7. A piece of paper 8 inches wide is folded as in Figure 1.17 so that one corner is placed on the opposite side. Express the length of the crease, L, in terms of the angle θ alone.

1.5.8. Let $P_1, P_2, \ldots, P_{12}$ be the successive vertices of a regular dodecagon (twelve sides). Are the diagonals $P_1 P_9, P_2 P_{11}, P_4 P_{12}$ concurrent?

1.5.9. Use algebra to support your answers to each of the following.

(a) A car travels from A to B at the rate of 40 miles per hour and then returns from B to A at the rate of 60 miles per hour. Is the average rate for the round trip more or less than 50 miles per hour?
(b) You are given a cup of coffee and a cup of cream, each containing the same amount of liquid. A spoonful of cream is taken from the cup and put into the coffee cup, then a spoonful of the mixture is put back into the cream cup. Is there now more or less cream in the coffee cup than coffee in the cream cup? (This problem has an elegant nonalgebraic solution based on the observation that the coffee in the cream cup has displaced an equal amount of cream which must be in the coffee cup.)
(c) Imagine that the earth is a smooth sphere and that a string is wrapped around it at the equator. Now suppose that the string is lengthened by six feet and the new length is evenly pushed out to form a larger circle just over the equator. Is the distance between the string and the surface of the earth more or less than one inch?

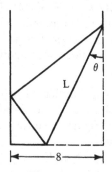

Figure 1.17.

1.5.10. A well-known theorem asserts that a prime $p > 2$ can be written as a sum of two perfect squares ($p = m^2 + n^2$, with m and n integers) if and only if p is one more than a multiple of 4. Assuming this result, show that:

(a) Every prime one more than a multiple of 8 can be written in the form $x^2 + 16y^2$, x and y integers.
(b) Every prime five more than a multiple of 8 can be written in the form $(2x + y)^2 + 4y^2$, x and y integers.

Additional Examples

1.1.10, 2.5.10, 3.2.15, 3.3.11, 3.3.28, 3.4.2, 3.4.4, 4.1.5, 6.4.2, 7.2.4, 8.1.15, 8.2.3, 8.2.17. Also, see Sections 2.5 (Recurrence Relations), 3.2 (Modular Arithmetic), 3.4 (Positional Notation), 8.3 (Vector Geometry), 8.4 (Complex Numbers in Geometry).

1.6. Exploit Symmetry

The presence of symmetry in a problem usually provides a means for reducing the amount of work in arriving at a solution. For example, consider the product $(a + b + c)(a^2 + b^2 + c^2 - ab - ac - bc)$. Since each factor is symmetrical in a, b, c (the expression remains unchanged whenever any pair of its variables are interchanged), the same will be true of the product. As a result, if a^3 appears in the product, so will b^3 and c^3. Similarly, if a^2b appears in the product, so will a^2c, b^2a, b^2c, c^2a, c^2b, and each will occur with the same coefficient, etc. Thus, a quick check shows the product will have the form

$$A(a^3 + b^3 + c^3) + B(a^2b + a^2c + b^2a + b^2c + c^2a + c^2b) + C(abc).$$

It is an easy matter to check that $A = 1$, $B = 0$, and $C = -3$.

1.6.1. Equilateral triangles ABK, BCL, CDM, DAN are constructed inside the square $ABCD$. Prove that the midpoints of the four segments KL, LM, MN, NK and the midpoints of the eight segments AK, BK, BL, CL, CM, DM, DN, AN are the twelve vertices of a regular dodecagon.

Solution. The twelve vertices are indicated in Figure 1.18 by heavy dots; two of these vertices are labeled a and b as shown.

Using the symmetry of the figure, it suffices to show that $\angle bOK = 15°$, $\angle aOb = 30°$, and $|aO| = |bO|$.

Note that AN is part of the perpendicular bisector of BK, and therefore $|KN| = |NB|$. Using symmetry it follows that MBN is an equilateral

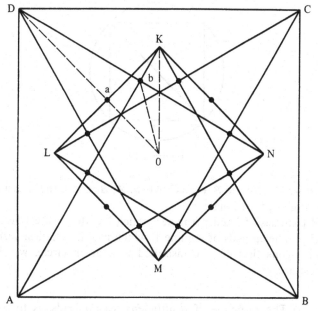

Figure 1.18.

triangle, say of side length s, and that $\angle CBN = 15°$. Now consider triangle DBN; note that Ob joins the midpoints of DB and DN, so Ob is parallel to BN and half its length. Thus $|Ob| = s/2$ and $\angle bOK = 15°$. From this it is easy to check that $\angle aOb = \angle DOK - \angle bOK = 45° - 15° = 30°$, and $|Oa| = |KN|/2 = s/2$.

The presence of symmetry in a problem also provides a clarity of vision which often enables us to see and discover relationships that might be more difficult to find by other means. For example, symmetry considerations alone suggest that the maximum value of xy, subject to $x + y = 1$, $x > 0$, $y > 0$, should occur when $x = y = \frac{1}{2}$ (x and y are symmetrically related). This is an example of the *principle of insufficient reason*, which can be stated briefly as follows: "Where there is no sufficient reason to distinguish, there can be no distinction." Thus, there is no reason to expect the maximum will occur when x is anything other than $\frac{1}{2}$, that is, closer to 0 or to 1. To verify this, let $x = \frac{1}{2} + e$. Then $y = \frac{1}{2} - e$, and, $xy = (\frac{1}{2} + e)(\frac{1}{2} - e) = \frac{1}{4} - e^2$. In this form it is clear that the maximum occurs when $e = 0$; that is, $x = y = \frac{1}{2}$.

The next problem offers several additional examples of this principle.

1.6.2.

(a) Of all rectangles which can be inscribed in a given circle, which has the greatest area?

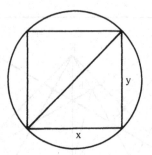

Figure 1.19.

(b) Maximize $\sin A + \sin B + \sin C$, where A, B, C are the measures of the three angles of a triangle.
(c) Of all triangles of fixed perimeter, which has the greatest area?
(d) Of all parallelepipeds of volume 1, which has the smallest surface area?
(e) Of all n-gons that can be inscribed in a given circle, which has the greatest area?

Solution. (a) The principle of insufficient reason leads us to suspect the rectangle of maximum area that can be inscribed in a circle is a square (Figure 1.19). To verify this, let x and y denote the length and width of the rectangle, and suppose without loss of generality that the units are chosen so that the diameter of the circle is unity. We wish to maximize xy subject to $x^2 + y^2 = 1$. It is equivalent to maximize $x^2 y^2$ subject to $x^2 + y^2 = 1$. But this is the same problem as that considered prior to this example; the maximum value occurs when $x^2 = y^2 = \frac{1}{2}$, that is, when the rectangle is a square.

(b) Notice that the sum, $\sin A + \sin B + \sin C$, is always positive (since each of the terms is positive), and it can be made arbitrarily small (in magnitude) by making A arbitrarily close to $180°$. There is no reason to expect the maximum will occur at any point other than $A = B = C = 60°$ (an equilateral triangle). A proof of this follows from the discussion in 2.4.1.

In a similar manner, we suspect the answers to (c), (d), and (e) are an equilateral triangle, a cube, and a regular n-gon. Proofs for these conjectures are given in 7.2.1, 7.2.12, and 2.4.1.

1.6.3. Evaluate

$$\int_0^{\pi/2} \frac{dx}{1 + (\tan x)^{\sqrt{2}}}.$$

Solution. Here is a problem that cannot be evaluated by the usual techniques of integration; that is to say, the integrand does not have an

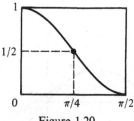

Figure 1.20.

antiderivative. However, the problem can be handled if we happen to notice that the integrand (Figure 1.20) is symmetric about the point $(\frac{1}{4}\pi, \frac{1}{2})$. To show this is so (it is not obvious), let $f(x) = 1/(1 + (\tan x)^{\sqrt{2}})$. It suffices to show that $f(x) + f(\pi/2 - x) = 1$ for all x, $0 \leqslant x \leqslant \pi/2$. Thus, we compute, for $r = \sqrt{2}$,

$$f(\pi/2 - x) + f(x) = \frac{1}{1 + \tan^r(\frac{1}{2}\pi - x)} + \frac{1}{1 + \tan^r x}$$

$$= \frac{1}{1 + \cot^r x} + \frac{1}{1 + \tan^r x}$$

$$= \frac{\tan^r x}{1 + \tan^r x} + \frac{1}{1 + \tan^r x}$$

$$= 1.$$

It follows from the symmetry just demonstrated that the area under the curve on $[0, \frac{1}{2}\pi]$ is one-half the area in the rectangle (see Figure 1.20); that is, the integral is $(\pi/2)/2 = \pi/4$.

Another way to take advantage of symmetry is in the choice of notation. Here are a couple of illustrations.

1.6.4. Let P be a point on the graph of $y = f(x)$, where f is a third-degree polynomial; let the tangent at P intersect the curve again at Q; and let A be the area of the region bounded by the curve and the segment PQ. Let B be the area of the region defined in the same way by starting with Q instead of P. What is the relationship between A and B?

Solution. We know that a cubic polynomial is symmetric about its inflection point (see 8.2.17). Since the areas of interest are unaffected by the choice of coordinate system, we will take the point of inflection as the origin. Therefore, we may assume the equation of the cubic is

$$f(x) = ax^3 + bx, \qquad a \neq 0$$

(see Figure 1.21).

Suppose x_0 is the abscissa of P. It turns out that the abscissa of Q is $-2x_0$. (We will not be concerned with the details of this straightforward

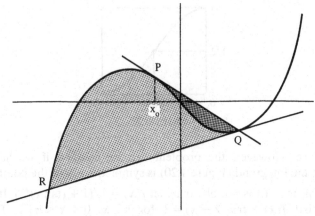

Figure 1.21.

computation. There is, indeed, a very elegant way to arrive at this fact, but it uses ideas found in Section 4.3 (see 4.3.7).)

A straightforward integration shows that the area A is equal to Kx_0^4, where K is independent of x_0. (Again, the details of this computation are not of concern here.)

We now can apply our previous conclusions to the point Q. The tangent at Q will intersect the curve at R, the abscissa of which evidently is $-2(-2x_0) = 4x_0$, and the area B is equal to $K(-2x_0)^4 = 16Kx_0^4 = 16A$.

1.6.5. Determine all values of x which satisfy

$$\tan x = \tan(x + 10°)\tan(x + 20°)\tan(x + 30°).$$

Solution. We will introduce symmetry by a simple change of variable. Thus, set $y = x + 15°$. The equation then is

$$\tan(y - 15°) = \tan(y - 5°)\tan(y + 5°)\tan(y + 15°),$$

which is equivalent to

$$\frac{\sin(y - 15°)\cos(y + 15°)}{\cos(y - 15°)\sin(y + 15°)} = \frac{\sin(y - 5°)\sin(y + 5°)}{\cos(y - 5°)\cos(y + 5°)}.$$

Using the identities

$$\sin A \cos B = \tfrac{1}{2}\big[\sin(A - B) + \sin(A + B)\big],$$

$$\sin A \sin B = \tfrac{1}{2}\big[\cos(A - B) - \cos(A + B)\big],$$

$$\cos A \cos B = \tfrac{1}{2}\big[\cos(A - B) + \cos(A + B)\big],$$

we get

$$\frac{\sin(-30°) + \sin 2y}{\sin(30°) + \sin 2y} = \frac{\cos(-10°) - \cos 2y}{\cos(-10°) + \cos 2y},$$

or equivalently,

$$\frac{2\sin 2y - 1}{2\sin 2y + 1} = \frac{\cos 10° - \cos 2y}{\cos 10° + \cos 2y}.$$

This simplifies to

$$\sin 4y = \cos 10°,$$

which implies that

$$4y = 80° + 360° k, \quad 100° + 360° k, \qquad k = 0, \pm 1, \pm 2, \ldots,$$
$$x = 5° + 90° k, \quad 10° + 90° k, \qquad k = 0, \pm 1, \pm 2, \ldots.$$

Problems

1.6.6.

(a) Exploit symmetry to expand the product

$$(x^2 y + y^2 z + z^2 x)(xy^2 + yz^2 + zx^2).$$

(b) If $x + y + z = 0$, prove that

$$\left(\frac{x^2 + y^2 + z^2}{2}\right)\left(\frac{x^5 + y^5 + z^5}{5}\right) = \frac{x^7 + y^7 + z^7}{7}.$$

(Substitute $z = -x - y$ and apply the binomial theorem. For another approach, see 4.3.9.)

1.6.7. The faces of each of the fifteen pennies, packed as exhibited in Figure 1.22, are colored either black or white. Prove that there exist three pennies of the same color whose centers are the vertices of an equilateral triangle. (There are many ways to exploit symmetry and create "without loss of generality" arguments.)

1.6.8. Make use of the principle of insufficient reason to minimize $x_1^2 + x_2^2 + \cdots + x_n^2$, subject to the condition that $0 < x_i < 1$, and $x_1 + x_2 + \cdots + x_n = 1$. Prove your conjecture. (For the proof, take $x_i = 1/n + e_i$.)

1.6.9. A point P is located in the interior of an equilateral triangle ABC. Perpendiculars drawn from P meet each of the sides in points D, E, and F, respectively. Where should P be located to make $PD + PE + PF$ a maximum? Where should P be located to make $PD + PE + PF$ a minimum?

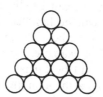

Figure 1.22.

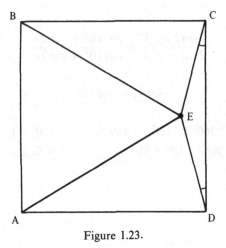

Figure 1.23.

Justify your answers. (Hint: It is helpful to reflect the figure about one of the sides. What happens to $PD + PE + PF$ as P moves parallel to the line of reflection?)

1.6.10. In Figure 1.23, $ABCD$ is a square, $\angle ECD = \angle EDC = 15°$. Show that triangle AEB is equilateral. (The key to this very beautiful problem is to create central symmetry. Specifically, add identical $15°$ angles on sides AB, BC, and AD (as on side CD) and create a diagram much like that constructed in 1.6.1.)

1.6.11. The product of four consecutive terms of an arithmetic progression of integers plus the fourth power of the common difference is always a perfect square. Verify this identity by incorporating symmetry into the notation.

Additional Examples

1.4.1, 8.1.4, 8.1.5, 8.1.8, 8.2.3.

1.7. Divide into Cases

It often happens that a problem can be divided into a small number of subproblems, each of which can be handled separately in a case-by-case manner. This is especially true when the problem contains a universal quantifier ("for all x . . . "). For example, the proof of a proposition of the form "for all integers . . . " might be carried out by arguing the even and odd cases separately. Similarly, a theorem about triangles might be proved by dividing it into three cases depending upon whether the triangle is acute,

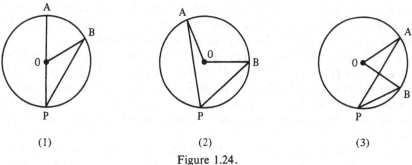

Figure 1.24.

right, or obtuse. Occasionally, the subproblems can be arranged hierarchically into subgoals, so that the first cases, once established, can be used to verify the succeeding stages. Such a procedure is called *hillclimbing*.

In the early stages of analysis, it is good to think about how a problem might be subdivided into a small number of (hopefully) simpler subproblems. The heuristic of this section is often given in the following form: "If you can't solve the problem, find a simpler related problem and solve it."

1.7.1. Prove that an angle inscribed in a circle is equal to one-half the central angle which subtends the same arc.

Solution. We are given a circle, say with center O, and an inscribed angle APB; some examples are shown in Figure 1.24. We are to prove that in all instances $\angle APB = \frac{1}{2}\angle AOB$. The three preceding figures represent three essentially different situations. Specifically, the center of the circle, O, is either inside $\angle APB$ (diagram 2), or outside $\angle APB$ (diagram 3), or on one of the rays of $\angle APB$ (diagram 1). We shall prove the theorem by considering each of these cases separately.

Case 1. Suppose the center O is on PA. Then $\angle AOB = \angle OPB + \angle OBP$ (exterior angle equals sum of opposite interior angles) $= 2\angle OPB$ ($\triangle OPB$ is isosceles) $= 2\angle APB$. The result follows.

Case 2. If O is interior to $\angle APB$ (diagram 2), extend line PO to cut the circle at D. We have just proved that $2\angle APD = \angle AOD$ and $2\angle DPB = \angle DOB$. Adding these equations gives the desired result.

Case 3. If O is exterior to $\angle APB$ (diagram 3), extend PO to cut the circle D. Then, using case 1, $2\angle DPB = \angle DOB$ and $2\angle DPA = \angle DOA$. Subtracting the second equation from the first yields the result. This completes the proof.

1.7.2. A real-valued function f, defined on the rational numbers, satisfies

$$f(x + y) = f(x) + f(y)$$

for all rational x and y. Prove that $f(x) = f(1) \cdot x$ for all rational x.

Solution. We will proceed in a number of steps. We will prove the result first for the positive integers, then for the nonpositive integers, then for the reciprocals of integers, and finally for all rational numbers.

Case 1 (positive integers). The result holds when $x = 1$. For $x = 2$, we have $f(2) = f(1 + 1) = f(1) + f(1) = 2f(1)$. For $x = 3$, $f(3) = f(2 + 1) = f(2) + f(1) = 2f(1) + f(1) = 3f(1)$. It is clear that this process can be continued, and that for any positive integer n, $f(n) = nf(1)$. (A formal proof can be given based on the principle of mathematical induction—see Chapter 2).

Case 2 (nonpositive integers). First, $f(0) = f(0 + 0) = f(0) + f(0)$. Subtract $f(0)$ from each side to get $0 = f(0)$; that is, $f(0) = 0 \cdot f(1)$. Now, $0 = f(0) = f(1 + (-1)) = f(1) + f(-1)$. From this, we see that $f(-1) = -f(1)$. Similarly, for any positive integer n, $f(n) + f(-n) = f(n + (-n)) = f(0) = 0$, so that $f(-n) = -nf(1)$.

Case 3 (reciprocals). For $x = \frac{1}{2}$, we proceed as follows: $f(1) = f(\frac{1}{2} + \frac{1}{2}) = f(\frac{1}{2}) + f(\frac{1}{2}) = 2f(\frac{1}{2})$. Divide by 2 to get $f(\frac{1}{2}) = f(1)/2$. For $x = \frac{1}{3}$, $f(1) = f(\frac{1}{3} + \frac{1}{3} + \frac{1}{3}) = f(\frac{1}{3}) + f(\frac{1}{3}) + f(\frac{1}{3}) = 3f(\frac{1}{3})$, or equivalently, $f(\frac{1}{3}) = f(1)/3$. In a similar way, for any positive integer n, $f(1/n) = f(1)/n$. For $x = -1/n$, we have $f(1/n) + f(-1/n) = f(1/n + (-1/n)) = f(0) = 0$, so $f(-1/n) = -f(1)/n$.

Case 4 (all rationals). Let n be an integer. Then $f(2/n) = f(1/n + 1/n) = f(1/n) + f(1/n) = 2f(1/n) = (2/n)f(1)$. Similarly, if m/n is any rational number, with m a positive integer and n an integer, then

$$f\left(\frac{m}{n}\right) = f\underbrace{\left(\frac{1}{n} + \cdots + \frac{1}{n}\right)}_{m \text{ times}} = \underbrace{f\left(\frac{1}{n}\right) + \cdots + f\left(\frac{1}{n}\right)}_{m \text{ times}}$$

$$= mf\left(\frac{1}{n}\right) = \frac{m}{n}f(1).$$

This establishes the result—a good example of hillclimbing.

1.7.3. Prove that the area of a lattice triangle is equal to $I + \frac{1}{2}B - 1$, where I and B denote respectively the number of interior and boundary lattice points of the triangle. (A lattice triangle is a triangle in the plane with lattice points as vertices.)

Solution. This is a special case of Pick's theorem (see 2.3.1). There are a number of ingenious proofs, each of which divide the set of lattice triangles into a few special types. One way to do this is to "circumscribe" about the triangle a rectangle with edges parallel to the coordinate axes. At least one vertex of the rectangle must coincide with a vertex of the triangle. Now it can be checked that every lattice triangle can be classified into one of the nonequivalent classes sketched in Figure 1.25.

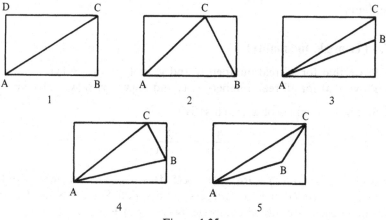

Figure 1.25.

In the first class are those right triangles whose legs are parallel to the coordinate axes. The second class includes acute-angled triangles one of whose sides is parallel to a coordinate axis. Such triangles are the "sum" of two triangles from the first class. In the third class are the obtuse triangles which have one side parallel to a coordinate axis. They are the "difference" of two triangles from the first class. The fourth and fifth classes cover those triangles having no sides parallel to the coordinate axes.

The proof of the result follows a hillclimbing pattern. To get started, let us consider the rectangle $ABCD$ in case 1. Suppose that line segments AB and AD contain a and b lattice points, respectively, not counting their endpoints. Then, with I and B the interior and boundary points of $ABCD$,

$$I + \tfrac{1}{2}B - 1 = ab + \tfrac{1}{2}(2a + 2b + 4) - 1$$
$$= ab + a + b + 1$$
$$= (a + 1)(b + 1)$$
$$= \text{Area } ABCD.$$

Now suppose that AB, BC, and AC contain a, b c lattice points, respectively, not counting their endpoints, and suppose that ABC contains i interior points. Then rectangle $ABCD$ has $2i + c$ interior points, and we have, with I and B the interior and boundary points of ABC,

$$I + \tfrac{1}{2}B - 1 = i + \tfrac{1}{2}(a + b + c + 3) - 1$$
$$= \tfrac{1}{2}(2i + a + b + c + 1)$$
$$= \tfrac{1}{2}\left[(2i + c) + \tfrac{1}{2}(2a + 2b + 4) - 1\right]$$
$$= \tfrac{1}{2}\text{Area } ABCD$$
$$= \text{Area } ABC.$$

The other cases can be handled in a similar way; we leave the details to the reader.

Problems

1.7.4 (Triangle inequality).

(a) Prove that for all real numbers x and y, $|x + y| \leqslant |x| + |y|$.
(b) Prove that for all real numbers x, y, and z, $|x - y| \leqslant |x - z| + |y - z|$.

1.7.5. Find all values of x which satisfy

$$\frac{3}{x - 1} < \frac{2}{x + 1}.$$

1.7.6. Let $S = \{i(3, 8) + j(4, -1) + k(5, 4) \mid i, j, k \text{ are integers}\}$, and $T = \{m(1, 5) + n(0, 7) \mid m, n \text{ are integers}\}$. Prove that $S = T$. (Note: Ordered pairs of integers are added componentwise: $(s, t) + (s', t') = (s + s', t + t')$, and $n(s, t) = (ns, nt)$.)

1.7.7. A real-valued function f, defined on the positive rational numbers, satisfies $f(x + y) = f(x)f(y)$ for all positive rational numbers x and y. Prove that $f(x) = [f(1)]^x$ for all positive rational x.

1.7.8. Determine $F(x)$ if, for all real x and y, $F(x)F(y) - F(xy) = x + y$.

Additional Examples

1.1.7, 2.5.11c, 2.5.12, 2.5.13, 2.6.3, 3.2.14, 3.2.15, 3.2.16, 3.2.17, 3.2.18, 3.4.1, 4.1.3, 4.1.4, 4.4.14, 4.4.29, 5.2.1, 5.3.14c, 6.5.4, 7.4.3, 7.6.2, 7.6.4, 7.6.10, 8.2.4. Some particularly nice examples which reduce to the study of very special cases are 3.3.8, 3.3.9, 3.3.21, 3.3.22, 3.3.26.

1.8. Work Backward

To work backward means to assume the conclusion and then to draw deductions from the conclusion until we arrive at something known or something which can be easily proved. After we arrive at the given or the known, we then reverse the steps in the argument and proceed forward to the conclusion.

This procedure is common in high-school algebra and trigonometry. For example, to find all real numbers which satisfy $2x + 3 = 7$, we argue as follows. Suppose that x satisfies $2x + 3 = 7$. Then, subtract 3 from each side of the equation and divide each side by 2, to get $x = 2$. Since each step in this derivation can be reversed, we conclude that 2 does indeed satisfy $2x + 3 = 7$ and is the only such number.

Often, in routine manipulations, such as in the previous example, an explicit rewriting of the steps is not done. However, it is important to be aware of what can, and what cannot, be reversed. For example, consider the equation $\sqrt{x+1} - \sqrt{x-1} = 2$. (Here, as usual, the square root is interpreted as the positive square root.) Write the equation in the form $\sqrt{x+1} = \sqrt{x-1} + 2$, and square each side to get $x + 1 = x - 1 + 4\sqrt{x-1} + 4$, or equivalently, $\sqrt{x-1} = -\frac{1}{2}$. Square a second time to get $x - 1 = \frac{1}{4}$, or $x = \frac{5}{4}$. We conclude that *if* there is a number x such that $\sqrt{x+1} - \sqrt{x-1} = 2$, it has to equal $\frac{5}{4}$. However, $\frac{5}{4}$ does not satisfy the original equation. The reason for this is that the steps are not all reversible. Thus, in this example, we proceed from $\sqrt{x-1} = -\frac{1}{2}$ to $x - 1 = \frac{1}{4}$. When this is reversed, however, the argument goes from $x - 1 = \frac{1}{4}$ to $\sqrt{x-1} = \frac{1}{2}$.

1.8.1. Let α be a fixed real number, $0 < \alpha < \pi$, and let

$$F(\theta) = \frac{\sin\theta + \sin(\theta + \alpha)}{\cos\theta - \cos(\theta + \alpha)}, \qquad 0 \leqslant \theta \leqslant \pi - \alpha.$$

Show that F is a constant. (This problem arose in 1.2.1.)

Solution. Suppose that F is a constant. Then $F(\theta) = F(0)$ for all θ, $0 \leqslant \theta \leqslant \pi - \alpha$. That is,

$$\frac{\sin\theta + \sin(\theta + \alpha)}{\cos\theta - \cos(\theta + \alpha)} = \frac{\sin\alpha}{1 - \cos\alpha}, \tag{1}$$

$$\left[\sin\theta + \sin(\theta + \alpha)\right]\left[1 - \cos\alpha\right] = \sin\alpha\left[\cos\theta - \cos(\theta + \alpha)\right], \tag{2}$$

$$\sin\theta + \sin(\theta + \alpha) - \sin\theta\cos\alpha - \sin(\theta + \alpha)\cos\alpha$$
$$= \sin\alpha\cos\theta - \sin\alpha\cos(\theta + \alpha), \tag{3}$$

$$\sin\theta + \sin(\theta + \alpha) - \left[\sin\theta\cos\alpha + \sin\alpha\cos\theta\right]$$
$$- \left[\sin(\theta + \alpha)\cos\alpha - \sin\alpha\cos(\theta + \alpha)\right] = 0, \tag{4}$$

$$\sin\theta + \sin(\theta + \alpha) - \sin(\theta + \alpha) - \sin(\theta + \alpha - \alpha) = 0. \tag{5}$$

The last equation is an identity. For the *proof*, we must reverse these steps. The only questionable step is from (2) to (1): the proof is valid only if we do not divide by zero in going from (2) to (1). But $(1 - \cos\alpha) \neq 0$ since $0 < \alpha < \pi$, and $\cos\theta - \cos(\theta + \alpha) > 0$ since $0 \leqslant \theta < \theta + \alpha \leqslant \pi$. The proof therefore can be carried out; that is, starting with the known identity at (5), we can argue (via steps (4), (3), (2), (1)) that for all θ, $0 \leqslant \theta \leqslant \pi - \alpha$, $F(\theta) = \sin\alpha/(1 - \cos\alpha) = $ constant.

1.8.2. If a, b, c denote the lengths of the sides of a triangle, show that

$$3(ab + bc + ca) \leqslant (a + b + c)^2 \leqslant 4(ab + bc + ca).$$

Solution. Consider the leftmost inequality:

$$3(ab + bc + ca) \leqslant (a + b + c)^2,$$

$$3(ab + bc + ca) \leqslant a^2 + b^2 + c^2 + 2(ab + bc + ca),$$

$$ab + bc + ca \leqslant a^2 + b^2 + c^2,$$

$$a^2 + b^2 + c^2 - ab - bc - ca \geqslant 0,$$

$$2a^2 + 2b^2 + 2c^2 - 2ab - 2bc - 2ca \geqslant 0,$$

$$(a^2 - 2ab + b^2) + (b^2 - 2bc + c^2) + (c^2 - 2ca + a^2) \geqslant 0,$$

$$(a - b)^2 + (b - c)^2 + (c - a)^2 \geqslant 0.$$

This last inequality is true for all values of a, b, c. Now consider the right inequality:

$$(a + b + c)^2 \leqslant 4(ab + bc + ca),$$

$$a^2 + b^2 + c^2 + 2(ab + bc + ca) \leqslant 4(ab + bc + ca),$$

$$a^2 + b^2 + c^2 \leqslant 2(ab + bc + ca),$$

$$a^2 + b^2 + c^2 \leqslant a(b + c) + b(a + c) + c(b + a).$$

This final inequality is true, since the sum of any two sides of a triangle is larger than the remaining side. Thus, $a^2 \leqslant a(b + c)$, $b^2 \leqslant b(a + c)$, and $c^2 \leqslant c(b + a)$.

The steps in each of these arguments can be reversed, so the proof is complete.

1.8.3. Given: AOB is a diameter of the circle O; BM is tangent to the circle at B; CF is tangent to the circle at E and meets BM at C; the chord AE, when extended, meets BM at D. Prove that $BC = CD$. (See Figure 1.26.)

Solution. Suppose $BC = CD$. Then $CE = CD$, since $BC = CE$ (tangents from C to the circle at E and B are equal). Thus, $\angle CED = \angle CDE$ (base angles of an isosceles triangle are equal). We are led to consider the angles as labeled in Figure 1.26.

Now, $\angle d$ is complementary to $\angle a$ since $\triangle ABD$ is a right triangle, and $\angle e$ is complementary to $\angle c$ since $\angle BEA$ is a right angle (AOB is a diameter). Therefore, $\angle a = \angle c$. But we know that $\angle a = \angle c$, since they both cut off the equal arc BE on the circle O.

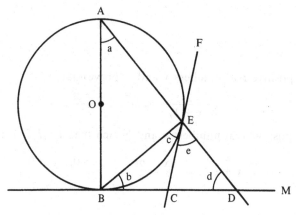

Figure 1.26.

The proof can now be completed by reversing these steps. Thus (omitting reasons), $\angle a = \angle c$, and therefore, $\angle e = \angle d$. Hence $CD = CE$, $CE = BC$, and therefore $BC = CD$.

1.8.4. In a round-robin tournament with n players $P_1, P_2, \ldots, P_n$, where $n > 1$, each player plays one game with each of the other players and rules are such that no ties can occur. Let W_r and L_r be the number of games won and lost, respectively, by player P_r. Show that

$$\sum_{r=1}^{n} W_r^2 = \sum_{r=1}^{n} L_r^2.$$

Solution. Suppose $\sum_{r=1}^{n} W_r^2 = \sum_{r=1}^{n} L_r^2$. Then,

$$\sum_{r=1}^{n} (W_r^2 - L_r^2) = 0,$$

$$\sum_{r=1}^{n} (W_r - L_r)(W_r + L_r) = 0.$$

But $W_r + L_r = n - 1$ for each r, so

$$(n - 1) \sum_{r=1}^{n} (W_r - L_r) = 0,$$

$$\sum_{r=1}^{n} (W_r - L_r) = 0,$$

$$\sum_{r=1}^{n} W_r = \sum_{r=1}^{n} L_r.$$

This last equation is true, since the total number of games won by the n players has to equal the total number of games lost. The proof follows on reversing the preceding argument.

Problems

1.8.5.

(a) Given positive real numbers x and y, prove that

$$\frac{2}{1/x + 1/y} \leqslant \sqrt{xy} \leqslant \frac{x+y}{2}.$$

(b) Given positive real numbers a and b such that $a + b = 1$, prove that

$$\frac{2}{a/x + b/y} \leqslant ax + by, \qquad x > 0, \quad y > 0.$$

1.8.6.

(a) If a, b, c are positive real numbers, and $a < b + c$, show that

$$\frac{a}{1+a} < \frac{b}{1+b} + \frac{c}{1+c}.$$

(b) If a, b, c are lengths of three segments which can form a triangle, show that the same is true for $1/(a + c)$, $1/(b + c)$, $1/(a + b)$.

1.8.7. Two circles are tangent externally at A, and a common external tangent touches them at B and C. The line segment BA is extended, meeting the second circle at D. Prove that CD is a diameter.

1.8.8. Consider the following argument. Suppose θ satisfies

$$\cot \theta + \tan 3\theta = 0.$$

Then, since

$$\tan(\alpha + \beta) = \frac{\tan \alpha + \tan \beta}{1 - \tan \alpha \tan \beta},$$

it follows that

$$\cot \theta + \frac{\tan \theta + \tan 2\theta}{1 - \tan \theta \tan 2\theta} = 0,$$
$$\cot \theta (1 - \tan \theta \tan 2\theta) + \tan \theta + \tan 2\theta = 0,$$
$$\cot \theta - \tan 2\theta + \tan \theta + \tan 2\theta = 0,$$
$$\cot \theta + \tan \theta = 0,$$
$$1 + \tan^2 \theta = 0,$$
$$\tan^2 \theta = -1.$$

Since this last equation cannot hold, the original equation does not have a solution (we don't need to reverse any steps because the final step doesn't yield any contenders). However, $\theta = \frac{1}{4}\pi$ does satisfy $\cot \theta + \tan 3\theta = 0$. What's wrong with the argument?

1.8.9. With Euclidean tools (straightedge and compass), inscribe a square in a given triangle so that one side of the square lies on a given side of the triangle. (Hint: Begin with the square and construct a triangle around it similar to the given triangle. Then use the fact that similar figures have proportional parts.)

Additional Examples

2.1.5, 7.1.1, 7.4.6. Also, see Section 2.2 (Induction) and Section 2.5 (Recursion).

1.9. Argue by Contradiction

To argue by contradiction means to assume the conclusion is not true and then to draw deductions until we arrive at something that is contradictory either to what is given (*the indirect method*) or to what is known to be true (*reductio ad absurdum*). Thus, for example, to prove $\sqrt{2}$ is irrational, we might assume it is rational and proceed to derive a contradiction. The method is often appropriate when the conclusion is easily negated, when the hypotheses offer very little substance for manipulation, or when there is a dearth of ideas about how to proceed.

As a simple example of this method of proof, consider the following argument which shows that the harmonic series diverges. Suppose on the contrary, that it converges—say to r. Then

$$r = 1 + \tfrac{1}{2} + \tfrac{1}{3} + \tfrac{1}{4} + \tfrac{1}{5} + \tfrac{1}{6} + \tfrac{1}{7} + \tfrac{1}{8} + \cdots$$
$$> \tfrac{1}{2} + \tfrac{1}{2} + \tfrac{1}{4} + \tfrac{1}{4} + \tfrac{1}{6} + \tfrac{1}{6} + \tfrac{1}{8} + \tfrac{1}{8} + \cdots$$
$$= 1 \;+\; \tfrac{1}{2} \;+\; \tfrac{1}{3} \;+\; \tfrac{1}{4} \;+\cdots$$
$$= r,$$

a contradiction. We are forced to conclude that the series diverges.

1.9.1. Given that a, b, c are odd integers, prove that equation $ax^2 + bx + c = 0$ cannot have a rational root.

Solution. Suppose p/q is a rational root, where (without loss of generality) p and q are not both even integers. We will first establish that neither p nor q is even. For suppose that p is even. From $a(p/q)^2 + b(p/q) + c = 0$ we find that $ap^2 + bpq + cq^2 = 0$. Since $ap^2 + bpq$ is even, cq^2 must be even, but this is impossible, since c and q are both odd. We get a similar contradiction if we suppose q is even. Therefore, both p and q are odd and $ap^2 + bpq + cq^2 = 0$. But this last equation states that the sum of three odd numbers is zero, an impossibility. Therefore, the equation has no rational root.

It is instructive to consider another proof of this result. The roots of $ax^2 + bx + c = 0$ are rational if and only if $b^2 - 4ac$ is a perfect square. So, suppose that $b^2 - 4ac = (2n + 1)^2$ for some integer n (by supposition, $b^2 - 4ac$ is odd, and therefore, if it is a square, it must be the square of an odd integer). Collecting multiples of 4 we have

$$b^2 - 1 = 4\big[n(n + 1) + ac\big].$$

Since either n or $n + 1$ is even, $n(n + 1) + ac$ is odd. Thus, the right side of the last equation is divisible by 4 but not by 8. However, the left side is divisible by 8, since $b^2 - 1 = (b - 1)(b + 1)$ and one of $b - 1$ and $b + 1$ is divisible by 4, while the other is divisible by 2. Therefore the displayed equation above cannot hold, and we have a contradiction. (In this proof, we have reached a contradiction by looking at how two numbers stand relative to multiples of 8, rather than multiples of 2 as in the first proof. We will return to a deeper consideration of this idea in Section 3.2.)

The next two sections contain additional illustrations of proof by contradiction.

Problems

1.9.2. In a party with 2000 persons, among any set of four there is at least one person who knows each of the other three. There are three people who are not mutually acquainted with each other. Prove that the other 1997 people know everyone at the party. (Assume that "knowing" is a symmetric relation; that is, if A knows B then B also knows A. What is the answer if "knowing" is not necessarily symmetric?)

1.9.3. Prove that there do not exist positive integers a, b, c, and n such that $a^2 + b^2 + c^2 = 2^n abc$. (From 1.4.3, we may assume that a and b are odd and c is even. How are the sides of the equation related to 4?)

1.9.4. Every pair of communities in a county are linked directly by exactly one mode of transportation: bus, train, or airplane. All three modes of transportation are used in the county; no community is served by all three modes, and no three communities are linked pairwise by the same mode.

Four communities can be linked according to these stipulations in the following way: bus, AB, BC, CD, DA; train, AC; airplane, BD.

(a) Give an argument to show that no community can have a single mode of transportation leading to each of three different communities.
(b) Give a proof to show that five communities cannot be linked in the required manner.

1.9.5. Let S be a set of rational numbers that is closed under addition and multiplication (that is, whenever a and b are members of S, so are $a + b$ and ab), and having the property that for every rational number r exactly one of the following three statements is true: $r \in S$, $-r \in S$, $r = 0$.

(a) Prove that 0 does not belong to S.
(b) Prove that all positive integers belong to S.
(c) Prove that S is the set of all positive rational numbers.

Additional Examples

1.5.10, 1.6.7, 3.2.1, 3.2.6, 3.2.11, 3.2.13, 3.2.15, 3.2.17, 3.2.18, 3.3.4, 3.3.11, 3.3.15, 3.3.28, 3.4.2, 4.1.3, 4.4.6, 5.4.1. Also, see Section 1.10 (Parity) and Section 1.11 (Extreme Cases).

1.10. Pursue Parity

The simple idea of parity—evenness and oddness—is a powerful problem-solving concept with a wide variety of applications. We will consider some examples in this section, and then generalize the idea in Section 3.2.

1.10.1. Let there be given nine lattice points in three-dimensional Euclidean space. Show that there is a lattice point on the interior of one of the line segments joining two of these points.

Solution. There are only eight different parity patterns for the lattice points: (even, even, even), (even, even, odd), . . . , (odd, odd, odd). Since there are nine given points, two of them have the same parity pattern. Their midpoint is a lattice point, and the proof is complete.

1.10.2. Place a knight on each square of a 7-by-7 chessboard. Is it possible for each knight to simultaneously make a legal move?

Solution. Assume a chessboard is colored in the usual checkered pattern. The board has 49 squares; suppose 24 of them are white and 25 are black.

Consider 25 knights which rest on the black squares. If they were to each make a legal move, they would have to move to 25 white squares. However, there are only 24 white squares available, therefore such a move cannot be made.

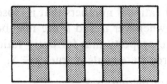

Figure 1.27.

1.10.3. Place a knight on a 4-by-n chessboard. Is it possible, in $4n$ consecutive knight moves, to visit each square of the board and return to the original square?

Solution. Before considering this problem, it is interesting to consider the same question for the 7-by-7 chessboard. Suppose that such a "closed tour" is attempted. On the first move the knight moves to a square of the opposite color; on the second move it returns to a square of the same color; and so forth. We see that after an odd number of moves the knight will occupy a square opposite in color from its original square. Now a closed tour of the 7-by-7 board requires 49 moves, an odd number. Therefore the knight cannot occupy its original square, and the closed tour is impossible.

Consider, now, the 4-by-n board. The argument for the 7-by-7 does not carry over to this case, because $4n$ is an even number. To handle this case, color the 4-by-n board in the manner indicated in Figure 1.27.

Notice that knight moves made from the white squares in the top and bottom rows lead to white squares in the second and third rows. Conversely, in a tour of the required type, knight moves from the inner two rows must necessarily be to the white squares in the outer two rows. This is because there are exactly n white squares in the outer two rows, and these can be reached only from the n white squares in the inner two rows. Therefore, the knight path can never move from the white squares to the black squares, and so such a closed tour is impossible.

1.10.4. Let n be an odd integer greater than 1, and let A by an n-by-n symmetric matrix such that each row and each column of A consists of some permutation of the integers $1, \ldots, n$. Show that each one of the integers $1, \ldots, n$ must appear in the main diagonal of A.

Solution. Off-diagonal elements occur in pairs because A is symmetric. Each number appears exactly n times, and this, together with knowing that n is odd, implies the result.

1.10.5. Let $a_1, a_2, \ldots, a_{2n+1}$ be a set of integers with the following property (P): if any of them is removed, the remaining ones can be divided into two sets of n integers with equal sums. Prove that $a_1 = a_2 = \cdots = a_{2n+1}$.

Solution. First, observe that all of the integers $a_1, \ldots, a_{2n+1}$ have the same parity. To see this, let $A = a_1 + \cdots + a_{2n+1}$. The claim follows after noting that for each i, $A - a_i$ is even (otherwise the remaining numbers could not be divided in the required manner).

Let a denote the smallest number among $a_1, \ldots, a_{2n+1}$, and for each i, let $b_i = a_i - a$. The problem is equivalent to showing that $b_i = 0$ for all i.

Now $b_1, b_2, \ldots, b_{2n+1}$ satisfy property (P). Since one of them is zero, it must be the case that they all are even. If they are not all zero, let k be the largest positive integer for which 2^k divides each of the b_i. For each i, let $c_i = b_i/2^k$. Then $c_1, c_2, \ldots, c_{2n+1}$ satisfy (P); however, they don't all have the same parity (since one of them is zero, and another is odd because of the choice of k). Therefore, all the b_i are zero and the proof is complete.

Problems

1.10.6.

(a) Remove the lower left corner square and the upper right corner square from an ordinary 8-by-8 chessboard. Can the resulting board be covered by 31 dominos? Assume each domino will cover exactly two adjacent squares of the board.

(b) Let thirteen points $P_1, \ldots, P_{13}$ be given in the plane, and suppose they are connected by the segments $P_1P_2, P_2P_3, \ldots, P_{12}P_{13}, P_{13}P_1$. Is it possible to draw a straight line which passes through the interior of each of these segments?

1.10.7.

(a) Is it possible to trace a path along the arcs of Figure 1.28(a) which traverses each arc once and only once? (Hint: Count the number of arcs coming out of each vertex.)

(b) Is it possible to trace a path along the lines of Figure 1.28(b) which passes through each juncture point once and only once? (Hint: Color the vertices in an alternating manner.)

1.10.8. Let $a_1, a_2, \ldots, a_n$ represent an arbitrary arrangement of the numbers $1, 2, \ldots, n$. Prove that, if n is odd, the product

$$(a_1 - 1)(a_2 - 2) \cdots (a_n - n)$$

is an even number.

1.10.9. Show that $(2^a - 1)(2^b - 1) = 2^{2^c} + 1$ is impossible in nonnegative integers a, b, and c. (Hint: Write the equation in the equivalent form $2^{a+b} - 2^a - 2^b = 2^{2^c}$ and investigate the possibilities for a, b, and c.)

1.10.10. Show that $x^2 - y^2 = a^3$ always has integral solutions for x and y whenever a is a positive integer.

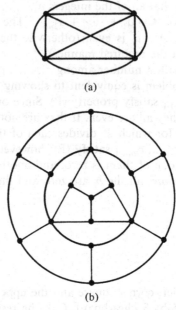

(a)

(b)

Figure 1.28.

Additional Examples

1.5.10, 1.9.1, 2.2.7, 3.2.13, 3.3.4, 3.3.20, 4.2.16(a), 4.3.4, 7.4.6. See Section 3.2 for a generalization of this method.

1.11. Consider Extreme Cases

In the beginning stages of problem exploration, it is often helpful to consider the consequences of letting the problem parameters vary from one extreme value to another. In this section we shall see that the existence of extreme positions are often the key to understanding existence results (problems of the sort "prove there is an x such that $P(x)$").

1.11.1. Given a finite number of points in the plane, not all collinear, prove there is a straight line which passes through exactly two of them.

Solution. If P is a point and L a line, let $d(P, L)$ denote the distance from P to L. Let S denote the set of positive distances $d(P, L)$ as P varies over the given points, and L varies over those lines which do not pass through P but

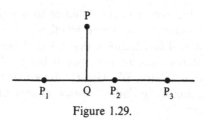

Figure 1.29.

which do pass through at least two of the given points. The set S is nonempty (because the given points are not all collinear) and finite (there are only a finite number of points and a finite number of lines which pass through at least two such points). Therefore S has a minimal element, say $d(P, M)$. We claim that M passes through exactly two of the given points.

Suppose that M passes through three of the given points, say P_1, P_2, and P_3. Let Q denote the point on M which is closest to P. At least two of the points P_1, P_2, P_3 lie on the same side of Q (one may equal Q), say P_2 and P_3 (see Figure 1.29). Suppose the points are labeled so that P_2 is closer to P than P_3. Now let N denote the line through P and P_3, and note that $d(P_2, N) < d(P, M)$, a contradiction to our choice of P and M. It follows that M can only pass through two of the given points.

1.11.2. Let A be a set of $2n$ points in the plane, no three of which are collinear. Suppose that n of them are colored red and the remaining n blue. Prove or disprove: There are n closed straight line segments, no two with a point in common, such that the endpoints of each segment are points of A having different colors.

Solution. If we disregard line crossings, there are a number of ways the given n red points can be paired with the given blue points by n closed straight line segments. Assign to each such pairing the total length of all the line segments in the configuration. Because there are only a finite number of such pairings, one of these configurations will have minimal total length. This pairing will have no segment crossings. (If segments $R_1 B_1$ and $R_2 B_2$ intersected, R_1, R_2 being red points and B_1, B_2 blue points, then we could reduce the total length of the configuration by replacing these segments with $R_1 B_2$ and $R_2 B_1$.) (For another solution, see 6.2.3.)

1.11.3. At a party, no boy dances with every girl, but each girl dances with at least one boy. Prove there are two couples bg and $b'g'$ which dance, whereas b does not dance with g' nor does g dance with b'.

Solution. Although not necessary, it may make the problem more understandable if we interpret the problem in matrix terms. Let the rows of a matrix correspond to the boys and the columns to the girls. Enter a 1 or 0 in the b-row and g-column according to whether b and g dance or don't dance with one another. The condition that no boy dances with every girl implies that (i) every row has at least one 0 entry. Similarly, (ii) every column has at least one 1 entry. We wish to prove that there are two rows, b and b', and two columns, g and g', whose entries at their intersection points have the pattern

$$
\begin{pmatrix} --- 1 --- 0 --- \\ --- 0 --- 1 --- \end{pmatrix} \quad \text{or} \quad \begin{pmatrix} --- 0 --- 1 --- \\ --- 1 --- 0 --- \end{pmatrix}
$$

Let h denote an arbitrary row. By (i) there is a 0 entry in this row, say in column k, and by (ii) there is a 1 entry in column k, say in row m:

$$
\begin{array}{c} \\ h \\ \\ m \end{array}
\begin{pmatrix} \overset{k}{} \\ --- 1 --- 0 --- \\ --- 0? --- 1 --- \end{pmatrix}
$$

Now, we're done if there is a column which contains a 1 in row h and a 0 in column m. In general, such a column may not exist. However, if h had been chosen in advance as a row with a maximal number of 1's, then such a column would have to exist, and the problem would be solved.

With this background, we can rewrite the solution in language independent of the matrix interpretation. Let b be a boy who dances with a maximal number of girls. Let g' be a girl with whom b does not dance, and b' a boy with whom g' dances. Among the partners of b, there must be at least one girl g who does not dance with b' (for otherwise b' would have more partners than b). The couples bg and $b'g'$ solve the problem.

1.11.4. Prove that the product of n successive integers is always divisible by $n!$.

Solution. First, notice that it suffices to prove the result for n successive positive integers. For the result is obviously true if one of the integers in the product is 0, whereas if all the integers are negative, it suffices to show that $n!$ divides their absolute value.

So suppose there are n successive positive integers whose product is not divisible by $n!$. Of all such numbers n, choose the smallest; call it N. Note that $N > 2$, since the product of two successive integers is always even. We are supposing, therefore, that there is a nonnegative integer m such that $(m + 1)(m + 2) \cdots (m + N)$ is not divisible by $N!$. Of all such numbers m, let M be the smallest. Note that $M > 0$, since $N!$ is divisible by $N!$. Thus, we are supposing that $(M + 1)(M + 2) \cdots (M + N)$ is not divisible by $N!$. Now,

$$(M + 1)(M + 2) \cdots (M + N - 1)(M + N)$$
$$= M\big[(M + 1)(M + 2) \cdots (M + N - 1)\big]$$
$$+ N\big[(M + 1)(M + 2) \cdots (M + N - 1)\big].$$

By our choice of M, $N!$ divides $M[(M + 1)(M + 2) \cdots (M + N - 1)]$. By our choice of N, $(N - 1)!$ divides $(M + 1)(M + 2) \cdots (M + N - 1)$, and consequently, $N!$ divides $N[(M + 1)(M + 2) \cdots (M + N - 1)]$. Combining, we see that $N!$ divides the right side of the last equation, contrary to our supposition. This contradiction establishes the result.

(A slick proof of this result is to recognize that the quotient $(m + 1)(m + 2) \cdots (m + n)/n!$ is equal to the binomial coefficient $\binom{n+m}{n}$, and is therefore an integer if m is an integer.)

Problems

1.11.5. Let $f(x)$ be a polynomial of degree n with real coefficients and such that $f(x) \geqslant 0$ for every real number x. Show that $f(x) + f'(x) + \cdots + f^{(n)}(x) \geqslant 0$ for all real x. ($f^{(k)}(x)$ denotes the kth derivative of $f(x)$.)

1.11.6. Give an example to show that the result of 1.11.1 does not necessarily hold for an infinite number of points in the plane. Where does the proof of 1.11.1 break down for the infinite case?

1.11.7. Show that there exists a rational number, c/d, with $d < 100$, such that

$$\left[\!\left[k\frac{c}{d} \right]\!\right] = \left[\!\left[k\frac{73}{100} \right]\!\right] \qquad \text{for} \quad k = 1, 2, 3, \ldots, 99.$$

1.11.8. Suppose that P_n is a statement, for $n = 1, 2, 3, \ldots$. Suppose further that

(i) P_1 is true, and
(ii) for each positive integer m, P_{m+1} is true if P_m is true.

Prove that P_n is true for all n. (Hint: Let S denote the set of all positive integers for which P_n is not true. Let m denote the smallest element in S, assuming that S is nonempty.)

Additional Examples

3.1.9, 3.3.11, 3.3.28, 4.4.7, 4.4.10, and the referrals given in 6.3.7. Also, see Sections 7.6 (The Squeeze Principle) and 6.2 (The Intermediate-Value Theorem) for examples which require consideration of "extremelike" cases.

1.12. Generalize

It may seem paradoxical, but it is often the case that a problem can be simplifed, and made more tractable and understandable, when it is generalized. This fact of life is well appreciated by mathematicians; in fact, abstraction and generalization are basic characteristics of modern mathematics. A more general setting provides a broader perspective, strips away nonessential features, and provides a whole new arsenal of techniques.

1.12.1. Evaluate the sum $\sum_{k=1}^{n} k^2/2^k$.

Solution. We will instead evaluate the sum $S(x) = \sum_{k=1}^{n} k^2 x^k$ and then calculate $S(\frac{1}{2})$. The reason for introducing the variable x is that we can now use the techniques of analysis. We know that

$$\sum_{k=1}^{n} x^k = \frac{1 - x^{n+1}}{1 - x}, \qquad x \neq 1.$$

Differentiating each side we get

$$\sum_{k=1}^{n} kx^{k-1} = \frac{(1-x)(-(n+1)x^n) + (1-x^{n+1})}{(1-x)^2}$$

$$= \frac{1 - (n+1)x^n + nx^{n+1}}{(1-x)^2}.$$

Multiplying each side of this equation by x, differentiating a second time, and multiplying the result by x yields

$$S(x) = \sum_{k=1}^{n} k^2 x^2 = \frac{x(1+x) - x^{n+1}(nx - n - 1)^2 - x^{n+2}}{(1-x)^3}.$$

It follows that

$$S(\tfrac{1}{2}) = \sum_{k=1}^{n} \frac{k^2}{2^k} = 6 - \frac{1}{2^{n-2}} \left(\tfrac{1}{2}n - n - 1 \right)^2 - \frac{1}{2^{n-1}}$$

$$= 6 - \left(\frac{n^2 + 4n + 6}{2^n} \right).$$

1.12.2. Evaluate the following determinant (Vandermonde's determinant):

$$\det \begin{bmatrix} 1 & a_1 & a_1^2 & \cdots & a_1^{n-1} \\ 1 & a_2 & a_2^2 & \cdots & a_2^{n-1} \\ \vdots & \vdots & \vdots & & \vdots \\ 1 & a_n & a_n^2 & \cdots & a_n^{n-1} \end{bmatrix}.$$

Solution. We will assume that $a_i \neq a_j$, $i \neq j$, for otherwise the determinant is zero. In order to more clearly focus on the main idea, consider the case $n = 3$:

$$\det \begin{bmatrix} 1 & a & a^2 \\ 1 & b & b^2 \\ 1 & c & c^2 \end{bmatrix}.$$

In this determinant, replace c by a variable x. Then, the determinant is a polynomial $P(x)$ of degree 2. Moreover, $P(a) = 0$ and $P(b) = 0$, since the corresponding matrix, with c replaced by a or b respectively, then has two identical rows, Therefore,

$$P(x) = A(x - a)(x - b)$$

for some constant A. Now, A is the coefficient of x^2, and, returning to the determinant, we find that this coefficient is

$$\det \begin{pmatrix} 1 & a \\ 1 & b \end{pmatrix}.$$

Thus, $A = b - a$, and the original 3-by-3 determinant is

$$P(c) = (b - a)[(c - a)(c - b)].$$

The general case is analogous. Let D_n denote the desired determinant (of order n). Replace a_n in the bottom row of the matrix by the variable x. The resulting determinant is a polynomial $P_n(x)$ of degree $n - 1$, which vanishes at $a_1, a_2, \ldots, a_{n-1}$. Hence, by the Factor Theorem (see Section 4.2),

$$P_n(x) = A(x - a_1)(x - a_2) \cdots (x - a_{n-1}),$$

where A is a constant. As before, A is the coefficient of x^n, and expanding along the bottom row makes it clear that $A = D_{n-1}$. That is,

$$D_n = P_n(a_n) = D_{n-1}[(a_n - a_1)(a_n - a_2) \cdots (a_n - a_{n-1})].$$

We can repeat the argument for D_{n-1}, etc. The final result will be

$$D_n = \prod_{k=2}^{n} \left[\prod_{i=1}^{k-1} (a_k - a_i) \right].$$

1.12.3. Given that $\int_0^\infty (\sin x)/x \, dx = \frac{1}{2}\pi$, evaluate $\int_0^\infty (\sin^2 x)/x^2 \, dx$.

Solution. We will evaluate the more general integral

$$I(a) = \int_0^\infty \frac{\sin^2 ax}{x^2} \, dx, \qquad a \geqslant 0,$$

by using a technique called parameter differentiation.

Differentiating each side of the previous equation with respect to a, we get

$$I'(a) = \int_0^\infty \frac{2 \sin ax \cos ax \cdot x}{x^2} \, dx$$

$$= \int_0^\infty \frac{\sin 2ax}{x} \, dx.$$

Now, with $y = 2ax$, we get $dy = 2a \, dx$, and

$$I'(a) = \int_0^\infty \frac{\sin y}{y} \, dy = \frac{1}{2}\pi.$$

Integrating each side gives

$$I(a) = \frac{1}{2}\pi a + C, \qquad C \text{ constant}.$$

Since $I(0) = 0$, we get $C = 0$. Thus $I(a) = \frac{1}{2}\pi a$, $a \geqslant 0$. Setting $a = 1$ yields $I(1) = \int_0^\infty (\sin^2 x)/x^2 \, dx = \frac{1}{2}\pi$. (Incidentally, the value of $\int_0^\infty (\sin x)/x \, dx$ can be found by evaluating a more general integral—an integral of a complex-valued function over a contour in the complex plane.)

Problems

1.12.4. By setting x equal to the appropriate values in the binomial expansion

$$(1 + x)^n = \sum_{k=1}^{n} \binom{n}{k} x^k$$

(or one of its derivatives, etc.) evaluate each of the following:

(a) $\displaystyle\sum_{k=1}^{n} k^2 \binom{n}{k}$, (b) $\displaystyle\sum_{k=1}^{n} 3^k \binom{n}{k}$,

(c) $\displaystyle\sum_{k=1}^{n} \frac{1}{k+1} \binom{n}{k}$, (d) $\displaystyle\sum_{k=1}^{n} (2k+1) \binom{n}{k}$.

1.12.5. Evaluate

$$\det \begin{bmatrix} 1 & a & a^2 & a^4 \\ 1 & b & b^2 & b^4 \\ 1 & c & c^2 & c^4 \\ 1 & d & d^2 & d^4 \end{bmatrix}.$$

(Replace d by a variable x; make use of the fact that the sum of the roots of a fourth-degree polynomial is equal to the coefficient of x^3 (see Section 4.3).)

1.12.6.

(a) Evaluate $\int_0^\infty (e^{-x}\sin x)/x\, dx$. (Consider $G(k) = \int_0^\infty (e^{-x}\sin kx)/x\, dx$ and use parameter differentiation.)

(b) Evaluate $\int_0^1 (x-1)/\ln x\, dx$. (Consider $H(m) = \int_0^1 (x^m - 1)/\ln x\, dx$ and use parameter differentiation.)

(c) Evaluate

$$\int_0^\infty \frac{\arctan(\pi x) - \arctan x}{x}\, dx.$$

(Consider $F(a) = \int_0^\infty (\arctan(ax) - \arctan x)/x\, dx$ and use parameter differentiation.)

1.12.7. Which is larger $\sqrt[3]{60}$ or $2 + \sqrt[3]{7}$? (Cubing each number leads to complications that are not easily resolved. Consider instead the more general problem: Which is larger, $\sqrt[3]{4(x+y)}$ or $\sqrt[3]{x} + \sqrt[3]{y}$, where $x, y \geqslant 0$? Take $x = a^3, y = b^3$.)

Additional Examples

1.4.2, 2.2.6, 2.2.7, 4.1.4, 5.1.3, 5.1.4, 5.1.9, 5.1.11, 5.4.4, 5.4.5, 5.4.6, 5.4.7, 6.9.2, 7.4.4. Also, see Section 2.4 (Induction and Generalization).

Chapter 2. Two Important Principles: Induction and Pigeonhole

Mathematical propositions come in two forms: universal propositions which state that something is true *for all* values of x in some specified set, and existential propositions which state that something is true for *some* value of x in some specified set. The former type are expressible in the form "For all x (in a set S), $P(x)$"; the latter type are expressible in the form "There exists an x (in the set S) such that $P(x)$," where $P(x)$ is a statement about x. In this chapter we will consider two important techniques for dealing with these two kinds of statements: (i) the principle of mathematical induction, for universal propositions, and (ii) the pigeonhole principle, for existential propositions.

2.1. Induction: Build on $P(k)$

Let a be an integer and $P(n)$ a proposition (statement) about n for each integer $n \geqslant a$. The principle of mathematical induction states that:

If

    (i) *$P(a)$ is true, and*
    (ii) *for each integer $k \geqslant a$, $P(k)$ true implies $P(k + 1)$ true,*
then $P(n)$ is true for all integers $n \geqslant a$.

Notice that the principle enables us, in two simple steps, to prove an *infinite* number of propositions (namely, $P(n)$ is true for all integers $n \geqslant a$). The method is especially suitable when a pattern has been established (see Section 1.1, "Search for a Pattern") for the first few special cases ($P(a)$,

$P(a + 1), P(a + 2), \dots)$. In this section we consider induction arguments which, in step (ii), proceed directly from the truth of $P(k)$ to the truth of $P(k + 1)$—that is, the truth of $P(k + 1)$ is "built on" an initial consideration of the truth of $P(k)$. This is in slight contrast to arguments (considered in the next section) which begin with a consideration of $P(k + 1)$.

2.1.1. Use mathematical induction to prove the binomial theorem:

$$(a + b)^n = \sum_{i=0}^{n} \binom{n}{i} a^i b^{n-i}, \quad n \text{ a positive integer.}$$

Solution. It is easy to check that the result holds when $n = 1$.

Assuming the result for the integer k (we will build on the truth of $P(k)$), multiply each side by $(a + b)$ to get

$$(a + b)^k (a + b) = \left[\sum_{i=0}^{k} \binom{k}{i} a^i b^{k-i} \right] (a + b)$$

$$= \sum_{i=0}^{k} \binom{k}{i} a^{i+1} b^{k-i} + \sum_{i=0}^{k} \binom{k}{i} a^i b^{k+1-i}.$$

In the first sum, make the change of variable $j = i + 1$, to get

$$= \sum_{j=1}^{k+1} \binom{k}{j-1} a^j b^{k+1-j} + \sum_{i=0}^{k} \binom{k}{i} a^i b^{k+1-i}$$

$$= \left[\sum_{j=1}^{k} \binom{k}{j-1} a^j b^{k+1-j} + a^{k+1} \right] + \left[\sum_{i=1}^{k} \binom{k}{i} a^i b^{k+1-i} + b^{k+1} \right]$$

$$= a^{k+1} + \left[\sum_{i=1}^{k} \left[\binom{k}{i-1} + \binom{k}{i} \right] a^i b^{k+1-i} \right] + b^{k+1}$$

$$= a^{k+1} + \sum_{i=1}^{k} \binom{k+1}{i} a^i b^{k+1-i} + b^{k+1}$$

$$= \sum_{i=0}^{k+1} \binom{k+1}{i} a^i b^{k+1-i},$$

where we have made use of the basic identity $\binom{k}{i-1} + \binom{k}{i} = \binom{k+1}{i}$ (see 2.5.2). This is the form for $P(k + 1)$, so by induction, the proof is complete.

2.1.2. Let $0 < a_1 < a_2 < \cdots < a_n$, and let $e_i = \pm 1$. Prove that $\sum_{i=1}^{n} e_i a_i$ assumes at least $\binom{n+1}{2}$ distinct values as the e_i range over the 2^n possible combinations of signs.

Solution. When $n = 1$, there are exactly 2 distinct values (a_1 and $-a_1$), and $\binom{2}{2} = 1$, so the result holds.

Suppose the result is true when $n = k$; that is, that $\sum_{i=1}^{k} e_i a_i$ assumes at least $\binom{k+1}{2}$ distinct values, and suppose $a_{k+1} > a_k$. To begin, generate $\binom{k+1}{2}$ sums by subtracting a_{k+1} from each of the distinct sums generated by $a_1, a_2, \ldots, a_k$ (the inductive assumption). We need to generate $\binom{k+2}{2} - \binom{k+1}{2} = k + 1$ additional sums. These can be found in the following manner: Let $S = \sum_{i=1}^{k} a_i$ and note that $S + a_{k+1}, S + (a_{k+1} - 2a_1), \ldots, S + (a_{k+1} - 2a_{k-1}), S + (a_{k+1} - 2a_k)$, are distinct and greater than each of the sums obtained above. (To see this, note that $S + (a_{k+1} - 2a_k) > S + (a_{k+1} - 2a_{k+1}) = S - a_{k+1}$). There are $k + 1$ numbers in this list, so the result follows by induction.

Mathematical induction is a method that can be tried on any problem of the form "Prove that $P(n)$ holds for all $n \geq a$." The clue is often signaled by the mere presence of the parameter n. But it should be noted that induction also applies to many problems where the quantification is over more general sets. For example, a proposition about all polynomials might be proved by inducting on the degree of the polynomial. A theorem about all matrices might be handled by inducting on the size of the matrix. Several results concerning propositions in symbolic logic are carried out by inducting on the number of logical connectives in the proposition. The list of unusual "inductive sets" could go on and on; we will be content to give just two examples here; other examples are scattered throughout the book (e.g. see the next four sections and the listings in the "Additional Examples").

2.1.3. If V, E, and F are, respectively, the number of vertices, edges, and faces of a connected planar map, then

$$V - E + F = 2.$$

Solution. Your intuitive understanding of the terms in this result are probably accurate, but to make certain, here are the definitions.

A *network* is a figure (in a plane or in space) consisting of finite, nonzero number of arcs, no two of which intersect except possibly at their endpoints. The endpoints of these arcs are called *vertices* of the network. A *path* in a network is a sequence of different arcs in the network that can be traversed continuously without retracing any arc. A network is *connected* if every two different vertices of the network are vertices of some path in the network. A *map* is a network, together with a surface which contains the network. If this surface is a plane the map is called a *planar map*. The arcs of a planar map are called *edges*. The *faces* of a planar map are the regions that are defined by the boundaries (edges) of the map (the "ocean" is counted as a face).

Figure 2.1 shows three examples of connected neworks. The first two are planar maps. In the first, $V = 4$, $E = 4$, $F = 2$; in the second, $V = 5$, $E = 6$, $F = 3$. The third network is not a planar map. However, if we should

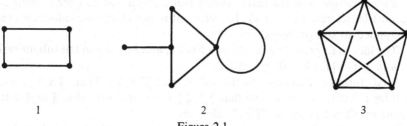

1   2   3

Figure 2.1.

flatten it onto a plane and place vertices at the intersection points, we
would have $V = 10$, $E = 20$, $F = 12$.

Now return to a consideration of the theorem. The key idea in the proof
of this result is to realize that connected planar maps can be built from a
single vertex by a sequence of the following constructions (each of which
leaves the map connected):

(i) Add a vertex in an existing edge (e.g. •————• becomes •—•—•).
(ii) Add an edge from a vertex back to itself (e.g. • becomes ◯).
(iii) Add an edge between the two existing vertices (e.g. ⎡ ⎤ becomes
⎣____⎦).
(iv) Add an edge and a vertex to an existing vertex (e.g. • becomes •——•).

We will induct on the number of steps required to construct the
connected planar map. If the network consists of a single point, then $V = 1$,
$F = 1$, $E = 0$, and $V - E + F = 2$.

Suppose the result holds when k steps are required in the construction.
The net change for each of the steps is given in the following table:

Operation	ΔV	ΔE	ΔF	$\Delta(V - E + F)$
(i)	$+1$	$+1$	0	0
(ii)	0	$+1$	$+1$	0
(iii)	0	$+1$	$+1$	0
(iv)	$+1$	$+1$	0	0

Since the quantity $V - E + F$ remains unchanged when the $(k + 1)$st step
is taken, the proof is complete by induction.

2.1.4. Given a positive integer n and a real number x, prove that

$$[\![x]\!] + \left[\!\left[x + \frac{1}{n} \right]\!\right] + \left[\!\left[x + \frac{2}{n} \right]\!\right] + \cdots + \left[\!\left[x + \frac{n-1}{n} \right]\!\right] = [\![nx]\!].$$

Solution. Although there is an integer parameter n in this problem, it will
not work to induct on n for a fixed x. Also, of course, we cannot induct on

x, since x ranges over the real numbers (for a given real number x, there is no next larger real number y). Therefore, it is not clear that induction can be applied to this problem.

The idea is to prove the result for a fixed n, and for all x in the subinterval $[k/n, (k+1)/n]$ for $k = 0, \pm 1, \pm 2, \ldots$.

First, suppose x belongs to the subinterval $[0, 1/n)$. Then $[\![x + i/n]\!] = 0$ for $i = 0, 1, \ldots, n-1$, so that $\sum_{i=0}^{n-1} [\![x + i/n]\!] = 0$. Also $[\![nx]\!] = 0$, so the result is true in the "first" subinterval.

Now suppose the result holds in the interval $[(k-1)/n, k/n)$, where k is a positive integer, and let x be any real number in this interval. Then

$$[\![x]\!] + \left[\!\left[x + \frac{1}{n} \right]\!\right] + \left[\!\left[x + \frac{2}{n} \right]\!\right] + \cdots + \left[\!\left[x + \frac{n-1}{n} \right]\!\right] = [\![nx]\!].$$

By adding $1/n$ to x (thereby getting an arbitrary number in $[k/n, (k+1)/n)$), each of the terms, except the final term, on the left side of the previous equation is "shifted" one term to the right, and the final term, $[\![x + (n-1)/n]\!]$, becomes $[\![x + 1]\!]$, which exceeds $[\![x]\!]$ by 1. Thus, replacing x by $x + 1/n$ increases the left side of the previous equation by 1.

At the same time, when x in $[\![nx]\!]$ is replaced by $x + 1/n$, the value is increased by 1. Since each side of the equation increases by 1 when x is replaced by $x + 1/n$, the result continues to hold for all numbers in the interval $[k/n, (k+1)/n)$.

By induction, the result is true for all positive values of x. A similar argument shows it is true for all negative values of x (replace x by $x - 1/n$).

The next example is a good illustration of "building" $P(k+1)$ from $P(k)$.

2.1.5. If $a > 0$ and $b > 0$, then $(n-1)a^n + b^n \geqslant na^{n-1}b$, n a positive integer, with equality if and only if $a = b$.

Solution. The result is true for $n = 1$; assume the result true for the integer k. To build $P(k+1)$, we must, to get the proper left side,

(i) multiply by a:

$$(k-1)a^{k+1} + b^k a \geqslant ka^k b,$$

(ii) add a^{k+1}:

$$ka^{k+1} + b^k a \geqslant ka^k b + a^{k+1},$$

(iii) subtract $b^k a$:

$$ka^{k+1} \geqslant ka^k b + a^{k+1} - b^k a,$$

(iv) add b^{k+1}:

$$ka^{k+1} + b^{k+1} \geqslant ka^k b + a^{k+1} - b^k a + b^{k+1}.$$

We are assuming that this inequality is an equality if and only if $a = b$. It only remains to show that $ka^kb + a^{k+1} - b^ka + b^{k+1} \geqslant (k+1)a^kb$ with equality if and only if $a = b$. To do this we work backwards:

$$ka^kb + a^{k+1} - b^ka + b^{k+1} \geqslant (k+1)a^kb,$$

$$-a^kb + a^{k+1} - b^ka + b^{k+1} \geqslant 0,$$

$$a^k(a-b) + b^k(b-a) \geqslant 0,$$

$$(a^k - b^k)(a-b) \geqslant 0,$$

and the final step is true ($a - b$ and $a^k - b^k$ have the same signs) with equality if and only if $a = b$. Thus, the proof follows by induction. (Note: This result is a special case of the arithmetic mean-geometric mean inequality; see Section 7.2.)

Problems

2.1.6.

(a) Use induction to prove that $1 + 1/\sqrt{2} + 1/\sqrt{3} + \cdots + 1/\sqrt{n} < 2\sqrt{n}$.

(b) Use induction to prove that $2! \, 4! \cdots (2n)! \geqslant ((n+1)!)^n$.

2.1.7. The Euclidean plane is divided into regions by drawing a finite number of straight lines. Show that it is possible to color each of these regions either red or blue in such a way that no two adjacent regions have the same color.

2.1.8. Prove that the equation $x^2 + y^2 = z^n$ has a solution in positive integers (x, y, z) for all $n = 1, 2, 3, \ldots$. (For a nice proof, divide into two cases: even n and odd n. For a noninductive proof, see 3.5.1.)

2.1.9. A group of n people play a round-robin tournament. Each game ends in either a win or a loss. Show that it is possible to label the players $P_1, P_2, P_3, \ldots, P_n$ in such a way that P_1 defeated P_2, P_2 defeated $P_3, \ldots, P_{n-1}$ defeated P_n.

2.1.10. Show that if a round robin tournament has an odd number of teams, it is possible for every team to win exactly half its games. (Hint: Assuming the result for $2n - 1$ teams, show how the same thing could happen when 2 teams are added.)

2.1.11. The following steps lead to another proof of the binomial theorem. We know that $(a + x)^n$ can be written as a polynomial of degree n, so there are constants $A_0, A_1, \ldots, A_n$ such that

$$(a + x)^n = A_0 + A_1x + A_2x^2 + \cdots + A_nx^n.$$

(a) Use induction to describe the equation which results upon taking the kth derivative of each side of this equation ($k = 1, 2, \ldots, n$).

(b) Evaluate A_k for $k = 0, 1, \ldots, n$ by setting $x = 0$ in the kth equation found in part (a).

2.1.12. Suppose that $f: R \to R$ is a function for which $f(2x - f(x)) = x$ for all x, and let r be a fixed real number.

(a) Prove that if $f(x) = x + r$, then $f(x - nr) = (x - nr) + r$ for all positive integers n.
(b) Prove that if f is a one-to-one function (i.e., $f(x) = f(y)$ implies $x = y$) then the property in (a) also holds for *all* integers n.

Additional Examples:

1.1.2, 1.1.8, 3.2.8, 6.5.13, 7.1.4.

2.2. Induction: Set Up $P(k + 1)$

In this section we consider induction arguments which begin with a direct assault on $P(k + 1)$ and which work backwards to exploit the assumed truth of $P(k)$. Theoretically, the arguments in this section could all be recast into the form of the previous section, and vice versa. However, from a practical standpoint, it is often much more convenient to think the one way rather than the other.

2.2.1. Prove that $n^5/5 + n^4/2 + n^3/3 - n/30$ is an integer for $n = 0, 1, 2, \ldots$.

Solution. The result is obviously true when $n = 0$. Assume the result holds for $n = k$. We need to prove that

$$\frac{(k + 1)^5}{5} + \frac{(k + 1)^4}{2} + \frac{(k + 1)^3}{3} - \frac{(k + 1)}{30}$$

is an integer. We expand,

$$\frac{k^5 + 5k^4 + 10k^3 + 10k^2 + 5k + 1}{5} + \frac{k^4 + 4k^3 + 6k^2 + 4k + 1}{2}$$

$$+ \frac{k^3 + 3k^2 + 3k + 1}{3} - \frac{k + 1}{30}$$

and recombine (to make use of $P(k)$):

$$\left[\frac{k^5}{5} + \frac{k^4}{2} + \frac{k^3}{3} - \frac{k}{30} \right]$$

$$+ \left[(k^4 + 2k^3 + 2k^2 + k) + (2k^3 + 3k^2 + 2k) + (k^2 + k) \right].$$

The first grouping is an integer by the inductive assumption, and the second grouping is an integer because it is a sum of integers. Thus, the proof follows by induction. (Notice how difficult it would have been to arrive at $P(k + 1)$ by starting from $P(k)$.)

2.2.2. Let $a, b, p_1, p_2, \ldots, p_n$ be real numbers with $a \neq b$. Define $f(x) = (p_1 - x)(p_2 - x)(p_3 - x) \cdots (p_n - x)$. Show that

$$
\det
\begin{pmatrix}
p_1 & a & a & a & \cdots & a & a \\
b & p_2 & a & a & \cdots & a & a \\
b & b & p_3 & a & \cdots & a & a \\
b & b & b & p_4 & \cdots & a & a \\
\vdots & \vdots & \vdots & \vdots & & \vdots & \vdots \\
b & b & b & b & \cdots & p_{n-1} & a \\
b & b & b & b & \cdots & b & p_n
\end{pmatrix}
= \frac{bf(a) - af(b)}{b - a}.
$$

Solution. This is similar to many determinant problems that can be worked by mathematical induction. When $n = 1$, we have $f(x) = p_1 - x$, and $\det(p_1) = p_1$, and

$$
\frac{bf(a) - af(b)}{b - a} = \frac{b(p_1 - a) - a(p_1 - b)}{b - a} = p_1,
$$

so the result holds.

Assume the result holds for $k - 1$, $k > 1$, and consider the case for k real numbers $p_1, \ldots, p_k$. (We begin by setting up the situation for $P(k)$ and plan to fall back on the truth of $P(k - 1)$ to complete the inductive step.) We wish to evaluate

$$
\det
\begin{pmatrix}
p_1 & a & a & a & \cdots & a & a \\
b & p_2 & a & a & \cdots & a & a \\
b & b & p_3 & a & \cdots & a & a \\
\vdots & \vdots & \vdots & \vdots & & \vdots & \vdots \\
b & b & b & b & \cdots & p_{k-1} & a \\
b & b & b & b & \cdots & b & p_k
\end{pmatrix}.
$$

Subtract the second column from the first (this does not change the determinant):

$$
\det
\begin{pmatrix}
p_1 - a & a & a & a & \cdots & a & a \\
b - p_2 & p_2 & a & a & \cdots & a & a \\
0 & b & p_3 & a & \cdots & a & a \\
0 & b & b & p_4 & \cdots & a & a \\
\vdots & \vdots & \vdots & \vdots & & \vdots & \vdots \\
0 & b & b & b & \cdots & p_{k-1} & a \\
0 & b & b & b & \cdots & b & a
\end{pmatrix},
$$

and expand down the first column to get

$$(p_1 - a)\det\begin{pmatrix} p_2 & a & \cdots & a & a \\ b & p_3 & \cdots & a & a \\ \vdots & \vdots & & \vdots & \vdots \\ b & b & \cdots & p_{k-1} & a \\ b & b & \cdots & b & p_k \end{pmatrix}$$

$$-(b - p_2)\det\begin{pmatrix} a & a & \cdots & a & a \\ b & p_3 & \cdots & a & a \\ \vdots & \vdots & & \vdots & \vdots \\ b & b & \cdots & p_{k-1} & a \\ b & b & \cdots & b & p_k \end{pmatrix}.$$

The latter two determinants (on $(k-1)$-by-$(k-1)$ matrices) are of the form for which we can apply the inductive assumption $P(k-1)$. To do this, we will need to introduce some notation. For the first determinant, set $F(x) = (p_2 - x)(p_3 - x) \cdots (p_k - x)$ and for the second, set $G(x) = (a - x)(p_3 - x) \cdots (p_k - x)$. Then, by the inductive assumption, the last expression equals

$$(p_1 - a)\left[\frac{bF(a) - aF(b)}{b - a}\right] - (b - p_2)\left[\frac{bG(a) - aG(b)}{b - a}\right].$$

But $G(a) = 0$ and $(p_1 - a)F(a) = f(a)$, and therefore we have

$$\frac{bf(a) - a(p_1 - a)(p_2 - b) \cdots (p_k - b) - a(a - b)(p_2 - b) \cdots (p_k - b)}{b - a},$$

$$\frac{bf(a) - a(p_2 - b) \cdots (p_k - b)[(p_1 - a) + (a - b)]}{b - a},$$

$$\frac{bf(a) - af(b)}{b - a}.$$

The result follows by induction.

Problems

2.2.3. Give a proof for the inductive step in 1.1.3.

2.2.4. For all x in the interval $0 \leqslant x \leqslant \pi$, prove that

$$|\sin nx| \leqslant n \sin x, \qquad n \text{ a nonnegative integer.}$$

2.2.5. Let Q denote the set of rational numbers. Find all functions f from Q to Q which satisfy the following two conditions: (i) $f(1) = 2$, and (ii) $f(xy) = f(x)f(y) - f(x + y) + 1$ for all x, y in Q.

2.2.6. If $a, b, c \geqslant 1$, prove that $4(abc + 1) \geqslant (1 + a)(1 + b)(1 + c)$. (Hint: Prove, more generally, that $2^{n-1}(a_1 a_2 \cdots a_n + 1) \geqslant (1 + a_1)(1 + a_2) \cdots (1 + a_n)$.)

2.2.7. Given a set of 51 integers between 1 and 100 (inclusive), show that at least one member of the set must divide another member of the set. (Hint: Prove, more generally, that the same property will hold whenever $n + 1$ integers are chosen from the integers between 1 and $2n$ (inclusive).) For a noninductive proof, see 2.6.1.

2.2.8. Criticize the proof given below for the following theorem:

> *An n-by-n matrix of nonnegative integers has the property that for any zero entry, the sum of the row plus the sum of the column containing that zero is at least n. Show that the sum of all elements of the array is at least $n^2/2$.*

Proof (?): The result holds for $n = 1$. Assume the result holds for $n = k - 1$, and consider a k-by-k matrix. If there are no zero entries, the result obviously holds. If $a_{ij} = 0$, the sum of row i and column j is at least k, by assumption, and the sum of the elements in the $(k - 1)$-by-$(k - 1)$ submatrix obtained by deleting row i and column j is at least $(k - 1)^2/2$ (by the inductive assumption). It follows that the sum of the elements in the k-by-k matrix is at least $(k - 1)^2/2 + k = (k^2 - 2k + 1)/2 + k = (k^2 + 1)/2 > k^2/2$. The result follows by induction.

Additional Examples

1.1.11, 1.12.2, 3.1.11, 4.2.21, 4.3.5$^-$, 4.3.24, 6.5.12, 6.6.1, 7.1.6, 7.1.13, 7.2.5, 7.3.5.

2.3. Strong Induction

Let a be an integer and $P(n)$ a proposition about n for each integer $n \geqslant a$. The strong form of mathematical induction states that:

If

    (i) *$P(a)$ is true, and*
    (ii) *for each integer $k \geqslant a$, $P(a), P(a + 1), \ldots, P(k)$ true implies $P(k + 1)$ true,*

then $P(n)$ is true for all integers $n \geqslant a$.

This differs from the previous induction principle in that we are allowed a stronger assumption in step (ii), namely, we may assume $P(a), P(a + 1), \ldots, P(k)$, instead of only $P(k)$, to prove $P(k + 1)$. Theoretically, the two forms of induction are equivalent, but in practice there are problems which are more easily worked with this stronger form.

2.3.1 (Pick's theorem). Prove that the area of a simple lattice polygon (a polygon with lattice points as vertices whose sides do not cross) is given by $I + \frac{1}{2}B - 1$, where I and B denote respectively the number of interior and boundary lattice points of the polygon.

Solution. We will induct on the number of sides in the polygon. The case of a triangle is given in 1.7.3. Consider, then, a simple lattice polygon P with k sides, $k > 3$. We first establish that such a polygon has an interior diagonal. This is clear if the polygon in convex (equivalently, if all the interior angles are less than 180°). So suppose the interior angle at some vertex, say V, is more than 180°. Then a ray emanating from V and sweeping the interior of the polygon must strike another vertex (otherwise the polygon encloses an infinite area), and this determines an interior diagonal D with V as one endpoint.

Suppose that our polygon P has I interior points and B boundary points. The interior diagonal D divides P into two simple lattice polygons P_1 and P_2 with I_1 and I_2 interior points respectively, and B_1 and B_2 boundary points respectively. Suppose there are x lattice points on D, excluding its endpoints. Then $B = B_1 + B_2 - 2 - 2x$, and $I = I_1 + I_2 + x$.

Now, let A, A_1, and A_2 denote the areas of P, P_1, and P_2 respectively. Then

$$
\begin{aligned}
A &= A_1 + A_2 \\
&= (I_1 + \tfrac{1}{2}B_1 - 1) + (I_2 + \tfrac{1}{2}B_2 - 1) \\
&= (I_1 + I_2) + \tfrac{1}{2}(B_1 + B_2) - 2 \\
&= (I_1 + I_2 + x) + \tfrac{1}{2}(B_1 + B_2 - 2x) - 2 \\
&= I + \tfrac{1}{2}(B + 2) - 2 \\
&= I + \tfrac{1}{2}B - 1.
\end{aligned}
$$

The result follows by induction.

Notice in this example that it is the first step of the induction argument which is the most difficult (done in 1.7.3); the inductive step (step (ii)) is conceptionally very simple.

Problems

2.3.2.

(a) Prove that every positive integer greater than one may be written as a
    product of prime numbers.
(b) Bertrand's postulate, once a postulate but now a known theorem, states

that for every number $x > 1$, there is a prime number between x and $2x$. Use this fact to show that every positive integer can be written as a sum of *distinct* primes. (For this result, assume that one is a prime.)

2.3.3.

(a) Show that every positive integer can be written as a sum of distinct Fibonacci numbers.
(b) Let $k \gg m$ mean that $k \geqslant m + 2$. Show that every positive integer n has a representation of the form $n = F_{k_1} + F_{k_2} + \cdots + F_{k_r}$, where F_{k_i} are Fibonacci numbers and $k_1 \gg k_2 \gg \cdots \gg k_r \gg 0$.
(c) Show that the representation in part (b) is unique.

Additional Examples

3.1.1, 3.1.2, 3.1.18, 3.5.5, 6.2.3.

2.4. Induction and Generalization

We have seen (in Section 1.12) that a problem is sometimes easier to handle when it is recast into a more general form. This is true also in induction problems. For example, it may happen that the original propositions $P(1), P(2), P(3), \ldots$ do not contain enough information to enable one to carry out the inductive step (step (ii)). In this case it is natural to reformulate the propositions into a stronger, more general form $Q(1), Q(2), \ldots$ (where $Q(n)$ implies $P(n)$ for each n), and to look again for an inductive proof.

2.4.1. If $A_1 + \cdots + A_n = \pi$, $0 < A_i \leqslant \pi$, $i = 1, \ldots, n$, then

$$\sin A_1 + \cdots + \sin A_n \leqslant n \sin \frac{\pi}{n}.$$

Solution. Let $P(k)$ be the statement of the theorem for a given k, and suppose $P(k)$ is true. For the inductive step, suppose $A_1 + \cdots + A_k + A_{k+1} = \pi$, $0 < A_i \leqslant \pi$, $i = 1, \ldots, k + 1$. In this form, it is not clear how to make use of $P(k)$. We might, for example, group A_k and A_{k+1} together, so that $A_1 + \cdots + A_{k-1} + (A_k + A_{k+1}) = \pi$, and then apply the inductive assumption to get

$$\sin A_1 + \cdots + \sin A_{k-1} + \sin(A_k + A_{k+1}) \leqslant k \sin \frac{\pi}{k}.$$

But now it is not at all clear that this implies $P(k + 1)$:

$$\sin A_1 + \cdots + \sin A_k + \sin A_{k+1} \leqslant (k + 1)\sin \frac{\pi}{k + 1}.$$

The requirement that the A_i's add to π seems too restrictive. Consider instead the following proposition $Q(n)$:

If $0 < A_i \leqslant \pi$, $i = 1, \ldots, n$, then

$$\sin A_1 + \cdots + \sin A_n \leqslant n \sin\left(\frac{A_1 + \cdots A_n}{n}\right).$$

(Note that $Q(n)$ implies $P(n)$.) Obviously, $Q(1)$ is true. Suppose that $Q(k)$ is true, and suppose that $0 < A_i \leqslant \pi$, $i = 1, \ldots, k + 1$. Then

$$\sin A_1 + \cdots + \sin A_k + \sin A_{k+1}$$

$$\leqslant k \sin\left(\frac{A_1 + \cdots + A_k}{k}\right) + \sin A_{k+1}$$

$$= (k + 1)\left[\frac{k}{k+1} \sin\left(\frac{A_1 + \cdots + A_k}{k}\right) + \frac{1}{k+1} \sin A_{k+1}\right]$$

$$\leqslant (k + 1)\left[\sin\left(\frac{k}{k+1}\left(\frac{A_1 + \cdots + A_k}{k}\right) + \frac{1}{k+1} A_{k+1}\right)\right]$$

$$= (k + 1)\sin\left(\frac{A_1 + \cdots + A_{k+1}}{k+1}\right).$$

(The inequality in the next to last step follows from the result in 1.2.12(b).) The result now follows by induction.

We are now able to prove the conjecture made in 1.6.2(e): The polygon of greatest area that can be inscribed in a circle is the regular polygon. To do this, suppose that $P_1, P_2, \ldots, P_n$, $n \geqslant 3$, are the successive vertices of an inscribed polygon (inscribed in a circle of radius r). Let O denote the center of the circle; let T_i denote the area of triangle $P_i O P_{i+1}$, $i = 1, \ldots, n$ (we set $P_{n+1} = P_1$); let $A_i = \angle P_i O P_{i+1}$ (Figure 2.2). Then

$$T_i = 2\left[\tfrac{1}{2}(r\cos\tfrac{1}{2}A_i)(r\sin\tfrac{1}{2}A_i)\right]$$

$$= r^2\cos\tfrac{1}{2}A_i\sin\tfrac{1}{2}A_i$$

$$= \tfrac{1}{2}r^2\sin A_i.$$

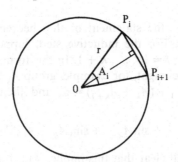

Figure 2.2.

The polygon of maximal area must satisfy $0 < A_i < \pi$ for each i. Thus, our preceding result shows that

$$\text{Area of polygon} = \sum_{i=1}^{n} T_i$$

$$= \sum_{i=1}^{n} \tfrac{1}{2} r^2 \sin A_i = \tfrac{1}{2} r^2 \sum_{i=1}^{n} \sin A_i$$

$$\leqslant \frac{n}{2} r^2 \sin\left[\frac{1}{n} \sum_{i=1}^{n} A_i \right]$$

$$= n\left[\tfrac{1}{2} r^2 \sin\left(\frac{2\pi}{n} \right) \right].$$

The right-hand side is the area of a regular n-gon, and this completes the proof.

2.4.2. Let $f(x) = (x^2 - 1)^{1/2}$, $x > 1$. Prove that $f^{(n)}(x) > 0$ for odd n and $f^{(n)}(x) < 0$ for even n.

Solution. We might expect to be able to express $f^{(k+1)}(x)$ in terms of $f^{(k)}(x)$. But a look at the first few derivatives makes this plan appear hopeless:

$$f'(x) = \frac{x}{(x^2 - 1)^{1/2}}, \qquad f''(x) = -\frac{1}{(x^2 - 1)^{3/2}},$$

$$f'''(x) = \frac{3x}{(x^2 - 1)^{5/2}}, \qquad f^{(iv)}(x) = -\frac{12x^2 + 3}{(x^2 - 1)^{7/2}},$$

$$f^{(v)}(x) = \frac{60x^3 + 45x}{(x^2 - 1)^{9/2}}, \qquad f^{(vi)}(x) = -\frac{360x^4 + 540x^2 + 45}{(x^2 - 1)^{11/2}}.$$

Consider instead the following reformulation: If $f(x) = (x^2 - 1)^{1/2}$, $x > 1$, then

$$f^{(n)}(x) = \frac{g_n(x)}{(x^2 - 1)^{(2n-1)/2}},$$

where $g_n(x)$ is a polynomial of degree $n - 2$, and

$$g_n(x) \text{ is } \begin{cases} \text{an odd function all of whose} \\ \quad \text{coefficients are nonnegative} & \text{if } n \text{ is odd,} \\ \text{an even function all of whose} \\ \quad \text{coefficients are nonpositive} & \text{if } n \text{ is even.} \end{cases}$$

This proposition can be established by induction (we omit the messy details), and this implies the original result.

2.4.3. Let F_i denote the ith term in the Fibonacci sequence. Prove that $F_{n+1}^2 + F_n^2 = F_{2n+1}$.

Solution. The result holds for $n = 1$, so suppose the result holds for the integer k. Then

$$F_{k+2}^2 + F_{k+1}^2 = (F_{k+1} + F_k)^2 + F_{k+1}^2$$

$$= F_{k+1}^2 + 2F_{k+1}F_k + F_k^2 + F_{k+1}^2$$

$$= (F_{k+1}^2 + F_k^2) + (2F_{k+1}F_k + F_{k+1}^2)$$

$$= F_{2k+1} + (2F_{k+1}F_k + F_{k+1}^2),$$

the last step by the inductive assumption.

We would be done if we could show $2F_{k+1}F_k + F_{k+1}^2 = F_{2k+2}$, for we could then continue the previous argument, $F_{2k+1} + (2F_{k+1}F_k + F_{k+1}^2)$ $= F_{2k+1} + F_{2k+2} = F_{2k+3}$, and this completes the inductive step. Therefore, it remains to prove that $2F_{k+1}F_k + F_{k+1}^2 = F_{2k+2}$. We proceed by induction. It is true for $n = 1$, and assuming it true for k, we have

$$2F_{k+2}F_{k+1} + F_{k+2}^2 = 2(F_{k+1} + F_k)F_{k+1} + F_{k+2}^2$$

$$= 2F_{k+1}^2 + 2F_{k+1}F_k + F_{k+2}^2$$

$$= (2F_{k+1}F_k + F_{k+1}^2) + (F_{k+1}^2 + F_{k+2}^2)$$

$$= F_{2k+2} + (F_{k+1}^2 + F_{k+2}^2).$$

But now we are back to the earlier problem: does $F_{k+2}^2 + F_{k+1}^2 = F_{2k+3}$? If so, $F_{2k+2} + (F_{k+1}^2 + F_{k+2}^2) = F_{2k+2} + F_{2k+3} = F_{2k+4}$ and the induction is complete. Thus, the problems are interrelated: the truth of the first depends upon the truth of the second, and conversely, the truth of the second depends upon the truth of the first.

We can resolve the difficulty by proving them both in the following manner. Consider the two propositions

$$P(n): \quad F_{n+1}^2 + F_n^2 = F_{2n+1},$$

$$Q(n): \quad 2F_{n+1}F_n + F_{n+1}^2 = F_{2n+2}.$$

$P(1)$ and $Q(1)$ are each true. The previous arguments show that $P(k)$ and $Q(k)$ imply $P(k+1)$, and that $P(k+1)$ and $Q(k)$ imply $Q(k+1)$. It follows that $P(k)$ and $Q(k)$ imply $P(k+1)$ and $Q(k+1)$, and the proof is complete.

2.4.4. Let $f(x) = a_1\sin x + a_2\sin 2x + \cdots + a_n\sin nx$, where $a_1, \ldots, a_n$ are real numbers and where n is a positive integer. Given that $|f(x)| \leqslant |\sin x|$ for all real x, prove that $|a_1 + 2a_2 + \cdots + na_n| \leqslant 1$.

Solution. Suppose we try inducting on the number of terms in $f(x)$. When $n = 1$, $f(x) = a_1 \sin x$, and since $|f(x)| \leqslant |\sin x|$, it follows that $|a_1| = |a_1 \sin(\pi/2)| = |f(\pi/2)| \leqslant |\sin(\pi/2)| = 1$.

Suppose the result holds for k, and consider the function

$$f(x) = a_1 \sin x + a_2 \sin 2x + \cdots + a_k \sin kx + a_{k+1} \sin(k+1)x,$$

for some choice of real numbers $a_1, a_2, \ldots, a_{k+1}$, and suppose that $|f(x)| \leqslant |\sin x|$ for all real x. Since $\sin(k+1)x = \sin kx \cos x + \sin x \cos kx$, we can write

$$f(x) = (a_1 + a_{k+1} \cos kx) \sin x + a_2 \sin 2x + \cdots$$
$$+ a_{k-1} \sin(k-1)x + (a_k + a_{k+1} \cos x) \sin kx.$$

We have now rewritten $f(x)$ as a sum of k terms, more or less of the type from which we can apply the induction assumption. The difficulty is that the coefficients of the sine terms in this expression are not constants; rather they contain functions of x. This suggests considering the following more general problem.

Let $a_1(x), \ldots, a_n(x)$ be differentiable functions of x, and let $f(x) = a_1(x) \sin x + a_2(x) \sin 2x + \cdots + a_n(x) \sin nx$. Given that $|f(x)| \leqslant |\sin x|$ for all real x, prove that

$$|a_1(0) + 2a_2(0) + \cdots + na_n(0)| \leqslant 1.$$

If we can prove this proposition, we will have solved the original problem also, because, taking $a_i(x) \equiv a_i$, a_i a constant, $i = 1, 2, \ldots, n$, for all x, we recover the original problem.

Again we proceed by induction. We are given $|a_1(x) \sin x| \leqslant |\sin x|$. As x approaches 0, $\sin x \neq 0$, so that for these x, $|a_1(x)| \leqslant 1$. Since $a_1(x)$ is continuous at $x = 0$, it follows that $|a_1(0)| \leqslant 1$. This implies that the result is true for the case $n = 1$.

Now suppose the result is true for $n = k$, and consider the function

$$f(x) = a_1(x) \sin x + a_2(x) \sin 2x + \cdots + a_{k+1}(x) \sin(k+1)x,$$

where $|f(x)| \leqslant |\sin x|$ and $a_i(x)$ are differentiable. As before, this can be rewritten in the equivalent form

$$f(x) = \left[a_1(x) - a_{k+1}(x) \cos kx \right] \sin x + a_2(x) \sin 2x + \cdots$$
$$+ a_{k-1}(x) \sin(k-1)x + \left[a_k(x) + a_{k+1}(x) \cos x \right] \sin kx.$$

We may now apply the inductive assumption, and conclude that

$$\left| \left[a_1(0) + a_{k+1}(0) \right] + 2a_2(0) + \cdots \right.$$
$$\left. + (k-1)a_{k-1}(0) + k \left[a_k(0) + a_{k+1}(0) \right] \right| \leqslant 1.$$

But this is the same as

$$|a_1(0) + 2a_2(0) + \cdots + ka_k(0) + (k+1)a_{k+1}(0)| \leqslant 1.$$

which is the desired form. (A noninductive proof of this result is given in 6.3.2.)

Problems

2.4.5. Let S denote an n-by-n lattice square, $n \geqslant 3$. Show that it is possible to draw a polygonal path consisting of $2n - 2$ segments which will pass through all of the n^2 lattice points of S.

2.4.6. Let $f_0(x) = 1/(1 - x)$, and define $f_{n+1}(x) = xf_n'(x)$. Prove that $f_{n+1}(x) > 0$ for $0 < x < 1$.

2.5. Recursion

In the second solution to 1.1.1, we let A_n denote the number of subsets of a set with n elements. We showed that $A_{n+1} = 2A_n$, $A_0 = 1$. This is an example of a recurrence relation. Even though we do not have an explicit formula for A_n (as the method of induction requires), the recurrence relation defines a "loop" or algorithm which shows us how to compute A_{n+1}. In this section we look at problems that can be reduced to equivalent problems with smaller parameters. The idea is to apply the reduction argument *recursively* until the parameters reach values for which the problem can be solved.

2.5.1 (Tower-of-Hanoi problem). Suppose n rings, with different outside diameters, are slipped onto an upright peg, the largest on the bottom, to form a pyramid (Figure 2.3). Two other upright pegs are placed sufficiently far apart. We wish to transfer all the rings, one at a time, to the second peg to form an identical pyramid. During the transfers, we are not permitted to place a larger ring on a smaller one (this necessitates using the third peg). What is the smallest number of moves necessary to complete the transfer?

Solution. Let M_n denote the minimal number of moves for a stack of n rings. Clearly $M_1 = 1$, so suppose $n > 1$. In order to get the largest ring on the bottom of the second peg, it is necessary to move the topmost $n - 1$ rings to the third peg. This will take a minimum of M_{n-1} moves (by our

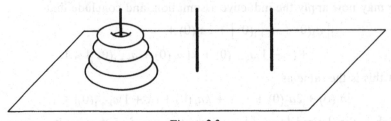

Figure 2.3.

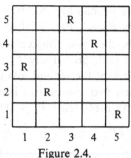

Figure 2.4.

choice of notation). One move is necessary to transfer the largest ring to the second peg, and then M_{n-1} moves are necessary to transfer the $n-1$ rings to the second peg. Thus

$$M_n = 2M_{n-1} + 1, \qquad M_1 = 1.$$

An easy induction, based on this recurrence, shows that $M_n = 2^n - 1$ ($M_{n+1} = 2M_n + 1 = 2[2^n - 1] + 1 = 2^{n+1} - 1$).

Let $a_1, a_2, \ldots, a_n$ be a permutation of $1, 2, \ldots, n$. We can interpret this permutation geometrically in the following way. Take an n by n chessboard, and for each i, place a rook in the ith column (from the left) and the a_ith row (from the bottom). For example, the permutation $3, 2, 5, 4, 1$ is represented in Figure 2.4. In this way we see that a permutation of $1, 2, \ldots, n$ corresponds to a placement of n "nonattacking" rooks on the n by n chessboard. This correspondence enables one to think of permutations geometrically and to use the language and imagery of nonattacking rooks on a chessboard.

2.5.2. Let Q_n denote the number of ways of placing n nonattacking rooks on the n-by-n chessboard so that the arrangement is symmetric about the diagonal from the lower left corner to the upper right corner. Show that

$$Q_n = Q_{n-1} + (n-1) Q_{n-2}.$$

Solution. A rook in the first column may or may not occupy the square in the lower left corner of the board. If it does, there are Q_{n-1} ways of placing the remaining $n-1$ rooks. If it doesn't, it can occupy any $n-1$ squares in the first column. Once it is placed, it uniquely determines the location of a symmetrically placed rook (symmetric with respect to the given diagonal) in the first row. The remaining $n-2$ rooks can be placed in Q_{n-2} ways. Putting these ideas together gives the result.

2.5.3. A coin is tossed n times. What is the probability that two heads will turn up in succession somewhere in the sequence of throws?

Solution. Let P_n denote the probability that two consecutive heads do not appear in n throws. Clearly $P_1 = 1$, $P_2 = \frac{3}{4}$. If $n > 2$, there are two cases.

If the first throw is tails, then two consecutive heads will not appear in the remaining $n - 1$ tosses with probability P_{n-1} (by our choice of notation). If the first throw is heads, the second toss must be tails to avoid two consecutive heads, and then two consecutive heads will not appear in the remaining $n - 2$ throws with probability P_{n-2}. Thus,

$$P_n = \tfrac{1}{2} P_{n-1} + \tfrac{1}{4} P_{n-2}, \qquad n > 2.$$

This recurrence can be transformed to a more familiar form by multiplying each side by 2^n:

$$2^n P_n = 2^{n-1} P_{n-1} + 2^{n-2} P_{n-2},$$

and setting $S_n = 2^n P_n$ for each n:

$$S_n = S_{n-1} + S_{n-2}.$$

This is the recurrence for the Fibonacci sequence (note that $S_n = F_{n+2}$). Thus, the probability we seek is $Q_n = 1 - P_n = 1 - F_{n+2}/2^n$.

Problems

2.5.5. Let P_n denote the number of regions formed when n lines are drawn in the Euclidean plane in such a way that no three are concurrent and no two are parallel. Show that $P_{n+1} = P_n + (n + 1)$.

2.5.6.

(a) Let E_n denote the determinant of the n-by-n matrix having -1's below the main diagonal (from upper left to lower right) and 1's on and above the main diagonal. Show that $E_1 = 1$ and $E_n = 2E_{n-1}$ for $n > 1$.

(b) Let D_n denote the determinant of the n-by-n matrix whose (i, j)th element (the element in the ith row and jth column) is the absolute value of the difference of i and j. Show that $D_n = (-1)^{n-1}(n-1)2^{n-2}$.

(c) Let F_n denote the determinant of the n-by-n matrix with a on the main diagonal, b on the superdiagonal (the diagonal immediately above the main diagonal—having $n - 1$ entries), and c on the subdiagonal (the diagonal immediately below the main diagonal—having $n - 1$ entries). Show that $F_n = aF_{n-1} - bcF_{n-2}$, $n > 2$. What happens when $a = b = 1$ and $c = -1$?

(d) Evaluate the n-by-n determinant A_n whose (i, j)th entry is $a^{|i-j|}$ by finding a recursive relationship between A_n and A_{n-1}.

2.5.7.

(a) Let $a_1, a_2, \ldots, a_n$ be positive real numbers and $A_n = (a_1 + \cdots + a_n)/n$. Show that $A_n \geqslant A_{n-1}^{(n-1)/n} a_n^{1/n}$ with equality if and only if $A_{n-1} = a_n$. (Hint: Apply the inequality of 2.1.5.)

(b) *Arithmetic-mean–geometric-mean inequality.* Using part (a), show that

$$\frac{a_1 + \cdots + a_n}{n} \geqslant (a_1 \cdots a_n)^{1/n}$$

with equality if and only if $a_1 = a_2 = \cdots = a_n$.

2.5.8. Two ping pong players, A and B, agree to play several games. The players are evenly matched; suppose, however, that whoever serves first has probability P of winning that game (this may be player A in one game, or player B in another). Suppose A serves first in the first game, but thereafter the loser serves first. Let P_n denote the probability that A wins the nth game. Show that $P_{n+1} = P_n(1 - P) + (1 - P_n)P$.

2.5.9. A gambling student tosses a fair coin and scores one point for each head that turns up and two points for each tail. Prove that the probability of the student scoring exactly n points at some time in a sequence of n tosses is $\frac{1}{3}[2 + (-\frac{1}{2})^n]$. (Hint: Let P_n denote the probability of scoring exactly n points at some time. Express P_n in terms of P_{n-1}, or in terms of P_{n-1} and P_{n-2}. Use this recurrence relation to give an inductive proof.)

2.5.10 (Josephus problem). Arrange the numbers $1, 2, \ldots, n$ consecutively (say, clockwise) about the circumference of a circle. Now, remove number 2 and proceed clockwise by removing every other number, among those that remain, until only one number is left. (Thus, for $n = 5$, numbers are removed in the order 2, 4, 1, 5, and 3 remains alone.) Let $f(n)$ denote the final number which remains. Show that

$$f(2n) = 2f(n) - 1,$$

$$f(2n + 1) = 2f(n) + 1.$$

(This problem is continued in 3.4.5.)

2.5.11.

(a) Let R_n denote the number of ways of placing n nonattacking rooks on the n-by-n chessboard so that the arrangement is symmetric about a 90° clockwise rotation of the board about the center. Show that

$$R_{4n} = (4n - 2)R_{4n-4},$$
$$R_{4n+1} = R_{4n},$$
$$R_{4n+2} = 0 = R_{4n+3}.$$

(b) Let S_n denote the number of ways of placing n nonattacking rooks on the n-by-n chessboard so that the arrangement is symmetric about the center of the board. Show that

$$S_{2n} = 2nS_{2n-2},$$
$$S_{2n+1} = S_{2n}.$$

(c) Let T_n denote the number of ways of placing n nonattacking rooks on the n-by-n chessboard so that the arrangement is symmetric about both diagonals. Show that

$$T_n = 2,$$
$$T_{2n+1} = T_{2n},$$
$$T_{2n} = 2T_{2n-2} + (2n - 2)T_{2n-4}.$$

2.5.12. A regular $(n + 2)$-gon is inscribed in a circle. Let T_n denote the number of ways it is possible to join its vertices in pairs so that the resulting segments do not intersect one another. If we set $T_0 = 1$, show that

$$T_n = T_0T_{n-1} + T_1T_{n-2} + T_2T_{n-3} + \cdots + T_{n-1}T_0.$$

(For a continuation of this problem, see 5.4.10.)

2.5.13. Let $a_1, a_2, \ldots, a_n$ be a permutation of the set $S_n = \{1, 2, \ldots, n\}$. An element i in S_n is called a fixed point of this permutation if $a_i = i$.

(a) A *derangement* of S_n is a permutation of S_n having no fixed points. Let g_n be the number of derangements of S_n. Show that

$$g_1 = 0, \qquad g_2 = 1,$$

and

$$g_n = (n - 1)(g_{n-1} + g_{n-2}), \qquad \text{for} \quad n > 2.$$

(Hint: a derangement either interchanges the first element with another or it doesn't.)

(b) Let f_n be the number of permutations of S_n with exactly one fixed point. Show that $|f_n - g_n| = 1$.

2.5.14. Suppose n men check in their hats as they arrive for dinner. As they leave, the hats are given back in a random order. What is the probability that no man gets back his own hat? (Hint: Let p_n denote this probability. Then $p_n = g_n/n!$, where g_n is as in 2.5.13. Let $C_n = p_n - p_{n-1}$. Use the recurrence relation found in 2.5.13(a) to show that $C_2 = \frac{1}{2}$, $C_n = -C_{n-1}/n$. Use this to show that $p_n = 1/2! - 1/3! + \cdots + (-1)^n/n!$. Then for large n, $p_n \approx 1/e$.)

2.5.15.

(a) Let $I_n = \int_0^{\pi/2} \sin^n x \, dx$. Find a recurrence relation for I_n.
(b) Show that

$$I_{2n} = \frac{1 \times 3 \times 5 \times \cdots \times (2n-1)}{2 \times 4 \times 6 \times \cdots \times 2n} \cdot \frac{\pi}{2}.$$

(c) Show that

$$I_{2n+1} = \frac{2 \times 4 \times 6 \times \cdots \times (2n-2)}{1 \times 3 \times 5 \times \cdots \times (2n-1)}.$$

Additional Examples

1.1.1 (Solution 2), 4.3.9, 5.3.5, 5.3.14, 5.3.15, 5.4.8, 5.4.9, 5.4.24, 5.4.25, 5.4.26. Closely related to induction and recursion are arguments based on "repeated arguments". Examples of what is meant here are 4.4.4, 4.4.17, the proof of the intermediate-value theorem in 6.1, 6.1.5, 6.1.6, 6.3.6, 6.8.10, and the heuristic for the arithmetic-mean–geometric-mean inequality given in Section 7.2.

2.6. Pigeonhole Principle

When a sufficiently large collection of objects is divided into a sufficiently small number of classes, one of the classes will contain a certain minimum number of objects. This is made more precise in the following self-evident proposition:

Pigeonhole Principle. *If $kn + 1$ objects ($k \geqslant 1$) are distributed among n boxes, one of the boxes will contain at least $k + 1$ objects.*

This principle, even when $k = 1$, is a very powerful tool for proving existence theorems. It takes some experience, however, to recognize when and how to use it.

2.6.1. Given a set of $n + 1$ positive integers, none of which exceeds $2n$, show that at least one member of the set must divide another member of the set.

Solution. This is the same as 2.2.7, where it was done by induction on n. However, the problem is really an existence problem for a given n, and it can be carried out very nicely by the pigeonhole principle, as we shall see.

Let the chosen numbers be denoted by $x_1, x_2, \ldots, x_{n+1}$, and for each i, write $x_i = 2^{n_i} y_i$, where n_i is nonnegative integer and y_i is odd. Let $T = \{y_i : i = 1, 2, \ldots, n+1\}$. Then T is a collection of $n+1$ odd integers, each less than $2n$. Since there are only n odd numbers less than $2n$, the pigeonhole principle implies that two numbers in T are equal, say $y_i = y_j$, $i < j$. Then

$$x_i = 2^{n_i} y_i \quad \text{and} \quad x_j = 2^{n_j} y_i.$$

If $n_i \leqslant n_j$, then x_i divides x_j; if $n_i > n_j$, then x_j divides x_i. This completes the proof.

2.6.2. Consider any five points P_1, P_2, P_3, P_4, P_5 in the interior of a square S of side length 1. Denote by d_{ij} the distance between the points P_i and P_j. Prove that at least one of the distances d_{ij} is less than $\sqrt{2}/2$.

Solution. Divide S into four congruent squares as shown in Figure 2.5. By the pigeonhole principle, two points belong to one of these squares (a point on the boundary of two smaller squares can be claimed by both squares). The distance between these points is less than $\sqrt{2}/2$.

2.6.3. Suppose that each square of a 4-by-7 chessboard, as shown below, is colored either black or white. Prove that in any such coloring, the board must contain a rectangle (formed by the horizontal and vertical lines of the board), such as the one outlined in the Figure 2.6, whose distinct corner squares are all the same color.

Solution. Such a rectangle exists even on a 3-by-7 board. The color configurations of the columns each must be of one of the types shown in Figure 2.7.

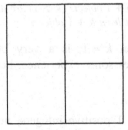

Figure 2.5.

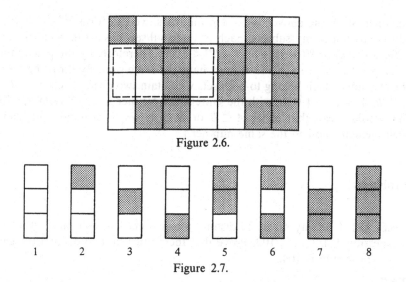

Figure 2.6.

1   2   3   4   5   6   7   8

Figure 2.7.

Suppose one of the columns is of type 1. We are done if any of the remaining six columns are of type 1, 2, 3, or 4. So suppose each of the other columns is of type 5, 6, 7, or 8. Then, by the pigeonhole principle, two of these six columns must have the same type and we are done.

The same argument applies if one of the columns is of type 8.

So suppose none of the columns are of type 1 or type 8. Then we have seven columns but only six types. By the pigeonhole principle, two columns have the same type and the proof is complete.

2.6.4. Prove that there exist integers a, b, c not all zero and each of absolute value less than one million, such that

$$|a + b\sqrt{2} + c\sqrt{3}| < 10^{-11}.$$

Solution. Let S be the set of 10^{18} real numbers $r + s\sqrt{2} + t\sqrt{3}$ with each of r, s, t in $\{0, 1, 2, \ldots, 10^6 - 1\}$, and let $d = (1 + \sqrt{2} + \sqrt{3})10^6$. Then each x in S is in the interval $0 \leqslant x < d$. Partition this interval into $10^{18} - 1$ equal subintervals, each of length $e = d/(10^{18} - 1)$. By the pigeonhole principle, two of the 10^{18} numbers of S must be in the same subinterval. Their difference, $a + b\sqrt{2} + c\sqrt{3}$, gives the desired a, b, c, since $e < 10^7/10^{18} = 10^{-11}$.

2.6.5. Given any set of ten natural numbers between 1 and 99 inclusive (decimal notation), prove that there are two disjoint nonempty subsets of the set with equal sums of their elements.

Solution. With the chosen set of ten numbers, we can form $2^{10} - 1 = 1023$ (different) nonempty subsets. Each of these subsets has a sum smaller than 1000, since even $90 + 91 + \cdots + 99 < 1000$. Therefore, by the pigeonhole principle, two subsets A and B must have the same sum. By throwing away the elements which belong to both sets we obtain two disjoint sets $X = A - A \cap B$, $Y = B - A \cap B$, with the same sum. (Neither X nor Y is empty, for this would mean that either $A \subset B$ or $B \subset A$, which is impossible, since their elements add to the same number.)

Problems

2.6.6. Let A be any set of 20 distinct integers chosen from the arithmetic progression $1, 4, 7, \ldots, 100$. Prove that there must be two distinct integers in A whose sum is 104.

2.6.7.

(a) Let S be a square region (in the plane) of side length 2 inches. Show that among any nine points in S, there are three which are the vertices of a triangle of area $\leqslant \frac{1}{2}$ square inch.

(b) Nineteen darts are thrown onto a dartboard which has the shape of a regular hexagon with side length one foot. Show that two darts are within $\sqrt{3}/3$ feet of each other.

2.6.8. Show that if there are n people at a party, then two of them know the same number of people (among those present).

2.6.9. Fifteen chairs are evenly placed around a circular table on which are name cards for fifteen guests. The guests fail to notice these cards until after they have sat down, and it turns out that no one is sitting in front of his own card. Prove that the table can be rotated so that at least two of the guests are simultaneously correctly seated.

2.6.10. Let X be any real number. Prove that among the numbers
$$X, 2X, \ldots, (n-1)X$$
there is one that differs from an integer by at most $1/n$.

2.6.11.

(a) Prove that in any group of six people there are either three mutual friends or three mutual strangers. (Hint: Represent the people by the vertices of a regular hexagon. Connect two vertices with a red line segment if the couple represented by these vertices are friends; otherwise connect them with a blue line segment. Consider one of the vertices, say A. At least three line segments emanating from A have the same color. There are two cases to consider.)

(b) Seventeen people correspond by mail with one another—each one with all the rest. In their letters only three topics are discussed. Each pair of correspondents deals with only one of the topics. Prove that there are at least three people who write to each other about the same topic.

2.6.12. Prove that no seven positive integers, not exceeding 24, can have sums of all subsets different.

Additional Examples

1.10.1, 3.2.1, 3.2.5, 3.2.19, 3.2.20, 3.3.24, 4.4.10.

Chapter 3. Arithmetic

In this chapter we consider problem-solving methods that are important in solving arithmetic problems. Perhaps the most basic technique is based on the fundamental theorem of arithmetic, which states that every integer can be written uniquely as a product of primes. The theoretical background necessary for the proof of this key theorem requires a discussion of the notion of divisibility. Therefore, we will begin the chapter by considering problems about greatest common divisors and least common multiples. Important to this understanding are the division algorithm and the Euclidean algorithm.

In the second section we introduce the technique of modular arithmetic (a generalization of the notion of parity), and see in it an efficient and effective method for many problems concerned with relationships between integers. In the last two sections we are again reminded of the importance of notation in solving problems, and we consider problems related to the representation of numbers: the positional notation for integers, and the rectangular, polar, and exponential notations for representing complex numbers.

3.1. Greatest Common Divisor

Given integers a and b, we say that a *divides* b, and we write $a \mid b$, if there is an integer q such that $b = qa$. On the basis of this definition it is easy to prove the following very useful result: If n divides two of the terms in the expression $a = b + c$, then n divides all three of the terms. (Note: In this chapter, unless otherwise stated, all variables are integer variables.)

If $a_1, \ldots, a_n$ are given integers, we will denote their greatest common divisor by $\gcd(a_1, \ldots, a_n)$, and their least common multiple by $\mathrm{lcm}(a_1, \ldots, a_n)$.

3.1.1. Find all functions f which satisfy the three conditions

(i) $f(x, x) = x$,
(ii) $f(x, y) = f(y, x)$,
(iii) $f(x, y) = f(x, x + y)$,

assuming that the variables and the values of f are positive integers.

Solution. A look at special cases leads us to suspect that $f(x, y) = \gcd(x, y)$. We will prove this by inducting on the sum $x + y$.

The smallest value for $x + y$ is 2, and this occurs when $x = y = 1$. By (i), $f(1, 1) = 1$, and also $\gcd(1, 1) = 1$, so our supposition is confirmed in this case.

Suppose that x and y are positive integers such that $x + y = k > 2$, and suppose the claim has been shown for all smaller sums. By (i) and (ii), there is no loss in generality in supposing that $x < y$. By (iii), $f(x, y) = f(x, x + (y - x)) = f(x, y - x)$. But by the inductive assumption, $f(x, y - x) = \gcd(x, y - x)$. The proof will be complete if we can show that $\gcd(x, y - x) = \gcd(x, y)$.

If $c \mid x$ and $c \mid y$, then $c \mid x$ and $c \mid y - x$. It follows that $\gcd(x, y) \leqslant \gcd(x, y - x)$. Similarly, if $c \mid x$ and $c \mid y - x$, then $c \mid x$ and $c \mid y$, and therefore $\gcd(x, y - x) \leqslant \gcd(x, y)$. Putting these together, we find that $\gcd(x, y - x) = \gcd(x, y)$, and the proof is complete.

The following result rests at the very foundation of number theory.

Division Algorithm. *If a and b are arbitrary integers, $b > 0$, there are unique integers q and r such that*

$$a = qb + r, \qquad 0 \leqslant r < b.$$

By repeated use of the division algorithm we can compute the greatest common divisor of two integers. To see how this goes, suppose that b_1 and b_2 are positive integers, with $b_1 > b_2$. By the division algorithm there are integers q and b_3 such that

$$b_1 = qb_2 + b_3, \qquad 0 \leqslant b_3 < b_2.$$

It is easy to check, using this equation, that $\gcd(b_1, b_2) = \gcd(b_2, b_3)$.

If $b_3 = 0$, then $\gcd(b_1, b_2) = b_2$. If $b_3 > 0$, we can repeat the procedure, using b_2 and b_3 instead of b_1 and b_2, to produce an integer b_4 such that $\gcd(b_2, b_3) = \gcd(b_3, b_4)$, $b_3 > b_4 \geqslant 0$.

By continuing in this way, we will generate a decreasing sequence of nonnegative integers

$$b_1 > b_2 > b_3 > \cdots$$

such that $\gcd(b_1, b_2) = \gcd(b_2, b_3) = \cdots = \gcd(b_i, b_{i+1})$, $i = 1, 2, 3, \ldots$. Since such a sequence cannot decrease indefinitely, there will be a first n such that $b_{n+1} = 0$. At this point $\gcd(b_1, b_2) = \gcd(b_n, b_{n+1}) = b_n$.

This procedure for finding $\gcd(b_1, b_2)$ is called the *Euclidean algorithm*.

Before giving an example of this algorithm, we will state and prove the major result of this section.

3.1.2. Given positive integers a and b, there are integers s and t such that

$$sa + tb = \gcd(a, b).$$

Solution. We will prove the result by inducting on the number of steps required by the Euclidean algorithm to produce the greatest common divisor of a and b. (Another proof is outlined in 3.1.9.)

Suppose $a > b$. If only one step is required, there is an integer q such that $a = bq$, and in this case $\gcd(a, b) = b$. Also, in this case, $\gcd(a, b) = b = a + (1 - q)b$, so set $s = 1$, $t = 1 - q$, and the proof is complete.

Assume the result has been proved for all pairs of positive integers which require less than k steps, and assume that a and b are integers that require k steps, $k > 1$. By the division algorithm, there are integers q and r such that

$$a = qb + r, \qquad 0 < r < b.$$

The greatest common divisor of b and r can be computed by the Euclidean algorithm in $k - 1$ steps, so by the inductive assumption, there are integers c and d such that

$$cb + dr = \gcd(b, r).$$

From these last two equations, it follows that

$$\begin{aligned}
\gcd(a, b) &= \gcd(b, r) \\
&= cb + dr \\
&= cb + d(a - qb) \\
&= da + (c - dq)b,
\end{aligned}$$

and the proof is complete upon setting $s = d$ and $t = c - dq$.

The steps of this proof will be clarified by an example.

3.1.3. Find integers x and y such that

$$754x + 221y = \gcd(754, 221).$$

Solution. We first apply the steps of the Euclidean algorithm to find the greatest common divisor of 754 and 221. We find that

$$754 = 3 \times 221 + 91,$$
$$221 = 2 \times 91 + 39,$$
$$91 = 2 \times 39 + 13,$$
$$39 = 3 \times 13.$$

This shows that $\gcd(754, 221) = 13$.

To find the desired integers x and y, we proceed "backwards" through the steps of the Euclidean algorithm (this was the essence of the inductive proof given above):

$$13 = 91 - 2 \times 39$$
$$= 91 - 2(221 - 2 \times 91)$$
$$= 5 \times 91 - 2 \times 221$$
$$= 5(754 - 3 \times 221) - 2 \times 221$$
$$= 5 \times 754 - 17 \times 221.$$

Thus, one solution is $x = 5$ and $y = -17$.

The following result is often useful.

3.1.4. The equation $ax + by = c$, a, b, c integers, has a solution in integers x and y if and only if $\gcd(a, b)$ divides c. Moreover, if (x_0, y_0) is an integer solution, then for each integer k, the values

$$x' = x_0 + bk/d,$$
$$y' = y_0 - ak/d, \qquad d = \gcd(a, b),$$

are also a solution, and all integer solutions are of this form.

Solution. For the first part, it is clear that $\gcd(a, b)$ must divide c, since $\gcd(a, b)$ divides $ax + by$. Therefore, $\gcd(a, b) \mid c$ is a necessary condition for the existence of a solution. On the other hand, if c is a multiple of $\gcd(a, b)$, say $c = \gcd(a, b) \times q$, we can find an integer solution in the following manner. We know there are integers s and t such that $sa + tb = \gcd(a, b)$. So set $x = sq$ and $y = tq$. Then $ax + by = asq + btq = (as + tb)q = \gcd(a, b)q = c$.

A straightforward calculation shows that (x', y'), as given, gives a solution, provided (x_0, y_0) is a solution. To show all integer solutions have this form we argue geometrically as follows (Figure 3.1).

Note that the problem of solving $ax + by = c$ in integers x and y is equivalent to the problem of finding the lattice points that lie on the straight line $ax + by = c$. Suppose that (x_0, y_0) is a lattice point on the line $ax + by = c$; that is, suppose that

$$ax_0 + by_0 = c.$$

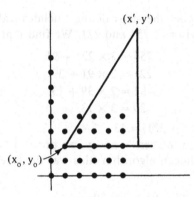

Figure 3.1.

The result is easy to prove if $b = 0$, so suppose here that $b \neq 0$. If (x', y') is any other lattice point in the plane, then (x', y') will be on the line $ax + by = c$ if and only if

$$\frac{y' - y_0}{x' - x_0} = -\frac{a}{b} = -\frac{a/d}{b/d}, \qquad \text{where} \quad d = \gcd(a, b).$$

Since a/d and b/d are relatively prime, this equation will hold if and only if there is an integer k such that

$$y' - y_0 = -(a/d)k,$$
$$x' - x_0 = (b/d)k.$$

It follows that all integer solutions of $ax + by = c$ are given by the equations

$$x' = x_0 + bk/d,$$
$$y' = y_0 - ak/d,$$

k an integer, $d = \gcd(a, b)$.

3.1.5. Prove that the fraction $(21n + 4)/(14n + 3)$ is irreducible for every natural number n.

Solution. We need to prove that $14n + 3$ and $21n + 4$ are relatively prime for all n. Our preceding discussion shows that we will be done if we can prove that there exist integers s and t such that

$$s(21n + 4) + t(14n + 3) = 1,$$

or equivalently,

$$7n(3s + 2t) + (4s + 3t) = 1.$$

This equation will hold for all n if we can find integers s and t which satisfy

$$3s + 2t = 0,$$
$$4s + 3t = 1.$$

It is straightforward to see that these equations are satisfied by $s = -2$ and $t = 3$, and this completes the proof.

3.1.6. The measure of a given angle is $180°/n$, where n is a positive integer not divisible by 3. Prove that the angle can be trisected by Euclidean means (straightedge and compass).

Solution. We do not expect this problem to have anything to do with numbers, and yet, what is the significance of the condition that n is not divisible by 3? This means 3 and n are relatively prime, so there are integers s and t such that

$$ns + 3t = 1.$$

We wish to construct an angle of $60°/n$. When we multiply each side of the last equation by $60°/n$, we get

$$60°s + (180°/n)t = 60°/n.$$

But now observe that the left side of this equation describes how to construct $60°/n$. This is because we can construct a $60°$ angle, we are given the angle $180°/n$, the integers s and t can be found, and therefore we can construct $60°s + (180°/n)t$.

Problems

3.1.7. If $\gcd(a, b) = 1$, prove that

(a) $\gcd(a - b, a + b) \leqslant 2$,
(b) $\gcd(a - b, a + b, ab) = 1$,
(c) $\gcd(a^2 - ab + b^2, a + b) \leqslant 3$.

3.1.8. The algebraic sum of any number of irreducible fractions whose denominators are relatively prime to each other cannot be an integer. That is, if $\gcd(a_i, b_i) = 1$, $i = 1, \ldots, n$, and $\gcd(b_i, b_j) = 1$ for $i \neq j$, show that

$$\frac{a_1}{b_1} + \frac{a_2}{b_2} + \cdots + \frac{a_n}{b_n}$$

is not an integer.

3.1.9. Let S be a nonempty set of integers such that

(i) the difference $x - y$ is in S whenever x and y are in S, and
(ii) all multiples of x are in S whenever x is in S.

(a) Prove that there is an integer d in S such that S consists of all multiples of d. (Hint: Consider the smallest positive integer in S.)
(b) Show that part (a) applies to the set $\{ma + nb \mid m$ and n are positive integers$\}$, and show that the resulting d is $\gcd(a, b)$.

3.1.10.

(a) Prove that any two successive Fibonacci numbers F_n, F_{n+1}, $n > 2$, are relatively prime.
(b) Given that $T_1 = 2$, and $T_{n+1} = T_n^2 - T_n + 1$, $n > 0$, prove that T_n and T_m are relatively prime whenever $n \neq m$.

3.1.11. For positive integers $a_1, \ldots, a_n$, prove there exist integers $k_1, \ldots, k_n$ such that $k_1 a_1 + \cdots + k_n a_n = \gcd(a_1, \ldots, a_n)$.

3.1.12. Prove that $(a + b)/(c + d)$ is irreducible if $ad - bc = 1$.

3.1.13. Prove that $\gcd(a_1, \ldots, a_m)\gcd(b_1, \ldots, b_n) = \gcd(a_1 b_1, a_2 b_2, \ldots, a_m b_n)$, where the parentheses on the right include all mn products $a_i b_j$, $i = 1, \ldots, m, j = 1, \ldots, n$.

3.1.14. When Mr. Smith cashed a check for x dollars and y cents, he received instead y dollars and x cents, and found that he had two cents more than twice the proper amount. For how much was the check written?

3.1.15. Find the smallest positive integer a for which
$$1001x + 770y = 1{,}000{,}000 + a$$
is possible, and show that it has then 100 solutions in positive integers.

3.1.16. A man goes to a stream with a 9-pint container and a 16-pint container. What should he do to get 1 pint of water in the 16-pint container? (Hint: Find integers s and t such that $9s + 16t = 1$.)

3.1.17. There is more than one integer greater than 1 which, when divided by any integer k such that $2 \leqslant k \leqslant 11$, has a remainder of 1. What is the difference between the two smallest such integers?

3.1.18. Let b be an integer greater than one. Prove that for every nonnegative integer N, there is a unique nonnegative integer n and unique integers a_i, $i = 0, 1, \ldots, n, 0 \leqslant a_i < b$, such that $a_n \neq 0$ and
$$N = a_n b^n + a_{n-1} b^{n-1} + \cdots + a_2 b^2 + a_1 b + a_0.$$
(The result is immediate for $N < b$, so assume $N \geqslant b$. Use induction.)

Additional Examples

3.2.4, 3.2.21, 3.3.11, 3.3.19, 3.3.28, 4.1.9, 4.2.1, 4.2.2, 4.2.4, corollary (iii) of Lagrange's theorem in Section 4.4, 4.4.5, 4.4.6, 4.4.8.

3.2. Modular Arithmetic

The parity of an integer tells us how that number stands relative to the number 2. Specifically, a number is even or odd according to whether its remainder when divided by 2 is zero or one respectively. This formulation of parity makes it natural to generalize the idea in the following manner.

Given an integer $n \geqslant 2$, divide the set of integers into "congruence" classes according to their remainders when they are divided by n; that is to say, two integers are put into the same congruence class if they have the same remainders when they are divided by n. For example, for $n = 4$, the integers are divided into four sets identified with the possible remainders $0, 1, 2, 3$. For an arbitrary $n \geqslant 2$, there will be n congruence classes, labeled $0, 1, 2, \ldots, n - 1$.

Two integers x and y are said to be *congruent modulo n*, written

$$x \equiv y \pmod{n},$$

if they each give the same remainder when they are divided by n (or, equivalently, and more conveniently in practice, if $x - y$ is divisible by n).

It is easy to prove that

(i) $x \equiv x \pmod{n}$,
(ii) $x \equiv y \pmod{n}$ implies $y \equiv x \pmod{n}$, and
(iii) $[x \equiv y \pmod{n}$ and $y \equiv z \pmod{n}]$ imply $x \equiv z \pmod{n}$.

These properties mean that congruence has the same characteristics as equality, and we often think of congruence as a kind of equality (in fact we sometimes read $x \equiv y \pmod{n}$ as "x equals y modulo n").

3.2.1. Prove that any subset of 55 numbers chosen from the set $\{1, 2, 3, 4, \ldots, 100\}$ must contain two numbers differing by 9.

Solution. There are nine congruence classes modulo 9: $0, 1, 2, 3, 4, 5, 6, 7, 8$. By the (generalized) pigeonhole principle, seven numbers from the chosen 55 are in the same congruence class (if each congruence class had six or less, this would account for at most 54 of the 55 elements). Let $a_1, \ldots, a_7$ be these numbers, and suppose they are labeled so that $a_1 < a_2 < a_3 < \cdots < a_7$. Since $a_{i+1} \equiv a_i \pmod{9}$, $a_{i+1} - a_i \in \{9, 18, \ldots\}$. We claim that $a_{i+1} - a_i = 9$ for some i. For if not, then for each i, $a_{i+1} - a_i \geqslant 18$, and this would mean that $a_7 - a_1 \geqslant 6 \times 18 = 108$. But this is impossible, since $a_7 - a_1 < 100$. Thus, two of the elements (among $a_1, \ldots, a_7$) differ by 9.

The real power of congruences is a consequence of the following easily proved property.

Modular Arithmetic. *If $x \equiv y$ (mod n) and $u \equiv v$ (mod n) then*

$$x + u \equiv y + v \text{ (mod } n),$$

and

$$x \cdot u \equiv y \cdot v \text{ (mod } n).$$

This result allows us to perform arithmetic by working solely with the "remainders" modulo n. For example, since

$$17 \equiv 5 \text{ (mod 12)} \quad \text{and} \quad 40 \equiv 4 \text{ (mod 12)},$$

we know that

$$17 + 40 \equiv 5 + 4 = 9 \text{ (mod 12)}$$

and

$$17 \times 40 \equiv 5 \times 4 \equiv 8 \text{ (mod 12)}.$$

Let n be a positive integer, $n > 1$, and let $Z_n = \{0, 1, 2, \ldots, n-1\}$. Observe that if x and y are elements of Z_n, there are unique elements r, s, t in Z_n such that

$$x - y \equiv r \text{ (mod } n),$$

$$x + y \equiv s \text{ (mod } n),$$

$$x \cdot y \equiv t \text{ (mod } n).$$

The set Z_n together with these operators of subtraction, addition, and multiplication is called the set of integers modulo n. In this system, computations are carried out as usual, except the result is always reduced (modulo n) to an equivalent number in the set Z_n.

3.2.2. Let $N = 22 \times 31 + 11 \times 17 + 13 \times 19$. Determine (a) the parity of N; (b) the units digit of N; (c) the remainder when N is divided by 7. (Of course, the idea is to make these determinations without actually computing N.)

Solution. For part (a), 22×31 is even, since 22 is even, 11×17 is odd, and 13×19 is odd, so the sum is even + odd + odd, and this is even. Notice that this reasoning is equivalent to computing modulo 2:

$$22 \times 31 + 11 \times 17 + 13 \times 19 \equiv 0 \times 1 + 1 \times 1 + 1 \times 1 \equiv 1 + 1 \equiv 0 \text{ (mod 2)}.$$

For part (b), we need only keep track of the units digit: 22×31 has a units digit of 2, 11×17 has a units digit of 7, and 13×19 has a units digit of 7. Therefore, the units digit of N is the units digit of $2 + 7 + 7$, or 6. Here again, this analysis is equivalent to computing N modulo 10:

$$22 \times 31 + 11 \times 17 + 13 \times 19 \equiv 2 \times 1 + 1 \times 7 + 3 \times 9 \text{ (mod 10)}$$

$$\equiv 2 + 7 + 7 \equiv 6 \text{ (mod 10)}.$$

Whereas parts (a) and (b) can be done without an awareness of modular arithmetic, it is not so apparent what should be done in part (c). The point of the example is that part (c) can be handled as a natural extension of the modular approach used in the previous cases. We work modulo 7:

$$22 \times 31 + 11 \times 17 + 13 \times 19 \equiv 1 \times 3 + 4 \times 3 + (-1) \times 5 \ (\text{mod } 7)$$

$$\equiv 3 + 5 - 5 \equiv 3 \ (\text{mod } 7).$$

Thus N is 3 more than a multiple of 7. (As a check: $N = 1116 = 159 \times 7 + 3$.)

3.2.3. What are the last two digits of 3^{1234}?

Solution. We work modulo 100. There are many way to build up to 3^{1234}. For example, $3^2 \equiv 9 \ (\text{mod } 100)$, $3^4 \equiv 81 \ (\text{mod } 100)$, $3^8 \equiv 81 \times 81 \equiv 61 \ (\text{mod } 100)$, $3^{10} \equiv 9 \times 61 \equiv 49 \ (\text{mod } 100)$, $3^{20} \equiv 49 \times 49 \equiv 1 \ (\text{mod } 100)$. Since $1234 = 20 \times 61 + 14$, we have $3^{1234} = (3^{20})^{61}(3)^{14} \equiv 3^{14} \equiv 3^4 3^{10} \equiv 81 \times 49 \equiv 69 \ (\text{mod } 100)$. The last two digits are thus seen to be 69.

3.2.4. Show that some positive multiple of 21 has 241 as its final three digits.

Solution. We must prove that there is a positive integer n such that

$$21n \equiv 241 \ (\text{mod } 1000).$$

Since 21 and 1000 are relatively prime, there are integers s and t such that

$$21s + 1000t = 1.$$

Multiply each side of this equation by 241, and rearrange in the form

$$21(241s) - 241 = -241 \times 1000t.$$

In congruence notation, the last equation means that

$$21 \times 241s \equiv 241 \ (\text{mod } 1000).$$

If s is positive, we are done, for we can set $n = 241s$. If s is not positive, let $n = 241s + 1000k$, where k is an integer large enough to make n positive (by choosing k in the appropriate manner, we may even assume that n is between 0 and 1000). It follows that

$$21n \equiv 21(241s + 1000k) \equiv 21 \times 241s \equiv 241 \ (\text{mod } 1000).$$

3.2.5. Prove that for any set of n integers, there is a subset of them whose sum is divisible by n.

Solution. Let $x_1, x_2, \ldots, x_n$ denote the given integers, and let

$$y_1 = x_1,$$
$$y_2 = x_1 + x_2,$$
$$\vdots$$
$$y_n = x_1 + x_2 + \cdots + x_n.$$

If $y_i \equiv 0 \pmod{n}$ for some i, we're done, so suppose this is not the case. Then we have n numbers $y_1, \ldots, y_n$, and $n - 1$ congruence classes modulo n (namely, $1, 2, \ldots, n - 1$), so by the pigeonhole principle, two of the y_i's must be congruent to one another modulo n. Suppose $y_i \equiv y_j \pmod{n}$, with $i < j$. Then

$$x_{i+1} + \cdots + x_j \equiv y_j - y_i \equiv 0 \pmod{n},$$

and the proof is complete.

In the preceding example, we made use of the fact that n divides a if and only if $a \equiv 0 \pmod{n}$. By means of this correspondence, problems concerning divisibility can be translated directly into the language of modular arithmetic.

3.2.6. Prove that if $2n + 1$ and $3n + 1$ are both perfect squares, then n is divisible by 40.

Solution. It is enough to show that n is divisible by both 5 and 8. This is equivalent to showing that $n \equiv 0 \pmod{5}$ and $n \equiv 0 \pmod{8}$.

Consider modulo 5. The table below shows that a square number is either 0, 1, or -1 modulo 5:

$x \pmod 5$	0	1	2	3	4
$x^2 \pmod 5$	0	1	-1	-1	1

Thus, $2n + 1$ and $3n + 1$ must be either 0, 1, or -1 modulo 5. There are nine cases to consider: $2n + 1$ can be 0, 1, or -1 modulo 5, and $3n + 1$ can be 0, 1, or -1. Some thought however reduces the number of cases to just two, as we shall see. Suppose that $2n + 1 \equiv a \pmod{5}$ and $3n + 1 \equiv b \pmod{5}$, $a, b \in \{0, 1, -1\}$.

Case 1. $a \neq b$. In this case, we add the last two equations to get

$$2 \equiv a + b \pmod{5}.$$

But this equation cannot hold for our choices of a and b, therefore this case can never occur.

Case 2. a = b. In this case, subtract the first equation from the second to get

$$n \equiv b - a \pmod{5}.$$

In this case n is divisible by 5 (which is part of what we wanted to prove).

Now consider modulo 8. In this case, the table shows that a square is either 0, 1, or 4 modulo 8:

$x \pmod 8$	0	1	2	3	4	5	6	7
$x^2 \pmod 8$	0	1	4	1	0	1	4	1

Again, there are nine cases, depending on the values of $2n + 1$ and $3n + 1$ modulo 8. These nine cases can be reduced to two exactly as in the modulo 5 case, and the argument in each case is exactly the same. We conclude that 8 divides n, and the proof is complete.

In congruence arithmetic, the operations of addition, subtraction, and multiplication behave as in ordinary arithmetic (except everything is taken with respect to the modulus under consideration). What about division?

We say that *a divides b modulo n* if there is an integer c such that $a \cdot c \equiv b \pmod n$. If there is an integer c such that $a \cdot c \equiv 1 \pmod n$, then c is called the (multiplicative) *inverse* of a, sometimes denoted by a^{-1}. Note that if a has an inverse, the equation $ax \equiv b \pmod n$ can be solved by simply multiplying each side of the equation by a^{-1}; $x = a^{-1}b \pmod n$.

An important theoretical fact is that an integer a has a multiplicative inverse with respect to modulo n arithmetic if and only if a and n are relatively prime (see 3.2.21).

As a special case of the result of the previous paragraph, consider the case in which the modulus n is a prime number, say p. In this case, each of $1, 2, \ldots, p - 1$ is relatively prime to p, so they all have multiplicative inverses. In fact, the numbers $Z_p = \{0, 1, 2, \ldots, p - 1\}$ can be added, subtracted, multiplied, and divided (by nonzero elements), and they form a field (see Section 4.4).

3.2.7. Prove that the expressions

$$2x + 3y, \qquad 9x + 5y$$

are divisible by 17 for the same set of integral values of x and y.

Solution. It suffices to show that

$$2x + 3y \equiv 0 \pmod{17} \qquad \text{if and only if} \qquad 9x + 5y \equiv 0 \pmod{17}.$$

The plan is to multiply each side of the left congruence by a suitable constant so as to transform it into the congruence on the right. So we ask: does there exist a constant c such that $c(2x + 3y) = 9x + 5y$ (mod 17)? For this to be possible, it is necessary that $2c = 9$ (mod 17). Since 2 is relatively prime to 17, it has an inverse. It turns out that $2^{-1} = 9$, and therefore, $c = 9 \times 9 = 81 \equiv 13$ (mod 17). Therefore, $2x + 3y \equiv 0$ (mod 17) implies

$$13(2x + 3y) \equiv 0 \text{ (mod 17)},$$

$$26x + 39y \equiv 0 \text{ (mod 17)},$$

$$9x + 5y \equiv 0 \text{ (mod 17)}.$$

Conversely, multiply each side of $9x + 5y \equiv 0$ (mod 17) by 4 to get $2x + 3y \equiv 0$ (mod 17).

The next example is a theoretical result which not only is interesting from a conceptional point of view, but also has many applications throughout mathematics.

3.2.8 (Chinese remainder theorem). If m and n are relatively prime integers greater than one, and a and b are arbitrary integers, there exists an integer x such that

$$x \equiv a \text{ (mod } m),$$

$$x \equiv b \text{ (mod } n).$$

More generally, if $m_1, m_2, \ldots, m_k$ are (pairwise) relatively prime integers greater than one, and $a_1, a_2, \ldots, a_k$ are arbitrary integers, there exists an integer x such that

$$x \equiv a_i \text{ (mod } m_i), \qquad i = 1, 2, \ldots, k.$$

Solution. Consider the n numbers $a, a + m, a + 2m, \ldots, a + (n - 1)m$. Each of these is congruent to a modulo m. Moreover, no two of them are congruent modulo n. For, if $a + im \equiv a + jm$ (mod n), $0 \leqslant i < j < n$, then $(i - j)m \equiv 0$ (mod n). But m and n are relatively prime, so this last congruence can hold only if n divides $i - j$. However, $i - j$ cannot be a multiple of n because of the restrictions on i and j. Therefore, $i = j$. It follows that the numbers $a, a + m, \ldots, a + (n - 1)m$ are congruent in some order to the numbers $0, 1, 2, \ldots, n - 1$ modulo n. Therefore, for some $i, a + mi \equiv b$ (mod n). The proof of the first part is established upon setting $x = a + mi$.

The more general statement can be proved in a similar way, using induction on k. (Let $c = m_1 \cdots m_{k-1}$, and consider $a, a + c, a + 2c, \ldots, a + (m_k - 1)c$, where a is chosen by the inductive hypothesis so that $a \equiv a_i$ (mod m_i), $i = 1, 2, \ldots, k - 1$. Then $a + ic \equiv a_i$ (mod m_i), $i = 1, \ldots, k - 1$, and no two of the numbers are congruent modulo m_k, etc.)

3.2.9. Do there exist 1,000,000 consecutive integers each of which contains a repeated prime factor?

Solution. Let $p_1, p_2, \ldots, p_{1,000,000}$ denote 1,000,000 distinct prime numbers. Then p_i^2 and p_j^2 are relatively prime if $i \neq j$, so by the Chinese remainder theorem, there is an integer x such that

$$x \equiv -k \ (\text{mod } p_k^2), \qquad k = 1, 2, \ldots, 10^6.$$

It follows that $x + k$ is divisible by p_k^2 (i.e., $x + k$ has a repeated prime factor), and the answer to the question is yes: take the consecutive integers $x + 1, x + 2, x + 3, \ldots, x + 1,000,000$.

3.2.10. A lattice point $(x, y) \in Z^2$ is *visible* if $\gcd(x, y) = 1$. Prove or disprove: Given a positive integer n, there exists a lattice point (a, b) whose distance from every visible point is $\geq n$.

Solution. We will look at a very special case first, but the pattern for the general case is a simple generalization which will be clear. Begin by choosing nine distinct primes $p_1, p_2, \ldots, p_9$. We now look for a lattice point (a, b) such that

$$a - 1 \equiv 0 \ (\text{mod } p_1 p_2 p_3),$$
$$a \equiv 0 \ (\text{mod } p_4 p_5 p_6), \qquad (1)$$
$$a + 1 \equiv 0 \ (\text{mod } p_7 p_8 p_9),$$

and

$$b + 1 \equiv 0 \ (\text{mod } p_1 p_4 p_7),$$
$$b \equiv 0 \ (\text{mod } p_2 p_5 p_8), \qquad (2)$$
$$b - 1 \equiv 0 \ (\text{mod } p_3 p_6 p_9).$$

Geometrically, (a, b) is a point characterized by the following diagram:

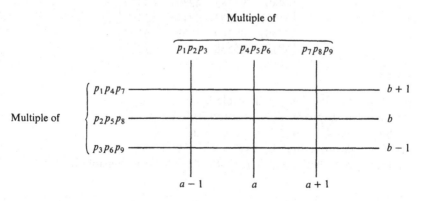

Since $p_1p_2p_3$, $p_4p_5p_6$, $p_7p_8p_9$ are relatively prime, the Chinese remainder theorem says that an integer a exists which satisfies equations (1). Similarly, since $p_1p_4p_7$, $p_2p_5p_8$, $p_3p_6p_9$ are relatively prime, an integer b exists which satisfies (2). By the way a and b are chosen, it is clear that the eight lattice points closest to (a, b) are invisible. Take, for instance, the point $(a, b + 1)$, which has the form $(k_1p_4p_5p_6, k_2p_1p_4p_7)$ for some integers k_1 and k_2. Since p_4 is a common factor of the coordinates, this point is invisible. A similar argument applies to the other seven closest lattice points.

The general case can be handled in exactly the same way, and we leave this as Problem 3.2.26.

Problems

3.2.11. Prove that any subset of 55 numbers chosen from the set $\{1, 2, 3, \ldots, 100\}$ must contain numbers differing by 10, 12, and 13, but need not contain a pair differing by 11.

3.2.12. The elements of a determinant are arbitrary integers. Determine the probability that the value of the determinant is odd. (Hint: Work modulo 2.)

3.2.13.

(a) Determine whether the following matrix is singular or nonsingular:

$$\begin{bmatrix} 54401 & 57668 & 15982 & 103790 \\ 33223 & 26563 & 23165 & 71489 \\ 36799 & 37189 & 16596 & 46152 \\ 21689 & 55538 & 79922 & 51237 \end{bmatrix}.$$

(Hint: A matrix A is nonsingular if $\det A \neq 0$. Examine the parity of the determinant of the given matrix; that is, compute its determinant modulo 2.)

(b) Determine whether the following matrix is singular or nonsingular:

$$\begin{bmatrix} 64809 & 91185 & 42391 & 44350 \\ 61372 & 26563 & 23165 & 71489 \\ 82561 & 39189 & 16596 & 46152 \\ 39177 & 55538 & 79922 & 51237 \end{bmatrix}.$$

3.2.14

(a) Show that $2^{2x+1} + 1$ is divisible by 3.
(b) Prove or disprove: $2^x \equiv 2^y \pmod{n}$ if $x \equiv y \pmod{n}$.
(c) Show that $4^{3x+1} + 2^{3x+1} + 1$ is divisible by 7.
(d) If $n > 0$, prove that 12 divides $n^4 - 4n^3 + 5n^2 - 2n$.
(e) Prove that $(2903)^n - (803)^n - (464)^n + (261)^n$ is divisible by 1897.

3.2.15.

(a) Prove that no prime three more than a multiple of four is a sum of two squares. (Hint: Work modulo 4.)

(b) Prove that the sequence (in base-10 notation)

$$11, 111, 1111, 11111, \ldots$$

contains no squares.

(c) Prove that the difference of the squares of any two odd numbers is exactly divisible by 8.

(d) Prove that $2^{70} + 3^{70}$ is divisible by 13.

(e) Prove that the sum of two odd squares cannot be a square.

(f) Determine all integral solutions of $a^2 + b^2 + c^2 = a^2 b^2$. (Hint: Analyze modulo 4.)

3.2.16.

(a) If $x^3 + y^3 = z^3$ has a solution in integers x, y, z, show that one of the three must be a multiple of 7.

(b) If n is a positive integer greater than 1 such that $2^n + n^2$ is prime, show that $n \equiv 3 \pmod 6$.

(c) Let x be an integer one less than a multiple of 24. Prove that if a and b are positive integers such that $ab = x$, then $a + b$ is a multiple of 24.

(d) Prove that if $n^2 + m$ and $n^2 - m$ are perfect squares, then m is divisible by 24.

3.2.17. Let S be a set of primes such that $a, b \in S$ (a and b need not be distinct) implies $ab + 4 \in S$. Show that S must be empty. (Hint: One approach is to work modulo 7.)

3.2.18. Prove that there are no integers x and y for which

$$x^2 + 3xy - 2y^2 = 122.$$

(Hint: Use the quadratic equation to solve for x; then look at the discriminant modulo 17. Can it ever be a perfect square?)

3.2.19. Given an integer n, show that an integer can always be found which contains only the digits 0 and 1 (in the base 10 notation) and which is divisible by n.

3.2.20. Show that if n divides a single Fibonacci number, then it will divide infinitely many Fibonacci numbers.

3.2.21. Suppose that a and n are integers, $n > 1$. Prove that the equation $ax \equiv 1 \pmod n$ has a solution if and only if a and n are relatively prime.

3.2.22. Let a, b, c, d be fixed integers with d not divisible by 5. Assume that m is an integer for which

$$am^3 + bm^2 + cm + d$$

is divisible by 5. Prove that there exists an integer n for which

$$dn^3 + cn^2 + bn + a$$

is also divisible by 5.

3.2.23. Prove that $(21n - 3)/4$ and $(15n + 2)/4$ cannot both be integers for the same positive integer n.

3.2.24.

(a) Do there exist n consecutive integers for which the jth integer, $1 \leqslant j \leqslant n$, has a divisor which does not divide any other member of the sequence?
(b) Do there exist n consecutive integers for which the jth integer, $1 \leqslant j \leqslant n$, has at least j divisors, none of which divides any other member of the sequence?

3.2.25. Let $m_0, m_1, \ldots, m_r$ be positive integers which are pairwise relatively prime. Show that there exist $r + 1$ consecutive integers $s, s + 1, \ldots, s + r$ such that m_i divides $s + i$ for $i = 0, 1, \ldots, r$.

3.2.26. Complete the proof of 3.2.10.

Additional Examples

3.3.11, 3.4.3, 3.4.9, 4.1.3, 4.2.4, 4.2.14, 4.3.4, 4.3.5, 4.4.6, 4.4.7, 4.4.8, 4.4.9, 4.4.19, 4.4.20, 4.4.21, 4.4.22, 4.4.23, 4.4.24, 4.4.29, 4.4.30, 4.4.31.

3.3. Unique Factorization

One of the most useful and far-reaching results at the heart of elementary number theory is the fact that every natural number greater than one can be factored uniquely (up to the order of the factors) into a product of prime numbers. More precisely, every natural number n can be represented in one and only one way in the form

$$n = p_1^{a_1} p_2^{a_2} \cdots p_k^{a_k}$$

where $p_1, p_2, \ldots, p_k$ are different prime numbers and $a_1, a_2, \ldots, a_k$ are positive integers. Here are some easily proved, but very useful, consequences.

3.3.1. All the divisors of

$$n = p_1^{a_1} p_2^{a_2} \cdots p_k^{a_k}$$

are of the form

$$m = p_1^{b_1} p_2^{b_2} \cdots p_k^{b_k}, \qquad 0 \leqslant b_i \leqslant a_i, \quad i = 1, \ldots, k,$$

and every such number is a divisor of n. It follows that n has exactly $(a_1 + 1)(a_2 + 1) \cdots (a_k + 1)$ distinct divisors.

3.3.2. An integer $n = p_1^{a_1} p_2^{a_2} \cdots p_k^{a_k}$ is a perfect square if and only if a_i is even for each i, a perfect cube if and only if each a_i is a multiple of three, and so forth.

3.3.3. Let $a, b, \ldots, g$ be a finite number of positive integers. Suppose their unique factorizations are

$$a = p_1^{a_1} p_2^{a_2} \cdots p_k^{a_k}, \quad b = p_1^{b_1} p_2^{b_2} \cdots p_k^{b_k}, \quad \ldots, \quad g = p_1^{g_1} p_2^{g_2} \cdots p_k^{g_k},$$

where $a_1, \ldots, a_k, b_1, \ldots, b_k, \ldots, g_1, \ldots, g_k$ are nonnegative integers (some may be zero). Then

$$\gcd(a, b, \ldots, g) = p_1^{m_1} p_2^{m_2} \cdots p_k^{m_k},$$

and

$$\operatorname{lcm}(a, b, \ldots, g) = p_1^{M_1} p_2^{M_2} \cdots p_k^{M_k},$$

where $m_i = \min\{a_i, b_i, \ldots, g_i\}$ and $M_i = \max\{a_i, b_i, \ldots, g_i\}$ for each $i = 1, 2, \ldots, k$. From this it easily follows that

$$\gcd(a, b)\operatorname{lcm}(a, b) = ab.$$

3.3.4. Use unique factorization to show that $\sqrt{2}$ is irrational.

Solution. Suppose there are integers r and s such that $\sqrt{2} = r/s$. Then $2s^2 = r^2$. But this equation cannot hold (by unique factorization), for on the left side, the prime 2 is raised to an odd power, and on the right side, 2 is raised to an even power (2 occurs an even number of times (perhaps zero) in s^2 and r^2). This contradiction implies that $\sqrt{2}$ must be irrational.

3.3.5. Find the smallest positive integer n such that $n/2$ is a perfect square, $n/3$ is a perfect cube, and $n/5$ is a perfect fifth power.

Solution. Since n is divisible by 2, 3, and 5, we may assume it has the form $n = 2^a 3^b 5^c$. Then $n/2 = 2^{a-1} 3^b 5^c$, $n/3 = 2^a 3^{b-1} 5^c$, $n/5 = 2^a 3^b 5^{c-1}$. The conditions are such that $a - 1$ must be even, and a must be a multiple of both 3 and 5. The smallest such a is $a = 15$. Similarly, the smallest values for b and c are $b = 10$ and $c = 6$. Thus $n = 2^{15} 3^{10} 5^6$ is the smallest such positive integer.

3.3.6. Prove there is one and only one natural number n such that $2^8 + 2^{11} + 2^n$ is a perfect square.

Solution. Suppose $m^2 = 2^8 + 2^{11} + 2^n$. Then

$$2^n = m^2 - 2^8 - 2^{11}$$
$$= m^2 - 2^8(1 + 2^3)$$
$$= m^2 - (3 \times 2^4)^2$$
$$= (m - 48)(m + 48).$$

Because of unique factorization, there are nonnegative integers s and t such that

$$m - 48 = 2^s, \qquad m + 48 = 2^t, \qquad s + t = n.$$

Thus $m = 2^s + 48$, $m = 2^t - 48$, so that

$$2^s + 48 = 2^t - 48,$$
$$2^t - 2^s = 96,$$
$$2^s(2^{t-s} - 1) = 2^5 \times 3.$$

Since $2^{t-s} - 1$ is odd, unique factorization implies that $2^{t-s} - 1 = 3$. It follows that $s = 5$, $t = 7$, and $n = 12$.

3.3.7. Let n be a given positive integer. How many solutions are there in ordered positive-integer pairs (x, y) to the equation

$$\frac{xy}{x + y} = n?$$

Solution. Write the equation in the form

$$xy = n(x + y),$$
$$xy - nx - ny = 0,$$
$$(x - n)(y - n) = n^2.$$

Since we want positive integer solutions, it must be the case that $x > n$ and $y > n$ ($0 < x < n$ and $0 < y < n$ imply $(x - n)(y - n) < n^2$).

Suppose the prime factorization of n is $p_1^{a_1} p_2^{a_2} \cdots p_k^{a_k}$. Then $n^2 = p_1^{2a_1} p_2^{2a_2} \cdots p_k^{2a_k}$. Each divisor of n^2 determines a solution, and therefore the number of such solutions is $(2a_1 + 1)(2a_2 + 1) \cdots (2a_k + 1)$.

3.3.8. Let r and s be positive integers. Derive a formula for the number of ordered quadruples (a, b, c, d) of positive integers such that

$$3^r 7^s = \text{lcm}(a, b, c) = \text{lcm}(a, b, d) = \text{lcm}(a, c, d) = \text{lcm}(b, c, d).$$

Solution. In view of the result of 3.3.3, it is apparent that each of a, b, c, and d must have the form $3^m 7^n$ with m in $\{0, 1, \ldots, r\}$ and n in $\{0,$

$1, \ldots, s\}$. Also, m must be r for at least two of the four numbers, and n must be s for at least two of the four numbers. There are $\binom{4}{2}r^2$ allowable ways of choosing the m's in which exactly two m's will equal r; there are $\binom{4}{3}r$ allowable ways in which exactly three m's will equal r; there are $\binom{4}{4}$ allowable ways in which all m's equal r. Putting this together, there are

$$\binom{4}{4} + \binom{4}{3}r + \binom{4}{2}r^2$$

choices of allowable m's. Similarly, there are

$$\binom{4}{4} + \binom{4}{3}s + \binom{4}{2}s^2$$

allowable n's. The desired number is therefore $(1 + 4r + 6r^2)(1 + 4s + 6s^2)$.

3.3.9. Given positive integers x, y, z, prove that

$$(x, y)(x, z)(y, z)[x, y, z]^2 = [x, y][x, z][y, z](x, y, z)^2,$$

where $(a, \ldots, g)$ and $[a, \ldots, g]$ denote $\gcd(a, \ldots, g)$ and $\operatorname{lcm}(a, \ldots, g)$ respectively.

Solution. Because of unique factorization, it suffices to show that for each prime p, the power of p on the left side (in its prime factorization) is equal to the power of p on the right side. So suppose $x = p^a r$, $y = p^b s$, and $z = p^c t$, for integers r, s, t, each relatively prime to p. We may assume (because of symmetry, and by relabeling if necessary) that $a \leqslant b \leqslant c$. Then the power of p in the unique factorization of $[x, y, z]^2$ is $2c$; the powers of p in (x, y), (x, z), and (y, z) are a, a, and b respectively. Hence the power of p on the left side is $2a + b + 2c$.

In the same manner, the power of p on the right side is $b + c + c + 2a = 2a + b + 2c$. Thus, by our earlier remarks, the proof is complete.

3.3.10. Show that 1000! ends with 249 zeros.

Solution. Write $1000! = 2^a 5^b r$, where r is an integer relatively prime to 10. Clearly, $a \geqslant b$, and the number of zeros at the end of 1000! will equal b. Thus, we must find b.

Every fifth integer in the sequence $1, 2, 3, 4, 5, 6, \ldots, 1000$ is divisible by 5; there are $[\![1000/5]\!] = 200$ multiples of 5 in the sequence. Every 25th integer in the sequence is divisible by 25, so each of these will contribute an additional factor; there are $[\![1000/25]\!] = 40$ of these. Every 125th integer in the sequence is divisible by 125, and each of these will contribute an additional factor; there are $[\![1000/125]\!] = 8$ of these. Every 625th integer will contribute an additional factor; there are $[\![1000/625]\!] = 1$ of these.

Thus, $b = [\![1000/5]\!] + [\![1000/25]\!] + [\![1000/125]\!] + [\![1000/625]\!]$
$= 200 + 40 + 8 + 1 = 249$.

In exactly the same manner, the highest power of p in $n!$ is given by the (finite) sum

$$[\![n/p]\!] + [\![n/p^2]\!] + [\![n/p^3]\!] + \cdots .$$

3.3.11. Prove that there are an infinite number of primes of the form $6n - 1$.

Solution. First, notice that if p is a prime number larger than 3, then either $p \equiv 1 \pmod 6$ or $p \equiv -1 \pmod 6$. [If $p \equiv 2 \pmod 6$, for example, then $p = 6k + 2$ for some k, which implies that p is even, a contradiction. A similar argument works for $p \equiv 3 \pmod 6$ or $p \equiv 4 \pmod 6$.]

Now suppose there are only a finite number of primes of the form $6n - 1$. Consider the number $N = p! - 1$, where p is the *largest* prime of the form $6n - 1$. Write N as a product of primes, say

$$N = p! - 1 = p_1 p_2 \cdots p_m . \tag{1}$$

Observe that each of the primes p_k is larger than p. For, if $p_k \leqslant p$ then equation (1) shows that p_k divides 1, an impossibility. Since p is the largest prime congruent to -1 modulo 6, it follows that $p_k \equiv 1 \pmod 6$ for each k.

If we now consider equation (1) modulo 6, we find that

$$p! - 1 \equiv 1 \pmod 6,$$

or equivalently,

$$p! \equiv 2 \pmod 6.$$

But this is clearly impossible, since $p! \equiv 0 \pmod 6$. Therefore, there must be an infinite number of primes of the form $6n - 1$.

Problems

3.3.12. In a certain college of under 5000 total enrollment, a third of the students were freshmen, two-sevenths were sophomores, a fifth were juniors and the rest seniors. The history department offered a popular course in which were registered a fortieth of all the freshmen in college, a sixteenth of all the sophomores, and a ninth of all the juniors, while the remaining third of the history class were all seniors. How many students were there in the history class?

3.3.13. Find the smallest number with 28 divisors.

3.3.14. Given distinct integers a, b, c, d such that

$$(x - a)(x - b)(x - c)(x - d) - 4 = 0$$

has an integral root r, show that $4r = a + b + c + d$.

3.3.15.

(a) Prove that $\sqrt[3]{72}$ is irrational.
(b) Prove that there is no set of integers m, n, p except $0, 0, 0$ for which
$m + n\sqrt{2} + p\sqrt{3} = 0$.

3.3.16. Given positive integers a, b, c, d such that $a^3 = b^2$, $c^3 = d^2$, and $c - a = 25$, determine a, b, c, and d.

3.3.17. Prove that if ab, ac, and bc are perfect cubes for some positive integers a, b, c, then a, b, and c must also be perfect cubes.

3.3.18. A changing room has n lockers numbered 1 to n, and all are locked. A line of n attendants $P_1, P_2, \ldots, P_n$ file through the room in order. Each attendant P_k changes the condition of those lockers (and only those) whose numbers are divisible by k: if such a locker is unlocked, P_k will lock it; if it is locked, P_k will unlock it. Which lockers are unlocked after all n attendants have passed through the room? What is the situation if each attendant performs the same operation, but they file through in some other order?

3.3.19. The geometry of the number line makes it clear that among any set of n consecutive integers, one of them is divisible by n. This fact is frequently useful, as it is for example in the following problems.

(a) Prove that if one of the numbers $2^n - 1$ and $2^n + 1$ is prime, $n > 2$, then the other number is composite.
(b) What is the largest number N for which you can say that $n^5 - 5n^3 + 4n$ is divisible by N for every integer n?
(c) Prove that every positive integer has a multiple whose decimal representation involves all ten digits.

3.3.20. For each positive integer n, let $H_n = 1 + 1/2 + \cdots + 1/n$. Show that for $n > 1$, H_n is not an integer. (Hint: Suppose H_n is an integer. Multiply each side of the equality by $\operatorname{lcm}(1, 2, \ldots, n)$, and show that the left side of the resulting identity is even whereas the right side is odd.)

3.3.21. If $\gcd(a, b) = 1$, then show that

(i) $\gcd((a + b)^m, (a - b)^m) \leq 2^m$, and
(ii) $\gcd(a^m + b^m, a^m - b^m) \leq 2$.

3.3.22. For positive integers $a, \ldots, g$, let $(a, \ldots, g)$ and $[a, \ldots, g]$ denote the $\gcd(a, \ldots, g)$ and $\operatorname{lcm}(a, \ldots, g)$ respectively. Prove that

(a) $xyz = (xy, xz, yz)[x, y, z]$,
(b) $(x[y, z]) = [(x, y), (x, z)]$,
(c) $[x, (y, z)] = ([x, y], [x, z])$,
(d) $([x, y], [x, z], [y, z]) = [(x, y), (x, z), (y, z)]$,
(e) $[x, y, z](x, y)(x, z)(y, z) = xyz(x, y, z)$,
(f) $(x, y) = (x + y, [x, y])$.

3.3.23. Let m be divisible by $1, 2, \ldots, n$. Show that the numbers $1 + m(1 + i)$, $i = 0, 1, 2, \ldots, n$, are pairwise relatively prime.

3.3.24. The prime factorizations of $r + 1$ positive integers ($r \geqslant 1$) together involve only r primes. Prove that there is a subset of these integers whose product is a perfect square.

3.3.25.

(a) Determine all positive rational solutions of $x^y = y^x$.
(b) Determine all positive rational solutions of $x^{x+y} = (x + y)^y$.

3.3.26. Suppose that $a^2 + b^2 = c^2$, a, b, c integers. Assume $\gcd(a, b) = \gcd(a, c) = \gcd(b, c) = 1$. Prove that there exist integers u and v such that $c - b = 2u^2$, $c + b = 2v^2$, $\gcd(u, v) = 1$. Conclude that $a = 2uv$, $b = v^2 - u^2$, $c = v^2 + u^2$. (Hint: By examination modulo 4, it is not the case that a and b are both odd; neither are they both even. So without loss of generality, a is even and b is odd.)

Conversely, show that if u and v are given, then the three numbers a, b, c given by the above formulas satisfy $a^2 + b^2 = c^2$.

3.3.27. Find all sets of three perfect squares in arithmetic progression. (Hint: Suppose $a < b < c$ and $b^2 - a^2 = c^2 - b^2$, or equivalently, $a^2 + c^2 = 2b^2$. Let $s = (c + a)/2$, $t = (c - a)/2$. Show that $s^2 + t^2 = b^2$. Now apply the result of 3.3.26.)

3.3.28.

(a) Suppose there are only a finite number of primes of the form $6n - 1$; call them $p_1, \ldots, p_k$. Reach a contradiction by considering $N = (p_1 \cdots p_k)^2 - 1$.
(b) Prove that there are an infinite number of primes of the form $4n - 1$.

Additional Examples

1.10.9, 1.10.10, 2.6.1, 3.1.4, 3.4.8, 4.1.3, 4.2.3, 4.2.16b, 4.4.9, 5.2.1, 5.2.4, 5.2.6, 5.2.9, 5.2.14, 5.2.15, 5.2.16, 5.2.17.

3.4. Positional Notation

We will assume a familiarity with the positional system of representing real numbers. Namely, if b is an integer greater than one (called the base), each real number x can be expressed (uniquely) in the positional form

$$x = A_n A_{n-1} \ldots A_1 A_0 . a_1 a_2 a_3 \ldots .$$

where $A_0, \ldots, A_n, a_1, a_2, \ldots$ (called the digits) are integers, $0 \leqslant A_i < b$, $0 \leqslant a_j < b$, and there is no integer m such that $a_k = b - 1$ for all $k > m$. This representation is used to denote the sum of the series

$$A_n b^n + A_{n-1} b^{n-1} + \cdots + A_1 b + A_0 + a_1 b^{-1} + a_2 b^{-2} + \cdots.$$

3.4.1. Let C denote the class of positive integers which, when written in base 3, do not require the digit 2. Show that no three integers in C are in arithmetic progression.

Solution. Let d denote the common difference for an arbitrary arithmetic progression of three positive integers, and suppose that when d is written in base 3 notation its first nonzero digit, counting from the right, occurs in the kth position. Now, let a be an arbitrary positive integer, and write it in base 3 notation. The following table gives the kth digit of each of the integers a, $a + d$, and $a + 2d$, depending upon the kth digit of d and a:

Then the kth digit of	If the kth digit of d is 1 and kth digit of a is			kth digit of d is 2 and kth digit of a is		
	0	1	2	0	1	2
a	0	1	2	0	1	2
a + d	1	2	0	2	0	1
a + 2d	2	0	1	1	2	0

In every case, one of a, $a + d$, $a + 2d$ has a 2 in the kth digit, which means the corresponding number does not belong to C.

3.4.2. Does $[\![x]\!] + [\![2x]\!] + [\![4x]\!] + [\![8x]\!] + [\![16x]\!] + [\![32x]\!] = 12345$ have a solution?

Solution. Suppose that x is such a number. It is an easy matter to show that $195 < x < 196$ (since $63 \times 195 = 12{,}285 < 12{,}345 < 12{,}348 = 63 \times 196$). Now, write the fractional part of x in base-2 notation (the $a, b, c, \ldots$ are either 0 or 1):

$$x = 195 + .abcdef \ldots.$$

Then

$$2x = 2 \times 195 + a.bcdef\ldots,$$
$$4x = 4 \times 195 + ab.cdef\ldots,$$
$$8x = 8 \times 195 + abc.def\ldots,$$
$$16x = 16 \times 195 + abcd.ef\ldots,$$
$$32x = 32 \times 195 + abcde.f\ldots.$$

In this form we see that

$$[\![\, x \,]\!] = 195,$$
$$[\![\, 2x \,]\!] = 2 \times 195 + a,$$
$$[\![\, 4x \,]\!] = 4 \times 195 + 2a + b,$$
$$[\![\, 8x \,]\!] = 8 \times 195 + 4a + 2b + c,$$
$$[\![\, 16x \,]\!] = 16 \times 195 + 8a + 4b + 2c + d,$$
$$[\![\, 32x \,]\!] = 32 \times 195 + 16a + 8b + 4c + 2d + e.$$

Adding, we find that $[\![\, x \,]\!] + [\![\, 2x \,]\!] + [\![\, 4x \,]\!] + [\![\, 8x \,]\!] + [\![\, 16x \,]\!] + [\![\, 32x \,]\!]$
$= 63 \times 195 + 31a + 15b + 7c + 3d + e$. The problem is therefore reduced to finding a, b, c, d, e, each 0 or 1, such that $31a + 15b + 7c + 3d + e = 60$. But this equation cannot hold under the restrictions on a, b, c, d, e, since $31a + 15b + 7c + 3d + e \leqslant 31 + 15 + 7 + 3 + 1 = 57 < 60$. Therefore, there can be no such x.

When an integer is written in decimal notation (base 10), it is possible to determine very easily if it is divisible by 2 or 5. There are other divisibility tests that are easy to apply. For example: An integer N is divisible by 4 if and only if its last two digits are divisible by 4. To see this, write N in base 10:

$$N = (a_n 10^n + \cdots + a_2 10^2) + (a_1 10 + a_0)$$

and note that $a_n 10^n + \cdots + a_2 10^2$ is always divisible by 4. Thus, $4 \mid N$ if and only if $4 \mid (a_1 10 + a_0)$.

One of the most striking and useful divisibility tests is that an integer is divisible by 9 if and only if the sum of its digits (in decimal notation) is divisible by 9. To see why this is so, notice that $10 \equiv 1 \pmod 9$, and therefore, by the properties of modular arithmetic, $10^2 \equiv 1 \pmod 9$, $10^3 \equiv 1 \pmod 9$, and so forth. It follows that

$$N = a_n 10^n + \cdots + a_1 10 + a_0 \equiv a_n + \cdots + a_1 + a_0 \pmod 9.$$

A similar proof shows that an integer is divisible by 3 if and only if the sum of its digits is divisible by 3. As an application of this test, suppose we ask: for what digits x is $4324x98765223$ divisible by 3? We simply need to add the digits modulo 3, and choose x that will make the sum congruent to zero modulo 3. In this case, the sum of the digits is $1 + x$ modulo 3, so the number is divisible by 3 if and only if $x = 2$, 5, or 8.

3.4.3. When 4444^{4444} is written in decimal notation, the sum of its digits is A. Let B be the sum of the digits of A. Find the sum of the digits of B. (A and B are written in decimal notation.)

Solution. Let $N = 4444^{4444}$. Then $N < (10^5)^{4444} = 10^{22220}$, which means that when N is written in decimal notation, it will have less than 22,220 digits. Since each of the digits of N must be less than or equal to 9, we are certain that $A < 22,200 \times 9 = 199,980$.

In a similar manner, A has at most 6 digits, so that the sum of the digits of A must be less than 54 ($= 6 \times 9$); that is, $B < 54$.

Of the positive integers less than 54, the number with the largest sum of digits is 49, and this sum equals 13. Let C denote the sum of the digits of B. We have just seen that $C \leqslant 13$.

From our reasoning preceding the problem, we know that

$$N \equiv A \equiv B \equiv C \ (\text{mod } 9),$$

so let us calculate the congruence class of C by calculating the congruence class of N. First, $4444 = 9 \times 493 + 7$, and therefore $4444 \equiv 7 \ (\text{mod } 9)$. Also, $7^3 \equiv 1 \ (\text{mod } 9)$. Since $4444 = 3 \times 1481 + 1$, we have

$$4444^{4444} \equiv 7^{4444} \ (\text{mod } 9)$$

$$\equiv 7^{3 \times 1481} \times 7 \ (\text{mod } 9)$$

$$\equiv 7 \ (\text{mod } 9).$$

Thus, $C \equiv 7 \ (\text{mod } 9)$ and $C \leqslant 13$. The only number which can satisfy both of these requirements is $C = 7$, and the problem is solved.

3.4.4. An (ordered) triple (x_1, x_2, x_3) of positive irrational numbers with $x_1 + x_2 + x_3 = 1$ is called *balanced* if each $x_i < \frac{1}{2}$. If a triple is not balanced, say if $x_j > \frac{1}{2}$, one performs the following "balancing act":

$$B(x_1, x_2, x_3) = (x_1', x_2', x_3'),$$

where $x_i' = 2x_i$ if $i \neq j$ and $x_j' = 2x_j - 1$. If the new triple is not balanced, one performs the balancing act on it. Does continuation of this process always lead to a balanced triple after a finite number of performances of the balancing act?

Solution. Write x_1, x_2, x_3 in base 2 notation in the manner described at the beginning of the section, say

$$x_1 = .a_1 a_2 a_3 \ldots ,$$
$$x_2 = .b_1 b_2 b_3 \ldots ,$$
$$x_3 = .c_1 c_2 c_3 \ldots ,$$

where a_i, b_i, c_i are each 0 or 1.

To say that each $x_i < \frac{1}{2}$ is to say that a_1, b_1, and c_1 are each equal to zero. Notice that the balancing act consists of moving the "decimal" point one place to the right and then disregarding the integer part. Thus, for example, if x_1, x_2, x_3 were not balanced, the representations (base 2) of x_1', x_2', x_3' are given by

$$x_1' = .a_2 a_3 a_4 \ldots,$$
$$x_2' = .b_2 b_3 b_4 \ldots,$$
$$x_3' = .c_2 c_3 c_4 \ldots.$$

Many examples can be given to show that the process need not terminate in a balanced triple. For example, define x_1, x_2, x_3 (using the earlier notation) by

$$a_i = \begin{cases} 1 & \text{if } i \text{ is a perfect square,} \\ 0 & \text{otherwise,} \end{cases}$$

$$b_i = \begin{cases} 1 & \text{if } i \text{ is one more than a perfect square,} \\ 0 & \text{otherwise,} \end{cases}$$

$$c_i = \begin{cases} 1 & \text{if } a_i + b_i = 0, \\ 0 & \text{otherwise,} \end{cases}$$

that is,

$$x_1 = .100100001000000100 \ldots,$$
$$x_2 = .010010000100000010 \ldots,$$
$$x_3 = .001001110011111001 \ldots.$$

Each of x_1, x_2, and x_3 is irrational (rational numbers are those which correspond to periodic "decimal" representations), and their sum is 1 (since $x_1 + x_2 + x_3 = \frac{1}{2} + \frac{1}{4} + \frac{1}{8} + \cdots = 1$). Repeated applications of the balancing act will never transform x_1, x_2, x_3 into a balanced triple (because, in every case, one of a_i, b_i, c_i is equal to 1).

3.4.5 (Continuation of 2.5.10). Suppose f is a function on the positive integers which satisfies

$$f(2k) = 2f(k) - 1,$$
$$f(2k + 1) = 2f(k) + 1.$$

Let a be an arbitrary positive integer whose binary representation is given by

$$a = a_n a_{n-1} \ldots a_2 a_1 a_0 \quad (= a_n 2^n + a_{n-1} 2^{n-1} + \cdots + a_1 2 + a_0).$$

Show that

$$f(a) = b_n 2^n + b_{n-1} 2^{n-1} + \cdots + b_1 2 + b_0,$$

where

$$b_i = \begin{cases} 1 & a_i = 1, \\ -1 & \text{if } a_i = 0. \end{cases}$$

(The idea is to replace each of the 0's in the binary sum for a with -1's; for example, for $n = 10$, $f(1010_2) = 1\bar{1}1\bar{1}_2$ (the $\bar{1}$'s stand for -1's) $= 8 - 4 + 2 - 1 = 5$.

Solution. We will induct on the number of digits in the binary representation of a.

The result is true for $a = 1$, so suppose it holds whenever a has fewer than $k + 1$ digits. Now consider an integer a with $k + 1$ digits (in base 2), say

$$a = a_k a_{k-1} \ldots a_2 a_1 a_0.$$

If $a_0 = 0$, then $a = 2(a_k a_{k-1} \ldots a_1)$, $f(a) = 2f(a_k \ldots a_1) - 1 = 2[b_k 2^{k-1} + \cdots + b_2 2 + b_1] - 1 = b_k 2^k + \cdots + b_2 2^2 + b_1 2 + b_0$, and the result holds. If $a_0 = 1$, then $a = 2(a_k a_{k-1} \ldots a_1) + 1$, $f(a) = 2f(a_k \ldots a_1) + 1 = 2(b_k 2^{k-1} + \cdots + b_1) + 1 = b_k 2^k + \cdots + b_1 2 + b_0$, and again the result holds.

This is a nice application of number representations. Notice how simple it is to compute: $f(25) = f(11001_2) = 11\bar{1}\bar{1}1_2 = 16 + 8 - 4 - 2 + 1 = 19$.

In the next example, a special number representation allows us to investigate and understand a set of real numbers of central importance in advanced analysis.

3.4.6. Let K denote the subset of $[0, 1]$ which consists of all numbers having a ternary expansion

$$\sum_{n=1}^{\infty} \frac{a_n}{3^n}$$

in which $a_n = 0$ or 2. This is called the *Cantor set*. Show that K is the complement of the union of disjoint open intervals I_n, $n = 1, 2, 3, \ldots$, whose lengths add to 1.

Solution. First observe that none of the numbers in the interval $I_1 = (\frac{1}{3}, \frac{2}{3})$ are in K. This is because numbers in this interval have ternary representations of the form

$$(.1a_2 a_3 a_4 \ldots)_3.$$

Similarly, none of the numbers in the interval $I_2 = (\frac{1}{9}, \frac{2}{9})$ are in K, because these numbers have ternary representations of the form

$$(.01a_3 a_4 a_5 \ldots)_3.$$

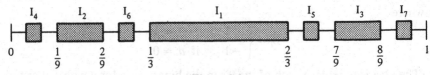

Figure 3.2.

Also, numbers in the interval $I_3 = (\frac{7}{9}, \frac{8}{9})$ have ternary representations of the form

$$(.21a_3a_4a_5 \cdots)_3,$$

so these are not in K. In the same manner, none of the intervals $I_4 = (\frac{1}{27}, \frac{2}{27})$, $I_5 = (\frac{19}{27}, \frac{20}{27})$, $I_6 = (\frac{7}{27}, \frac{8}{27})$, $I_7 = (\frac{25}{27}, \frac{26}{27})$ contain elements in K.

It is apparent that this process can be carried out systematically. Figure 3.2 and Table 3.1 help make the idea precise.

To find I_n (that is, X_n and Y_n) for an arbitrary positive integer n, write n in base 2 notation:

$$n = (a_k a_{k-1} \cdots a_2 a_1)_2$$

(i.e., $n = a_1 + 2a_2 + \cdots + 2^{k-1}a_k$, $a_i = 0$ or 1, $a_k \neq 0$), let $b_i = 2a_i$, $i = 1, 2, \ldots, k$, and set $I_n = (X_n, Y_n)$, where

$$X_n = \frac{b_1}{3} + \frac{b_2}{3^2} + \cdots + \frac{b_{k-1}}{3^{k-1}} + \frac{1}{3^k} = (.b_1 b_2 \cdots b_{k-1} 1)_3,$$

$$Y_n = \frac{b_1}{3} + \frac{b_3}{3^2} + \cdots + \frac{b_{k-1}}{3^{k-1}} + \frac{2}{3^k} = (.b_1 b_2 \cdots b_{k-1} 2)_3.$$

It is easy to see that X_n and Y_n are elements of K for each n (note that $X_n = b_1/3 + b_2/3^2 + \cdots + (b_{k-1})/(3^{k-1}) + \sum_{i=1}^{\infty}(2/3^{k+i})$, and that no elements in I_n are in K (the kth digit of every element of $I_n = (X_n, Y_n)$ is 1). From these facts it follows that the I_n's are disjoint.

Table 3.1. $I_n = (X_n, Y_n)$

n (base 10)	n (base 2)	X_n (base 3)	Y_n (base 3)	I_n (in fractional form)
1	1	0.1	0.2	$(\frac{1}{3}, \frac{2}{3})$
2	10	0.01	0.02	$(\frac{1}{9}, \frac{2}{9})$
3	11	0.21	0.22	$(\frac{7}{9}, \frac{8}{9})$
4	100	0.001	0.002	$(\frac{1}{27}, \frac{2}{27})$
5	101	0.201	0.202	$(\frac{19}{27}, \frac{20}{27})$
6	110	0.021	0.022	$(\frac{7}{27}, \frac{8}{27})$
7	111	0.221	0.222	$(\frac{25}{27}, \frac{26}{27})$
8	1000	0.0001	0.0002	$(\frac{1}{81}, \frac{2}{81})$
9	1001	0.2001	0.2002	$(\frac{55}{81}, \frac{56}{81})$

Also, the length of I_n is $1/3^k$, where $k = [\![\log_2 n]\!]$, and therefore

$$\sum_{n=1}^{\infty} I_n = \sum_{n=1}^{\infty} \frac{1}{3^{[\![\log_2 n]\!] + 1}} = \sum_{m=0}^{\infty} \left[\sum_{n=2^m}^{2^{m+1}-1} \left(\frac{1}{3^{[\![\log_2 n]\!] + 1}} \right) \right]$$

$$= \sum_{m=0}^{\infty} 2^m \left(\frac{1}{3^{m+1}} \right) = \frac{1}{3} \sum_{m=0}^{\infty} \left(\frac{2}{3} \right)^m = \frac{1}{3} \left(\frac{1}{1 - \frac{2}{3}} \right) = 1.$$

Our construction of the I_n's makes it clear that K is what remains after the intervals I_n are removed from $[0, 1]$, and the result is proved.

Problems

3.4.7. Prove that there does not exist an integer which is doubled when the initial digit is transferred to the end.

3.4.8. Find the smallest natural number n which has the following properties:

(i) its decimal representation has a 6 as its last digit, and
(ii) if the last digit 6 is erased and placed in front of the remaining digits, the resulting number is four times as large as the original number n.

3.4.9.

(a) Solve the following equation for the positive integers x and y:

$$(360 + 3x)^2 = 492y04.$$

(b) Devise a divisibility test for recognizing when a number is divisible by 11. (Hint: $10 \equiv -1 \pmod{11}$.)
(c) If $62ab427$ is a multiple of 99, find a and b.
(d) Find the probability that if the digits $0, 1, 2, \ldots, 9$ are placed in random order in the blank spaces of 5_383_8_2_936_5_8_203_9_3_76, the resulting number will be divisible by 396.

3.4.10. Given a two-pan balance and a system of weights of $1, 3, 3^2, 3^3,$ $3^4, \ldots$ pounds, show that one can weigh any integral number of pounds (weights can be put into either pan). (Hint: Show that any positive integer can be represented as sums and differences of powers of 3.)

3.4.11.

(a) Does the number $0.1234567891011121314\ldots$, which is obtained by writing successively all the integers, represent a rational number?
(b) Does the number $0.011010100010100\ldots$, where $a_n = 1$ if n is prime, 0 otherwise, represent a rational number?

3.4.12. Let $S = a_0 a_1 a_2 \ldots$, where $a_n = 0$ if there are an even number of 1's in the expression of n in base 2 and $a_n = 1$ if there are an odd number of

1's. Thus, $S = 01101001100\ldots$. Define $T = b_1b_2b_3\ldots$, where b_i is the number of 1's between the ith and the $(i + 1)$st occurrence of 0 in S. Thus, $T = 2102012\ldots$. Prove that T contains only three symbols $0, 1, 2$.

3.4.13. Show there is a one-to-one correspondence between the points of the closed interval $[0, 1]$ and the points of the open interval $(0, 1)$. Give an explicit description of such a correspondence.

Additional Examples

1.1.1 (Solution 5), 4.4.8, 5.2.5, 6.1.1, 6.1.4, 6.1.8, 6.2.13, 7.6.6.

3.5. Arithmetic of Complex Numbers

Recall that a complex number z can be written in several different forms:

$$\text{rectangular form:} \quad z = a + bi,$$

$$\text{polar form:} \quad z = r(\cos\theta + i\sin\theta),$$

$$\text{exponential form:} \quad z = re^{i\theta},$$

where a, b, r, and θ are related as in Figure 3.3, and $e^{i\theta} = \cos\theta + i\sin\theta$. The angle θ is the *argument* of z (determined only up to a multiple of 2π), and r is the *magnitude* (absolute value) of z; these are denoted by $\arg z$ and $|z|$ respectively. The numbers a and b are called the real part and the imaginary part of z respectively, and are denoted by $\text{Re}(z)$ and $\text{Im}(z)$.

If $z = a + bi$ and $w = c + di$, then $z + w = (a + c) + i(b + d)$ corresponds geometrically to the diagonal of the parallelogram having z and w as adjacent sides (see Figure 3.4).

If $z = re^{i\theta}$ and $w = se^{i\varphi}$, then $zw = rse^{i(\theta + \varphi)}$. Notice that $|zw| = rs = |z||w|$ and $\arg zw = \theta + \varphi = \arg z + \arg w$; that is, under multiplication, the absolute values multiply and the arguments add.

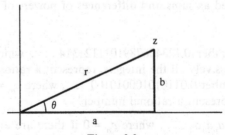

Figure 3.3.

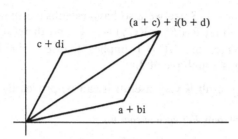

Figure 3.4.

3.5.1. If a, b, and n are positive integers, prove there exist integers x and y such that

$$\left(a^2 + b^2\right)^n = x^2 + y^2.$$

Solution. Let $z = a + bi$. Then $(a^2 + b^2)^n = (|z|^2)^n = |z|^{2n} = (|z|^n)^2$. But $z^n = x + iy$ for some integers x and y (because a and b are integers), so $(|z^n|)^2 = |x + iy|^2 = x^2 + y^2$, and the proof is complete.

3.5.2. Let n be an integer $\geqslant 3$, and let α, β, γ be complex numbers such that $\alpha^n = \beta^n = \gamma^n = 1$, $\alpha + \beta + \gamma = 0$. Show that n is a multiple of 3.

Solution. We may assume without loss of generality that $\alpha = 1$ (for if not, divide each side of $\alpha + \beta + \gamma = 0$ by α to get $1 + \beta/\alpha + \gamma/\alpha = 0$, and then set $\alpha_1 = 1$, $\beta_1 = \beta/\alpha$, $\gamma_1 = \gamma/\alpha$). We will assume that $0 \leqslant \arg \beta < \arg \gamma < 2\pi$.

Now, β and γ are of magnitude 1 (since $\beta^n = \gamma^n = 1$), so they lie on the unit circle (center $(0,0)$, radius 1). From the equation $\beta + \gamma = -1$, we can equate imaginary parts to see that $\mathrm{Im}(\beta + \gamma) = \mathrm{Im}(\beta) + \mathrm{Im}(\gamma) = 0$, or equivalently, $\mathrm{Im}(\beta) = -\mathrm{Im}(\gamma)$ (Figure 3.5). Equating real parts yields

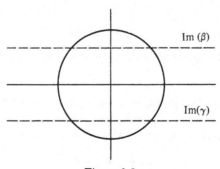

Figure 3.5.

$\mathrm{Re}(\beta) + \mathrm{Re}(\gamma) = -1$. Since we also have established that $|\beta| = |\gamma| = 1$, it must be the case that $\mathrm{Re}(\beta) = \mathrm{Re}(\gamma) = -\frac{1}{2}$, and therefore $\beta = e^{2\pi i/3}$, and $\gamma = e^{4\pi i/3}$. The fact that $\beta^n = 1$ implies that $e^{2\pi in/3} = 1$, and this can happen only if n is a multiple of 3.

The following result is very useful; it can be proved by induction.

de Moivre's Theorem. *For each integer n,*

$$(\cos\theta + i\sin\theta)^n = \cos n\theta + i\sin n\theta.$$

(*In exponential notation,* $(e^{i\theta})^n = e^{in\theta}$.)

3.5.3. Express $\cos 5\theta$ in terms of $\cos\theta$.

Solution. An efficient way to do this is to recognize that $\cos 5\theta$ is the real part of $e^{5i\theta}$. Then we can apply de Moivre's theorem:

$$\cos 5\theta + i\sin 5\theta = (\cos\theta + i\sin\theta)^5$$

$$= \cos^5\theta + 5\cos^4\theta\,(i\sin\theta) + 10\cos^3\theta\,(i^2\sin^2\theta)$$

$$+ 10\cos^2\theta\,(i^3\sin^3\theta) + 5\cos\theta\,(i^4\sin^4\theta) + i^5\sin^5\theta$$

$$= (\cos^5\theta - 10\cos^3\theta\sin^2\theta + 5\cos\theta\sin^4\theta)$$

$$+ i(\sin^5\theta - 10\sin^3\theta\cos^2\theta + 5\sin\theta\cos^4\theta).$$

Equating real and imaginary parts, we get

$$\cos 5\theta = \cos^5\theta - 10\cos^3\theta\sin^2\theta + 5\cos\theta\sin^4\theta,$$

$$\sin 5\theta = \sin^5\theta - 10\sin^3\theta\cos^2\theta + 5\sin\theta\cos^4\theta.$$

For the case of $\cos 5\theta$,

$$\cos 5\theta = \cos^5\theta - 10\cos^3\theta\,(1 - \cos^2\theta) + 5\cos\theta\,(1 - \cos^2\theta)^2$$

$$= 16\cos^5\theta - 20\cos^3\theta + 5\cos\theta.$$

3.5.4. Find constants $a_0, a_1, \ldots, a_6$ so that

$$\cos^6\theta = a_6\cos 6\theta + a_5\cos 5\theta + \cdots + a_1\cos\theta + a_0.$$

Solution. As in the last problem, we can do this very nicely by exploiting the relationship between trigonometric functions (especially the sine and cosine) and complex variables. In this case, write $\cos\theta$ in the form

$$\cos\theta = \frac{e^{i\theta} + e^{-i\theta}}{2}$$

and apply the binomial theorem to get

$$\cos^6\theta = \left(\frac{e^{i\theta} + e^{-i\theta}}{2}\right)^6$$

$$= \frac{1}{2^6}\left[(e^{i\theta})^6 + 6(e^{i\theta})^5(e^{-i\theta}) + 15(e^{i\theta})^4(e^{-i\theta})^2\right.$$

$$\left. + 20(e^{i\theta})^3(e^{-i\theta})^3 + 15(e^{i\theta})^2(e^{-i\theta})^4 + 6(e^{i\theta})(e^{-i\theta})^5 + (e^{-i\theta})^6\right]$$

$$= \frac{1}{2^6}\left[(e^{6i\theta} + e^{-6i\theta}) + 6(e^{4i\theta} + e^{-4i\theta}) + 15(e^{2i\theta} + e^{-2i\theta}) + 20\right]$$

$$= \frac{1}{2^6}\left[2\cos 6\theta + 2\times 6\cos 4\theta + 2\times 15\cos 2\theta + 20\right]$$

$$= \frac{1}{32}\left[\cos 6\theta + 6\cos 4\theta + 15\cos 20 + 10\right].$$

3.5.5. Let $G_n = x^n\sin nA + y^n\sin nB + z^n\sin nC$, where x, y, z, A, B, C are real and $A + B + C$ is an integral multiple of π. Prove that if $G_1 = G_2 = 0$, then $G_n = 0$ for all positive integral n.

Solution. A standard trick (similar to 3.5.3) is to recognize that G_n is the imaginary part of the expression

$$H_n = x^n e^{inA} + y^n e^{inB} + z^n e^{inC}.$$

Suppose that H_n is real for $n = 0, 1, \ldots, k$, and consider H_{k+1}. We have

$$H_1 H_k = H_{k+1} + H,$$

where

$$H = xe^{iA}y^k e^{ikB} + xe^{iA}z^k e^{ikC} + ye^{iB}x^k e^{ikA}$$

$$+ ye^{iB}z^k e^{ikC} + ze^{iC}x^k e^{ikA} + ze^{iC}y^k e^{ikB}$$

$$= xye^{i(A+B)}\left[y^{k-1}e^{i(k-1)B} + x^{k-1}e^{i(k-1)A}\right]$$

$$+ xze^{i(A+C)}\left[z^{k-1}e^{i(k-1)C} + x^{k-1}e^{i(k-1)A}\right]$$

$$+ yze^{i(B+C)}\left[y^{k-1}e^{i(k-1)B} + z^{k-1}e^{i(k-1)C}\right]$$

$$= xye^{i(A+B)}\left[H_{k-1} - z^{k-1}e^{i(k-1)C}\right]$$

$$+ xze^{i(A+C)}\left[H_{k-1} - y^{k-1}e^{i(k-1)B}\right]$$

$$+ yze^{i(B+C)}\left[H_{k-1} - x^{k-1}e^{i(k-1)A}\right]$$

$$= H_{k-1}\left[xye^{i(A+B)} + xze^{i(A+C)} + yze^{i(B+C)}\right] - xyze^{i(A+B+C)}H_{k-2}$$

$$= H_{k-1}K - xyze^{i(A+B+C)}H_{k-2},$$

where $K = xye^{i(A+B)} + xze^{i(A+C)} + yze^{i(B+C)}$.

Observe that $H_2 = H_1^2 + 2K$, and since H_1 and H_2 are real, by hypothesis, it must be the case that K is real. Also, by the inductive assumption, H_{k-1} and H_{k-2} are real. Because $A + B + C$ is a multiple of π, $e^{i(A+B+C)}$ is real. Putting these facts together, the formula of the last paragraph shows that H is real. Now since H_k is real, by the inductive assumption, and since $H_{k+1} = H_1 H_k - H$, it follows that H_{k+1} is real. Thus, the result of the problem follows by mathematical induction.

Problems

3.5.6.

(a) Given that $13 = 2^2 + 3^2$ and $74 = 5^2 + 7^2$, express $13 \times 74 = 962$ as a sum of two squares. (Hint: Let $z = 2 + 3i$, $w = 5 + 7i$, and use $|z|^2 |w|^2 = |zw|^2$.)
(b) Show that $4 \arctan \frac{1}{5} - \arctan \frac{1}{239} = \frac{1}{4}\pi$. (Hint: Consider $(5 - i)^4(1 + i)$.)

3.5.7. Suppose A is a complex number and n is a positive integer such that $A^n = 1$ and $(A + 1)^n = 1$. Prove that n is divisible by 6 and that $A^3 = 1$.

3.5.8. Show that

$$\binom{n}{1} - \binom{n}{3} + \binom{n}{5} - \binom{n}{7} + \cdots = 2^{n/2} \cos \frac{n\pi}{4} ,$$

and

$$\binom{n}{0} - \binom{n}{2} + \binom{n}{4} - \binom{n}{6} + \cdots = 2^{n/2} \sin \frac{n\pi}{4} .$$

(Hint: Consider $(1 + i)^n$.)

3.5.9. By considering possible magnitudes and arguments,

(a) find all values of $\sqrt[3]{-i}$;
(b) find which values of $(3 - 4i)^{-3/8}$ lie closest to the imaginary axis.

3.5.10.

(a) Prove that if $z = e^{i\theta}$ then $z - z^{-1} = 2i \sin \theta$ and $z^n - z^{-n} = 2i \sin n\theta$.
(b) Using part (a), express $\sin^{2n}\theta$ as a sum of sines whose angles are multiples of θ.

3.5.11. Show that

$$\tan n\theta = \frac{\binom{n}{1}\tan\theta - \binom{n}{3}\tan^3\theta + \cdots}{\binom{n}{0} - \binom{n}{2}\tan^2\theta + \cdots}$$

3.5.12.

(a) Prove that

$$\sum_{k=0}^{n}(-1)^{k+1}\binom{n}{k}\frac{\cos k\theta}{\cos^{k}\theta} = \begin{cases} 0 & \text{if } n \text{ is odd,} \\ (-1)^{1+n/2}\tan^{n}\theta & \text{if } n \text{ is even} \end{cases}$$

(Hint: Consider $i\tan\theta = -1 + (\cos\theta + i\sin\theta)/\cos\theta$.)

(b) Prove that

$$\sum_{k=0}^{n}(-1)^{k-1}\binom{n}{k}\cos^{k}\theta\cos k\theta = \begin{cases} (-1)^{(n+1)/2}\sin^{n}\theta\sin n\theta & \text{if } n \text{ is odd,} \\ (-1)^{1+n/2}\sin^{n}\theta\cos n\theta & \text{if } n \text{ is even.} \end{cases}$$

(Hint: Consider $-1 + \cos\theta[\cos\theta + i\sin\theta] = i\sin\theta[\cos\theta + i\sin\theta]$.)

3.5.13. Prove that

$$\frac{\cos n\theta}{\cos^{n}\theta} = 1 - \binom{n}{2}\tan^{2}\theta + \binom{n}{4}\tan^{4}\theta - \cdots.$$

3.5.14. Show that if $e^{i\theta}$ satisfies the equation $z^{n} + a_{n-1}z^{n-1} + \cdots + a_{1}z + a_{0} = 0$, where the a_{i} are real, then $a_{n-1}\sin\theta + a_{n-2}\sin 2\theta + \cdots + a_{1}\sin(n-1)\theta + a_{0}\sin n\theta = 0$.

Additional Examples

1.3.2, 4.2.10, 4.2.11, 4.2.13, 4.2.15, 4.2.17, 4.2.20, 4.2.22, 4.3.18, 5.2.2, 5.2.3, 5.2.11, 5.3.4, 5.3.10, 5.4.11, 5.4.28, 5.4.29. Also, see Section 8.4 (Complex Numbers in Geometry).

Chapter 4. Algebra

Algebra is one of the oldest branches of mathematics, and it continues to be one of the most active areas of mathematical research. The subject is still rich in new ideas, and it shows no signs of soon becoming exhausted or barren.

In high-school algebra one learns to manipulate equations and formulas into equivalent forms which are more understandable and interpretable. A large proportion of the problems in this book attest to the usefulness of this basic subject. One of the most important algebraic manipulations involves factorization of algebraic expressions. In the first section we will look at problems whose solution depends upon knowing some elementary factorization formulas.

The middle two sections are devoted to problems from classical algebra: namely, the study of polynomials. Much of this material once belonged to a branch of algebra called the theory of equations. The rudiments of this subject are now scattered throughout the high-school and college curriculum. In these sections we draw together the ideas of this subject that constitute essential knowledge for problem solving.

In the final section we introduce those topics which professional mathematicians think of when they speak of algebra. Here the emphasis is on formal systems and formal thinking. The subject contains a whole new world of concepts which generalize the classical ideas and methods. We introduce the most fundamental structures that make up the subject matter: groups, rings, and fields.

4.1. Algebraic Identities

In this section we will look at applications of some of the most basic factorization formulas, which include the following:

$$a^2 - b^2 = (a - b)(a + b),$$

$$a^2 + 2ab + b^2 = (a + b)^2,$$

$$a^2 + b^2 + c^2 + 2ab + 2ac + 2bc = (a + b + c)^2,$$

$$a^n - b^n = (a - b)(a^{n-1} + a^{n-2}b + \cdots + ab^{n-2} + b^{n-1}).$$

If n is an odd positive integer, we can replace b by $-b$ in the last formula and get a formula for the factorization of the sum of two perfect nth powers:

$$a^n + b^n = (a + b)(a^{n-1} - a^{n-2}b + \cdots - ab^{n-2} + b^{n-1}), \qquad n \text{ odd.}$$

4.1.1. Show that $n^4 - 20n^2 + 4$ is composite when n is any integer.

Solution. The idea is to try to factor the expression. If we proceed $n^4 - 20n^2 + 4 = (n^4 - 20n^2 + 100) - 96 = (n^2 - 10)^2 - 96$, we are stymied because 96 is not a perfect square. It does work, however, to argue that $n^4 - 20n^2 + 4 = (n^4 - 4n^2 + 4) - 16n^2 = (n^2 - 2)^2 - (4n)^2 = (n^2 - 2 - 4n) \times (n^2 - 2 + 4n)$. If we can show that neither of these factors equals ± 1, we are done.

Suppose $n^2 - 2 - 4n = 1$; or equivalently, $n^2 - 4n - 3 = 0$. By the quadratic formula, $n = 2 \pm \sqrt{7}$, and this is not an integer. Thus, if n is an integer, $n^2 - 2 - 4n$ is not equal to 1. A similar argument works for the other three cases.

4.1.2. Determine all solutions in real numbers x, y, z, w of the system

$$x + y + z = w,$$

$$\frac{1}{x} + \frac{1}{y} + \frac{1}{z} = \frac{1}{w}.$$

Solution. Some initial guesses lead us to suspect solutions only when one of x, y, z is equal to w and the other two are negatives of one another (for example, $x = w$, $y = -z$). Certainly, these are solutions, but how can we prove there are no others?

From the second equation,

$$\frac{yz + xz + xy}{xyz} = \frac{1}{w},$$

and this, together with the first equation yields

$$(x + y + z)(yz + xz + xy) = xyz.$$

This expands to

$$x^2y + x^2z + y^2x + y^2z + z^2x + z^2y + 2xyz = 0,$$

which in turn factors into

$$(x + y)(x + z)(y + z) = 0.$$

Our initial conjecture follows (i.e. one of $x + y, x + z, y + z$ equals zero, say $y = -z$, and thus $x = x + y + z = w$).

4.1.3.

(a) Find all pairs (m, n) of positive integers such that $|3^m - 2^n| = 1$.
(b) Find all pairs (m, n) of integers larger than 1 such that $|p^m - q^n| = 1$, where p and q are primes.

Solution. (a) When $m = 1$ or 2 we quickly find the solutions

$$(m, n) = (1, 1), (1, 2), (2, 3).$$

We will show there are no others.

Suppose that (m, n) is a solution of $|3^m - 2^n| = 1$, where $m > 2$ (and hence $n > 3$). Then $3^m - 2^n = 1$, or $3^m - 2^n = -1$.

Case 1. Suppose $3^m - 2^n = -1$, $n > 3$. Then $3^m \equiv -1 \pmod 8$. But this congruence cannot hold, since $3^m \equiv 1$ or $3 \pmod 8$, depending upon whether m is even or odd ($3 \equiv 3 \pmod 8$, $3^2 \equiv 1 \pmod 8$, $3^3 \equiv 3 \pmod 8$, $3^4 \equiv 1 \pmod 8$, ...).

Case 2. Suppose $3^m - 2^n = 1$, $n > 3$. Then $3^m \equiv 1 \pmod 8$, so m is even, say $m = 2k$, $k > 1$. Then $2^n = 3^{2k} - 1 = (3^k - 1)(3^k + 1)$. By unique factorization, $3^k + 1 = 2^r$ for some r, $r > 3$. But, by case 1, we know this cannot happen. This completes the proof of part (a).

(b) It is immediate that not both p and q are odd, for this would imply that $p^m - q^n$ is even. So suppose that $q = 2$. We will show, by using only the algebraic identities of this section, that the only solution is that found in part (a), namely $|3^2 - 2^3| = 1$.

Suppose m and n are larger than 1, and that $|p^m - 2^n| = 1$. It cannot be the case that m and n are both even, for if $m = 2r$ and $n = 2s$, then

$$1 = |p^m - 2^n| = |p^{2r} - 2^{2s}| = |p^r - 2^s||p^r + 2^s|,$$

and this is impossible (since $p^r + 2^s > 1$).

Suppose that m is odd. Then

$$2^n = p^m \pm 1 = (p \pm 1)(p^{m-1} \mp p^{m-2} + \cdots - p + 1),$$

and this is impossible, since the last factor on the right side of the equation is an odd number larger than 1.

Therefore, it must be the case that m is even and n is odd.

Suppose $m = 2^r k$, where k is odd and suppose $k > 1$. Then

$$2^n = p^m \pm 1 = \left(p^{2^r}\right)^k \pm 1 = \left(p^{2^r} \pm 1\right)\left(\left(p^{2^r}\right)^{k-1} \pm \cdots - \left(p^{2^r}\right) + 1\right),$$

and again the factor on the right is odd, a contradiction.

Therefore, $m = 2^r$ for some positive integer r and n is odd, and our equation has the form $|p^{2^r} - 2^n| = 1$. Either $p^{2^r} - 2^n = 1$ or $p^{2^r} - 2^n = -1$.

Case 1. If $p^{2^r} - 2^n = -1$, then

$$p^{2^r} = 2^n - 1 = (2 - 1)(2^{n-1} + 2^{n-2} + \cdots + 2 + 1) \equiv 3 \pmod 4),$$

but this is impossible, since for any odd integer x, $x^2 \equiv 1 \pmod 4$.

Case 2. If $p^{2^r} - 2^n = 1$, then $2^n = p^{2^r} - 1 = (p^{2^{r-1}} - 1)(p^{2^{r-1}} + 1)$. The only way both $p^{2^{r-1}} - 1$ and $p^{2^{r-1}} + 1$ could be powers of 2 is for $p^{2^{r-1}} - 1 = 2$ and $p^{2^{r-1}} + 1 = 4$. Adding these yields $p^{2^{r-1}} = 3$, and this implies that $p = 3$, $r = 1$, $m = 2$, and $n = 3$. This completes the proof.

4.1.4. Prove that there are no prime numbers in the infinite sequence of integers

$$10001, 100010001, 1000100010001, \ldots .$$

Solution. The terms of the sequence can be written as

$$1 + 10^4, 1 + 10^4 + 10^8, \ldots, 1 + 10^4 + \cdots + 10^{4n}, \ldots .$$

Consider, more generally, then, the sequence

$$1 + x^4, 1 + x^4 + x^8, \ldots, 1 + x^4 + \cdots + x^{4n}, \ldots$$

for an arbitrary integer x, $x > 1$.

If n is odd, say $n = 2m + 1$,

$$1 + x^4 + x^8 + \cdots + x^{4(2m+1)}$$
$$= (1 + x^4) + x^8(1 + x^4) + \cdots + x^{8m}(1 + x^4)$$
$$= (1 + x^4)(1 + x^8 + \cdots + x^{8m}).$$

Thus, if $m > 0$, the number is composite. For $m = 0$ and $x = 10$, we also get a composite number, since $10001 = 73 \times 137$.

Suppose n is even, say $n = 2m$. Then

$$1 + x^4 + \cdots + x^{4(2m)} = \frac{1 - (x^4)^{2m+1}}{1 - x^4}$$

$$= \left(\frac{1 - (x^{2m+1})^2}{1 - x^2}\right)\left(\frac{1 + (x^{2m+1})^2}{1 + x^2}\right)$$

$$= \left(1 + x^2 + \cdots + (x^2)^{2m}\right)$$

$$\times \left(1 - x^2 + \cdots + (x^2)^{2m}\right).$$

This factorization shows the number is composite.

Problems

4.1.5.

(a) If a and b are consecutive integers, show that $a^2 + b^2 + (ab)^2$ is a perfect square.
(b) If $2a$ is the harmonic mean of b and c (i.e., $2a = 2/(1/b + 1/c)$), show that the sum of the squares of the three numbers a, b, and c is the square of a rational number.
(c) If N differs from two successive squares between which it lies by x and y respectively, prove that $N - xy$ is a square.

4.1.6. Prove that there are infinitely many natural numbers a with the following property: The number $n^4 + a$ is not prime for any natural number n.

4.1.7. Supposing that an integer n is the sum of two triangular numbers,

$$n = \frac{a^2 + a}{2} + \frac{b^2 + b}{2},$$

write $4n + 1$ as the sum of two squares, $4n + 1 = x^2 + y^2$, and show how x and y can be expressed in terms of a and b.

Show that, conversely, if $4n + 1 = x^2 + y^2$, then n is the sum of two triangular numbers.

4.1.8. Let N be the number which when expressed in decimal notation consists of 91 ones:

$$N = \underbrace{111 \ldots 1}_{91}.$$

Show that N is a composite number.

4.1.9. Prove that any two numbers of the following sequence are relatively prime:

$$2 + 1, 2^2 + 1, 2^4 + 1, 2^8 + 1, \ldots, 2^{2^n} + 1, \ldots .$$

Show that this result proves that there are an infinite number of primes.

4.1.10. Determine all triplets of integers (x, y, z) satisfying the equation

$$x^3 + y^3 + z^3 = (x + y + z)^3.$$

Additional Examples

1.8.4, 1.12.7, 3.3.6, 4.2.5, 5.2.15, 5.3.7, 7.1.11. Also, see Section 5.2 (Geometric Series).

4.2. Unique Factorization of Polynomials

A *polynomial of degree n* (n a nonnegative integer) in the variable x is an expression of the form

$$a_n x^n + a_{n-1} x^{n-1} + \cdots + a_1 x + a_0,$$

where $a_0, a_1, \ldots, a_n$ are constants (called the *coefficients*), and $a_n \neq 0$. A polynomial all of whose coefficients are zero is called the *zero polynomial*; no degree is assigned to the zero polynomial. The coefficient a_n is called the *leading coefficient*; if it is equal to 1 we say the polynomial is a *monic polynomial*. Two polynomials are called (identically) *equal* if their coefficients are equal term for term, that is, their coefficients for the same power of the variable are equal.

If the coefficients of the polynomial $P(x)$ are integers, we say that $P(x)$ is a polynomial *over the integers*; similarly if the coefficients are rationals, we say the polynomial is *over the rationals*, and so forth.

In many respects polynomials are like integers. They can be added, subtracted, and multiplied; however, just as in the case of integers, when a polynomial divides another the result will be a quotient polynomial plus a remainder polynomial (more on this later). A polynomial F *divides* a polynomial G (exactly) if there is a polynomial Q such that $G = QF$ (that is, G is a multiple of F). A polynomial H is a *greatest common divisor* of polynomials F and G if and only if (1) H divides F and G and (2) if K is any other polynomial that divides F and G, then K divides H. It can be shown that H is unique up to a constant multiple.

Also, as in the case of integers, there is a division algorithm.

Division Algorithm for Polynomials. *If $F(x)$ and $G(x)$ are polynomials over a field K (for example, K might be the rationals, the reals, the complexes, the integers modulo p for p prime), there exist unique polynomials $Q(x)$ and $R(x)$ over the field K such that*

$$F(x) = Q(x)G(x) + R(x),$$

where $R(x) \equiv 0$ or $\deg R(x) < \deg G(x)$ (deg denotes degree).

Moreover, if K is an integral domain (such as the integers), the same result holds provided $G(x)$ is a monic polynomial.

As an example of the division algorithm for polynomials, let $F(x) = 3x^5 + 2x^2 - 5$ and $G(x) = 2x^3 + 6x + 3$. Then

$$
\begin{array}{r}
\frac{3}{2}x^2 - \frac{9}{2} \\
2x^3 + 6x + 1 \overline{\big)\ 3x^5 \qquad\quad + 2x^2 \qquad\quad - 5} \\
\underline{3x^5 + 9x^3 + \frac{3}{2}x^2} \\
-9x^3 + \frac{1}{2}x^2 \qquad\quad - 5 \\
\underline{-9x^3 \qquad\qquad - 27x - \frac{9}{2}} \\
\frac{1}{2}x^2 + 27x - \frac{1}{2}
\end{array}
$$

In this case, $Q(x) = \frac{3}{2}x^2 - \frac{9}{2}$, and $R(x) = \frac{1}{2}x^2 + 27x - \frac{1}{2}$. (This example should make it clear that the algorithm will work in the general case only if the coefficients come from a field; however, it also should be clear that an integral domain is sufficient if the divisor is monic.)

As in the case of the integers, the division algorithm can be used to find the greatest common divisor of two polynomials. Furthermore, as in the case of the integers, if F and G are polynomials (over a field K), there are polynomials S and T (over K) such that

$$\gcd(F, G) = SF + TG,$$

where $\gcd(F, G)$ denotes the greatest common divisor of F and G.

4.2.1. Find a polynomial $P(x)$ such that $P(x)$ is divisible by $x^2 + 1$ and $P(x) + 1$ is divisible by $x^3 + x^2 + 1$.

Solution. By the conditions of the problem, there are polynomials $S(x)$ and $T(x)$ such that

$$P(x) = (x^2 + 1)S(x),$$
$$P(x) + 1 = (x^3 + x^2 + 1)T(x).$$

It follows that $(x^2 + 1)S(x) = (x^3 + x^2 + 1)T(x) - 1$, or equivalently

$$(x^3 + x^2 + 1)T(x) - (x^2 + 1)S(x) = 1.$$

By our remarks preceding the example, $x^3 + x^2 + 1$ and $x^2 + 1$ are "relatively prime" and we can use the Euclidean algorithm for polynomials to find $S(x)$ and $T(x)$. Thus, we have

$$x^3 + x^2 + 1 = (x + 1)(x^2 + 1) + (-x),$$
$$x^2 + 1 = -x(-x) + 1,$$

and "working backwards," we have

$$1 = (x^2 + 1) + x(-x)$$
$$= (x^2 + 1) + x\left[(x^3 + x^2 + 1) - (x + 1)(x^2 + 1)\right]$$
$$= (x^2 + 1)\left[1 - x(x + 1)\right] + x\left[x^3 + x^2 + 1\right]$$
$$= (x^3 + x^2 + 1)x - (x^2 + 1)(x^2 + x - 1).$$

In this form, we find that we can take $S(x) = x^2 + x - 1$ and $T(x) = x$. It follows that

$$P(x) = (x^2 + 1)(x^2 + x - 1).$$

4.2.2. Prove that the fraction $(n^3 + 2n)/(n^4 + 3n^2 + 1)$ is irreducible for every natural number n.

Solution. We have

$$n^4 + 3n^2 + 1 = n(n^3 + 2n) + (n^2 + 1),$$

$$n^3 + 2n = n(n^2 + 1) + n,$$

$$n^2 + 1 = n(n) + 1,$$

$$n = n(1).$$

It follows that $\gcd(n^4 + 3n^2 + 1, n^3 + 2n) = 1$, and the proof is complete.

Let $F(x)$ be a polynomial over an integral domain D, and consider the polynomial equation $F(x) = 0$. If an element a of D is such that $F(a) = 0$, we say that a is a *root* of $F(x) = 0$, or that a is a *zero of* $F(x)$. The following very useful theorem is an easy application of the division algorithm.

Factor Theorem. *If $F(x)$ is a polynomial over an integral domain D, an element a of D is a root of $F(x) = 0$ if and only if $x - a$ is a factor of $F(x)$.*

By repeated application of the factor theorem, we can prove that there is a unique nonnegative integer m and a unique polynomial $G(x)$ over D such that

$$F(x) = (x - a)^m G(x),$$

where $G(a) \neq 0$. In this case, we say that a is a *zero of multiplicity* m.

The next two examples illustrate the use of the factor theorem.

4.2.3. Given the polynomial $F(x) = x^n + a_{n-1}x^{n-1} + \cdots + a_1 x + a_0$ with integral coefficients $a_0, a_1, \ldots, a_{n-1}$, and given also that there exist four distinct integers a, b, c, d such that $F(a) = F(b) = F(c) = F(d) = 5$, show that there is no integer k such that $F(k) = 8$.

Solution. Let $G(x) = F(x) - 5$. By the factor theorem, $x - a$, $x - b$, $x - c$, and $x - d$ are factors of $G(x)$, and we may write

$$G(x) = (x - a)(x - b)(x - c)(x - d)H(x),$$

where $H(x)$ is a polynomial with integer coefficients. If k is an integer such that $F(k) = 8$, then $G(k) = F(k) - 5 = 8 - 5 = 3$, or equivalently,

$$(k - a)(k - b)(k - c)(k - d)H(k) = 3.$$

The left side represents a product of five integers, and each of the integers $k - a, k - b, k - c, k - d$ must be distinct, since a, b, c, d are distinct. But this is impossible, since at most one of the numbers $k - a, k - b, k - c, k - d$ can equal ± 3, so the other three must be ± 1. Thus, such a k cannot be found.

4.2.4. Prove that if $F(x)$ is a polynomial with integral coefficients, and there exists an integer k such that none of the integers $F(1), F(2), \ldots, F(k)$ is divisible by k, then $F(x)$ has no integral zero.

Solution. It is equivalent to prove that if $F(x)$ has an integral zero, say r, then for any positive integer k, at least one of $F(1), F(2), \ldots, F(k)$ is divisible by k. So suppose $F(r) = 0$. By the factor theorem, we can write

$$F(x) = (x - r)G(x),$$

where $G(x)$ is a polynomial with integer coefficients. From the division algorithm for integers, there are integers q and s such that $r = qk + s$, $0 < s \leqslant k$ (note the inequalities on s). Substituting $s = r - qk$ into the equation above, we get

$$F(s) = (s - r)G(s) = -qkG(s).$$

This equation shows that $F(s)$ is divisible by k ($G(s)$ is an integer), and this completes the proof.

A simpler approach for this problem, based on modular arithmetic, is to observe that if $a \equiv b \pmod{k}$ then $F(a) \equiv F(b) \pmod{k}$. The result follows directly from the fact that for any given integer a, $F(a)$ is congruent to one of $F(1), \ldots, F(k)$ modulo k, and by assumption, none of these is divisible by k.

The unique-factorization theorem for integers states that every integer can be written uniquely as a product of primes. There is a similar theorem for polynomials: every polynomial over a field can be written uniquely as a product of irreducible polynomials (i.e., prime factors). In the case of the complex numbers, the irreducible factors are the first-degree (linear) polynomials. In the case of the real numbers, the irreducible polynomials are the linear polynomials and the quadratic polynomials with negative discriminant (that is, those of the form $ax^2 + bx + c$, where $b^2 - 4ac < 0$).

As in the case of integers, unique factorization is often a useful way of representing a polynomial. The next two examples illustrate the idea.

4.2.5. Prove that every polynomial over the complex numbers has a nonzero polynomial multiple whose exponents are all divisible by 1,000,000.

Solution. Let the given polynomial be represented by the unique factorization

$$P(x) = A(x - s_1)^{m_1}(x - s_2)^{m_2} \cdots (x - s_k)^{m_k},$$

where A is a constant, $s_1, \ldots, s_k$ are the roots of $P(x)$ of multiplicities $m_1, \ldots, m_k$ respectively. For any positive integer a (e.g., $a = 1{,}000{,}000$),

$(x^a - s_i^a)/(x - s_i)$ is a polynomial over the complex numbers (see Section 4.1). Set

$$Q(x) = x^a \left(\frac{x^a - s_1^a}{x - s_1} \right)^{m_1} \cdots \left(\frac{x^a - s_k^a}{x - s_k} \right)^{m_k}.$$

Then $Q(x)$ is a polynomial over the complex numbers, and

$$P(x)Q(x) = A(x - s_1)^{m_1} \cdots (x - s_k)^{m_k}$$

$$\times \left(x^a \left(\frac{(x^a - s_1^a)^{m_1}}{(x - s_1)^{m_1}} \right) \cdots \left(\frac{(x^a - s_k^a)^{m_k}}{(x - s_k)^{m_k}} \right) \right)$$

$$= Ax^a (x^a - s_1^a)^{m_1} \cdots (x^a - s_k^a)^{m_k}$$

is a polynomial all of whose exponents are divisible by a.

4.2.6. Let f be a polynomial with real coefficients. Show that all the zeros of f are real if and only if f^2 cannot be written as the sum of squares

$$f^2 = g^2 + h^2$$

where g and h are polynomials with real coefficients and $\deg g \neq \deg h$.

Solution. Suppose $f^2 = g^2 + h^2$, where g and h are polynomials with real coefficients, $\deg g \neq \deg h$, and suppose that all the zeros of f are real. Write f in factored form:

$$f(x) = A(x - a_1)^{m_1} \cdots (x - a_k)^{m_k},$$

where A is a nonzero real number.

From the equation

$$A^2 (x - a_1)^{2m_1} \cdots (x - a_k)^{2m_k} = (g(x))^2 + (h(x))^2$$

it follows that for each $i = 1, 2, \ldots, k$,

$$0 = (g(a_i))^2 + (h(a_i))^2.$$

Since $g(a_i)$ and $h(a_i)$ are both real numbers, it must be the case that $g(a_i) = 0$ and $h(a_i) = 0$. In fact, it follows that the multiplicity of these zeros is at least m_i. Thus, the factor theorem implies that there will be polynomials $g_1(x)$ and $h_1(x)$ with real coefficients such that $g(x) = f(x)g_1(x)$ and $h(x) = f(x)h_1(x)$. It follows that

$$1 = (g_1(x))^2 + (h_1(x))^2.$$

But this equation is impossible, because $\deg g_1 \neq \deg h_1$ (that is, not both of g_1 and h_1 are constants). This contradiction implies that f must have a zero that is not a real number.

Now suppose that not all the zeros of f are real numbers, and write f in factored form:

$$f(x) = A(x - a_1)^{m_1} \cdots (x - a_r)^{m_r}(x^2 + b_1 x + c_1)^{n_1} \cdots (x^2 + b_s x + c_s)^{n_s},$$

where A is a real number, $m_1, \ldots, m_r$ are nonnegative integers, s is a positive integer and $n_1, \ldots, n_s$ are positive integers, a_i, b_j, c_j are real numbers, and $b_j^2 - 4c_j \leqslant 0$ for $j = 1, \ldots, s$. We have

$$x^2 + b_j x + c_j = \left(x^2 + b_j x + \tfrac{1}{4}b_j^2\right) + \left(c_j - \tfrac{1}{4}b_j^2\right)$$

$$= \left(x + \tfrac{1}{2}b_j\right)^2 + \left(\tfrac{1}{2}\sqrt{4c_j - b_j^2}\right)^2,$$

which shows that each quadratic factor of f is a sum of squares. Replace each quadratic factor in the unique factorization of f^2 by its representation as a sum of squares. This yields an equation of the form

$$f^2(x) = A^2(x - a_1)^{2m_1} \cdots (x - a_r)^{2m_r}$$

$$\times \left(g_1^2(x) + h_1^2(x)\right)^{n_1} \cdots \left(g_s^2(x) + h_s^2(x)\right)^{n_s},$$

where $g_1, \ldots, g_s, h_1, \ldots, h_s$ are polynomials, deg $g_i = 1$, and deg $h_i = 0$.

The result now follows by repeated use of the fact that the product of a sum of two squares with another sum of two squares is itself expressible as a sum of two squares:

$$(f^2 + g^2)(h^2 + k^2) = (fh - gk)^2 + (fk + gh)^2.$$

Also, in this identity, if deg $f >$ deg g and deg $h >$ deg k, then deg$(fh - gk)$ $>$ deg$(fk + gh)$. Thus, we see that there are polynomials $g(x)$ and $h(x)$ with real coefficients, deg $g(x) \neq$ deg $h(x)$ such that $f^2 = g^2 + h^2$.

Problems

4.2.7. Find polynomials $F(x)$ and $G(x)$ such that

$$(x^8 - 1)F(x) + (x^5 - 1)G(x) = x - 1.$$

4.2.8. What is the greatest common divisor of $x^n - 1$ and $x^m - 1$?

4.2.9. Let $f(x)$ be a polynomial leaving the remainder A when divided by $x - a$ and the remainder B when divided by $x - b$, $a \neq b$. Find the remainder when $f(x)$ is divided by $(x - a)(x - b)$.

4.2.10. Show that $x^{4a} + x^{4b+1} + x^{4c+2} + x^{4d+3}$, a, b, c, d positive integers, is divisible by $x^3 + x^2 + x + 1$. (Hint: $x^3 + x^2 + x + 1 = (x^2 + 1)(x + 1)$.)

4.2.11. Show that each polynomial $(\cos \theta + x \sin \theta)^n - \cos n\theta - x \sin n\theta$ is divisible by $x^2 + 1$.

4.2.12. For what n is the polynomial $1 + x^2 + x^4 + \cdots + x^{2n-2}$ divisible by the polynomial $1 + x + x^2 + \cdots + x^{n-1}$?

4.2.13. A real number is called *algebraic* if it is a zero of a polynomial with integer coefficients.

(a) Show that $\sqrt{2} + \sqrt{3}$ is algebraic.
(b) Show that $\cos(\pi/2n)$ is algebraic for each positive integer n. (Hint: Use de Moivre's theorem to express $\cos nx$ as a polynomial in $\cos x$.)

4.2.14. If $P(x)$ is a monic polynomial with integral coefficients and k is any integer, must there exist an integer m for which there are at least k distinct prime divisors of $P(m)$? (Hint: First prove, by induction, that there are k distinct primes $q_1, \ldots, q_k$ and k integers $n_1, \ldots, n_k$ such that q_i divides $P(n_i)$ for $i = 1, \ldots, k$. Then prove that a prime q divides $P(n)$ if and only if q divides $P(n + sq)$ for all integers s. An affirmative answer follows from these facts together with an application of the Chinese Remainder Theorem.)

4.2.15.

(a) Factor $x^8 + x^4 + 1$ into irreducible factors (i) over the rationals, (ii) over the reals, (iii) over the complex numbers.
(b) Factor $x^n - 1$ over the complex numbers.
(c) Factor $x^4 - 2x^3 + 6x^2 + 22x + 13$ over the complex numbers, given that $2 + 3i$ is a zero.

4.2.16. Here are two results that are useful in factoring polynomials with integer coefficients into irreducibles.

Rational-Root Theorem. *If* $P(x) = a_n x^n + a_{n-1} x^{n-1} + \cdots + a_1 x + a_0$ *is a polynomial with integer coefficients, and if the rational number* r/s *(r and s relatively prime integers) is a root of* $P(x) = 0$, *then r divides a_0 and s divides a_n.*

Gauss' Lemma. *Let* $P(x)$ *be a polynomial with integer coefficients. If* $P(x)$ *can be factored into a product of two polynomials with rational coefficients, then* $P(x)$ *can be factored into a product of two polynomials with integer coefficients.*

(a) Let $f(x) = a_n x^n + a_{n-1} x^{n-1} + \cdots + a_1 x + a_0$ be a polynomial of degree n with integral coefficients. If a_0, a_n, and $f(1)$ are odd, prove that $f(x) = 0$ has no rational roots.
(b) For what integer a does $x^2 - x + a$ divide $x^{13} + x + 90$?

4.2.17.

(a) Suppose $f(x)$ is a polynomial over the real numbers and $g(x)$ is a divisor of $f(x)$ and $f'(x)$ which is irreducible or square free. Show that $(g(x))^2$ divides $f(x)$. (This fact can be used to check $f(x)$ for multiple roots.)

(b) Use the idea of part (a) to factor $x^6 + x^4 + 3x^2 + 2x + 2$ into a product of irreducibles over the complex numbers.

4.2.18. Determine all pairs of positive integers (m, n) such that $1 + x^n + x^{2n} + \cdots + x^{mn}$ is divisible by $1 + x + x^2 + \cdots + x^m$.

4.2.19.

(a) Let $F(x)$ be a polynomial over the real numbers. Prove that a is a zero of multiplicity m if and only if $F(a) = F'(a) = \cdots = F^{(m)}(a) = 0$ and $F^{(m+1)}(a) \neq 0$.
(b) The equation $f(x) = x^n - nx + n - 1 = 0$, $n > 1$, is satisfied by $x = 1$. What is the multiplicity of this root?

4.2.20. If $n > 1$, show that $(x + 1)^n - x^n - 1 = 0$ has a multiple root if and only if $n - 1$ is divisible by 6.

4.2.21. Let $P(x)$ be a polynomial with real coefficients, and assume that $P(x) \geqslant 0$ for all x. Prove that $P(x)$ can be expressed in the form $(Q_1(x))^2 + (Q_2(x))^2 + \cdots + (Q_n(x))^2$ where $Q_1(x), Q_2(x), \ldots, Q_n(x)$ are polynomials with real coefficients.

4.2.22.

(a) Set $\omega = \cos(2\pi/n) + i\sin(2\pi/n)$. Show that
$$x^{n-1} + x^{n-2} + \cdots + x + 1 = (x - \omega)(x - \omega^2) \cdots (x - \omega^{n-1}).$$
(b) Set $x = 1$ and take the absolute value of each side to show that
$$\sin\frac{\pi}{n}\sin\frac{2\pi}{n} \cdots \sin\frac{(n-1)\pi}{n} = \frac{n}{2^{n-1}}.$$

Additional Examples

1.12.2, 1.12.5, 6.5.13, 6.9.3.

4.3. The Identity Theorem

Let P be a nonzero polynomial of degree n over an integral domain D. According to the factor theorem, if a is a root of $P(x) = 0$, there is a polynomial Q of degree $n - 1$ such that $P(x) = (x - a)Q(x)$. Using this fact, an easy induction argument shows that P has at most n zeros.

The preceding observation has a very important corollary. Suppose that F and G are polynomials over a domain D, each of degree less than or equal to n, and suppose that F and G are equal for $n + 1$ distinct values. Then $F - G$ is a polynomial of degree less than $n + 1$ with $n + 1$ zeros. If $F - G$ is not the zero polynomial, we have a contradiction to the reasoning in the previous paragraph. Therefore, $F - G$ is the zero polynomial, and it follows that F equals G (coefficient for coefficient). (For another proof, see 6.5.10.)

Identity Theorem. *Suppose that two polynomials in x over an infinite integral domain are each of degree $\leqslant n$. If these polynomials have equal values for more than n distinct values of x, then the two polynomials are identical.*

4.3.1. Determine all polynomials $P(x)$ such that $P(x^2 + 1) = (P(x))^2 + 1$ and $P(0) = 0$.

Solution. We start by testing some cases:

$$P(1) = P(0^2 + 1) = (P(0))^2 + 1 = 1,$$

$$P(2) = P(1^2 + 1) = (P(1))^2 + 1 = 1 + 1 = 2,$$

$$P(5) = P(2^2 + 1) = (P(2))^2 + 1 = 4 + 1 = 5,$$

$$P(26) = P(5^2 + 1) = (P(5))^2 + 1 = 5^2 + 1 = 26.$$

In general, define $x_0 = 0$, and for $n > 0$ define $x_n = x_{n-1}^2 + 1$. Then an easy induction argument shows that $P(x_n) = x_n$. Thus, the polynomial $P(x)$ and the polynomial x are equal for an infinite number of integers, and therefore, by the identity theorem, $P(x) \equiv x$. That is, there is only one polynomial with the stated property, namely, $P(x) = x$.

4.3.2. Prove that if m an n are positive integers and $1 \leqslant k \leqslant n$, then

$$\sum_{r=0}^{k} \binom{m}{k-r} \binom{n}{r} = \binom{m+n}{k}.$$

Solution. We proved this identity in Chapter 1 (see 1.3.4) by using a counting argument. Here is another proof, based on the identity theorem. The technique is standard: the polynomials $(1 + x)^m (1 + x)^n$ and $(1 + x)^{m+n}$ are equal for all values of x. Therefore, by the identity theorem, their coefficients are equal; that is, for each k, the coefficient of x^k in $(1 + x)^m (1 + x)^n$ is equal to the coefficient of x^k in $(1 + x)^{m+n}$. It follows that

$$\sum_{r=0}^{k} \binom{m}{k-r} \binom{n}{r} = \binom{m+n}{k}.$$

4.3.3. For each positive integer n, show that the identity

$$(x + y)^n = \sum_{k=0}^{n} \binom{n}{k} x^k y^{n-k}, \qquad x, y \text{ positive integers}$$

implies the identity

$$(x + y)^n = \sum_{k=0}^{n} \binom{n}{k} x^k y^{n-k}, \qquad x, y \text{ real numbers.}$$

Solution. Let y_0 be an arbitrary but fixed positive integer, and let

$$P(x) = (x + y_0)^n, \qquad Q(x) = \sum_{k=0}^{n} \binom{n}{k} x^k y_0^{n-k}.$$

$P(x)$ and $Q(x)$ are polynomials in x, and they are equal whenever x is a positive integer. Therefore, by the identity theorem, $P(x)$ and $Q(x)$ are equal for all real numbers x.

Now, let x_0 be a fixed real number, and let

$$S(y) = (x_0 + y)^n \quad \text{and} \quad T(y) = \sum_{k=0}^{n} \binom{n}{k} x_0^k y^{n-k}.$$

$S(y)$ and $T(y)$ are polynomials in y, and since they are equal whenever y is a positive integer, it follows that $S(y) \equiv T(y)$ for all real numbers y. This completes the proof.

(Incidentally, the identity

$$(x + y)^n = \sum_{k=0}^{n} \binom{n}{k} x^k y^{n-k}, \qquad x, y \text{ positive integers}$$

can be proved neatly as follows. Let $S = \{1, 2, \ldots, n\}$; let A be a set with x elements and B be a set, disjoint from A, with y elements. Now, count, in two different ways, the number of functions from S to $A \cup B$. This, together with the preceding solution, constitutes another proof of the binomial theorem.)

4.3.4. Is $x^5 - x^2 + 1$ irreducible over the rationals?

Solution. By the rational-root theorem (see 4.2.16), the only possible rational zeros are ± 1, and neither of these is a zero. Therefore, if the polynomial is reducible, it must necessarily be the product of a quadratic and a cubic. So suppose

$$x^5 - x^2 + 1 = (x^2 + ax + b)(x^3 + cx^2 + dx + e).$$

By Gauss' lemma (see 4.2.16), we may assume that a, b, c, d, e are integers. Since these polynomials are equal for all x, their coefficients are equal; so, equating coefficients, we get the following equations:

$$a + c = 0,$$
$$b + ac + d = 0,$$
$$bc + ad + e = -1,$$
$$bd + ae = 0,$$
$$be = 1.$$

It is not difficult to show these equations cannot hold simultaneously. For example, the last equation shows that b and e are both odd. Thus, the fourth equation shows a and d have the same parity. Similarly, the first equation shows that a and c have the same parity. Therefore a, c, and d have the same parity. But then $ac + d$ is even, and the second equation cannot hold (b is odd). Therefore $x^5 - x^2 + 1$ is not reducible over the integers, or the rationals.

Another way to proceed with the problem is based on the following observation. If f, g, and h are polynomials over the integers and $f = gh$, then $\bar{f} \equiv \bar{g}\bar{h}(\bmod n)$, where $\bar{f}$, $\bar{g}$, and $\bar{h}$ are the polynomials formed from f, g, and h respectively by taking their coefficients modulo n. If f is reducible over the integers, then $\bar{f}$ is reducible over the integers taken modulo n. In the case at hand, the polynomial $x^5 - x^2 + 1$ transfoms to $x^5 + x^2 + 1$ (mod 2). The only irreducible quadratic polynomial over $Z_2 = \{0, 1\}$ is $x^2 + x + 1$ (the other quadratic polynomials and their factorizations modulo 2 are $x^2 = x \cdot x$, $x^2 + 1 = (x + 1)^2$, and $x^2 + x = x(x + 1)$). But $x^2 + x + 1$ does not divide $x^5 + x^2 + 1$ in Z_2 ($x^5 + x^2 + 1 = (x^3 + x^2)(x^2 + x + 1) + 1$ (mod 2)), and therefore, $x^5 + x^2 + 1$ is irreducible over Z_2. It follows that $x^5 - x^2 + 1$ is irreducible over the integers, and the rationals.

In the preceding discussion, we made use of the fact that polynomials over Z_n can be added, subtracted, and multiplied in the usual manner except that the arithmetic (on the coefficients) is done within Z_n (i.e. modulo n). If n is a prime number, say $n = p$, then Z_p is a field, so all the results concerning polynomials over fields (e.g., the factor theorem, the identity theorem) continue to hold. This is not the case if n is not a prime. For example, $2x^3 - 2x$, as a polynomial over Z_4, has four distinct zeros in Z_4, namely, 0, 1, 2, and 3, whereas it would have at most three if the arithmetic were carried out in a field.

Let p be a prime, and consider the binomial theorem modulo p

$$(1 + x)^p \equiv \sum_{k=0}^{n} \binom{p}{k} x^k \ (\bmod\ p),$$

where each side is regarded as a polynomial over Z_p. For $1 \leqslant k \leqslant p - 1$, we have $\binom{p}{k} \equiv 0 \ (\bmod\ p)$, since none of the factors in $k!(p - k)!$ divide the factor of p in $p!$. Thus, as polynomials over Z_p,

$$(1 + x)^p \equiv 1 + x^p \ (\bmod\ p).$$

More generally, for each positive integer n,

$$(1 + x)^{p^n} \equiv 1 + x^{p^n} \pmod{p}.$$

The argument is by induction. It is true for $n = 1$, and assuming it true for k, we have

$$(1 + x)^{p^{k+1}} \equiv \underbrace{(1 + x)^{p^k}(1 + x)^{p^k} \cdots (1 + x)^{p^k}}_{p \text{ times}} \pmod{p}$$

$$\equiv (1 + x^{p^k})(1 + x^{p^k}) \cdots (1 + x^{p^k}) \pmod{p}$$

$$\equiv (1 + x^{p^k})^p \pmod{p}$$

$$\equiv 1 + (x^{p^k})^p \pmod{p}$$

$$\equiv (1 + x^{p^{k+1}}) \pmod{p}.$$

By equating coefficients of x^i on each side, we find that

$$\binom{p^n}{i} \equiv 0 \pmod{p}, \qquad 1 \leqslant i < p^n.$$

4.3.5. Prove that the number of odd binomial coefficients in any finite binomial expression is a power of 2.

Solution. A conjecture, based on the examination of several special cases (see 1.1.9), is that the number of odd coefficients in $(1 + x)^n$ is 2^k, where k is the number of nonzero digits when n is expressed in binary notation.

An example will make it clear how the proof goes in the general case. Consider $n = 13$. In binary notation, $13 = 1101_2 = 8 + 4 + 1$. Therefore,

$$(1 + x)^{13} = (1 + x)^{8+4+1}$$

$$= (1 + x)^8(1 + x)^4(1 + x)$$

$$\equiv (1 + x^8)(1 + x^4)(1 + x) \pmod{2},$$

making use of the previously established result. From this we can see that there are eight odd binomial coefficients in $(1 + x)^{13}$. This is because when the right side in the preceding equation is expanded, $(1 + x^4)(1 + x)$ will have four terms, and $(1 + x^8)(1 + x^4 + x + x^5)$ will have eight terms. (In general, if $1 + x^n$ is multiplied by a polynomial $P(x)$ of degree smaller than n, the result will be a polynomial with twice as many nonzero coefficients as the corresponding number in $P(x)$.)

Consider the polynomial equation $x^2 + ax + b = 0$, and suppose its roots are r_1 and r_2. Then we can write

$$x^2 + ax + b = (x - r_1)(x - r_2)$$

$$= x^2 - (r_1 + r_2)x + r_1 r_2.$$

From this, using the identity theorem, it follows that

$$r_1 + r_2 = -a,$$
$$r_1 r_2 = b.$$

Similarly, if $x^3 + ax^2 + bx + c = 0$ has roots r_1, r_2, r_3 we have

$$x^3 + ax^2 + bx + c = (x - r_1)(x - r_2)(x - r_3)$$
$$= x^3 - (r_1 + r_2 + r_3)x^2$$
$$+ (r_1 r_2 + r_1 r_3 + r_2 r_3)x - r_1 r_2 r_3.$$

In this case,

$$r_1 + r_2 + r_3 = -a,$$
$$r_1 r_2 + r_1 r_3 + r_2 r_3 = b,$$
$$r_1 r_2 r_3 = -c.$$

In each case, we have expressed the coefficients of the polynomial equation in terms of the roots (in a rather patterned way). An induction argument shows this is true in general: specifically,

If $x^n + a_{n-1}x^{n-1} + \cdots + a_1 x + a_0 = 0$ has roots $r_1, r_2, \ldots, r_n$ then

$$S_1 = r_1 + r_2 + \cdots + r_n = -a_{n-1},$$
$$S_2 = r_1 r_2 + \cdots + r_1 r_n + r_2 r_3 + \cdots + r_2 r_n + \cdots + r_{n-1} r_n = a_{n-2},$$
$$S_3 = r_1 r_2 r_3 + r_1 r_2 r_4 + \cdots + r_2 r_3 r_4 + \cdots + r_{n-2} r_{n-1} r_n = -a_{n-3},$$

$$\vdots$$

$$S_n = r_1 r_2 \cdots r_n = (-1)^n a_0,$$

where S_i is the sum of all the products of the roots taken i at a time.

4.3.6. Consider all lines which meet the graph

$$y = 2x^4 + 7x^3 + 3x - 5$$

in four distinct points, say (x_i, y_i), $i = 1, 2, 3, 4$. Show that

$$\frac{x_1 + x_2 + x_3 + x_4}{4}$$

is independent of the line, and find its value.

Solution. Let $y = mx + b$ intersect the curve in four points (x_i, y_i), $i = 1, 2, 3, 4$. Then x_1, x_2, x_3, x_4 are the roots of the equation

$$mx + b = 2x^4 + 7x^3 + 3x - 5,$$

or equivalently, of

$$x^4 + \tfrac{7}{2}x^3 + \left(\frac{3 - m}{2} \right)x + \left(\frac{-5 - b}{2} \right) = 0.$$

It follows from our earlier remarks that $(x_1 + x_2 + x_3 + x_4)/4 = (-\tfrac{7}{2})/4 = -\tfrac{7}{8}$, and this is independent of m and b.

4.3.7. Let P be a point on the graph of $f(x) = ax^3 + bx$, and let the tangent at P intersect the curve $y = f(x)$ again at Q. Let the x-coordinate of P be x_0. Show that the x-coordinate of Q is $-2x_0$.

Solution. The straightforward approach is to write the equation of the tangent to the curve $y = f(x)$ at P, say $y = T(x)$, and to solve $y = T(x)$ and $y = f(x)$ simultaneously to find Q.

Another approach is to argue as follows. We recognize that solving $y = T(x)$ and $y = f(x)$ simultaneously is the same as finding the roots of $f(x) - T(x) = 0$. Now x_0 is a double root (that is, of multiplicity 2) of this equation, since $T(x)$ is tangent to $y = f(x)$ at x_0. What we seek is the third root, denoted by x_1. We know that the sum of the roots, $2x_0 + x_1$, is equal to $-k/a$, where k is the coefficient of the x^2 term. But the coefficient of the x^2 term is 0, so it follows that $x_1 = -2x_0$.

4.3.8. Let x_1 and x_2 be the roots of the equation

$$x^2 - (a + d)x + (ad - bc) = 0.$$

Show that x_1^3 and x_2^3 are the roots of

$$y^2 - (a^3 + d^3 + 3abc + 3bcd)y + (ad - bc)^3 = 0.$$

Solution. We know that

$$x_1 + x_2 = a + d,$$
$$x_1 x_2 = ad - bc.$$

Since $(x_1 + x_2)^3 = x_1^3 + 3x_1^2 x_2 + 3x_1 x_2^2 + x_2^3$, we have

$$x_1^3 + x_2^3 = (x_1 + x_2)^3 - 3x_1^2 x_2 - 3x_1 x_2^2$$
$$= (a + d)^3 - 3x_1 x_2 (x_1 + x_2)$$
$$= (a + d)^3 - 3(ad - bc)(a + d)$$
$$= (a + d)[a^2 + 2ad + d^2 - 3ad + 3bc]$$
$$= (a + d)(a^2 - ad + d^2 + 3bc)$$
$$= a^3 + d^3 + 3abc + 3bcd.$$

Furthermore,

$$x_1^3 x_2^3 = (ad - bc)^3,$$

and the proof is complete.

4.3.9. Let a, b, c be real numbers such that $a + b + c = 0$. Prove that

$$\frac{a^5 + b^5 + c^5}{5} = \left(\frac{a^3 + b^3 + c^3}{3}\right)\left(\frac{a^2 + b^2 + c^2}{2}\right).$$

Solution. Here is a very clever solution based on the ideas of this section. Let $A = ab + ac + bc$ and $B = abc$. Then a, b, c are roots of the equation

$$x^3 + Ax - B = 0.$$

For each positive integer n, let $T_n = a^n + b^n + c^n$. Then,

$$T_0 = 3,$$

$$T_1 = 0,$$

$$T_2 = (a + b + c)^2 - 2(ab + ac + bc) = -2A.$$

For $n \geqslant 0$, $T_{n+3} = -AT_{n+1} + BT_n$ (substitute a, b, c into $x^{n+3} = -Ax^{n+1} + Bx^n$ and add), and this gives

$$T_3 = -AT_1 + BT_0 = 3B,$$

$$T_4 = -AT_2 + BT_1 = 2A^2,$$

$$T_5 = -AT_3 + BT_2 = -5AB.$$

It follows that

$$\frac{T_5}{5} = -AB = \frac{T_3}{3} \cdot \frac{T_2}{2}.$$

4.3.10. Show that the polynomial equation with real coefficients

$$P(x) = a_n x^n + a_{n-1} x^{n-1} + \cdots + a_3 x^3 + x^2 + x + 1 = 0$$

cannot have all real roots.

Solution. Let $r_1, r_2, \ldots, r_n$ denote the roots of $P(x) = 0$. None of $r_1, \ldots, r_n$ is zero. Divide each side of $P(x) = 0$ by x^n and set $y = 1/x$, to get

$$Q(y) \equiv y^n + y^{n-1} + y^{n-2} + a_3 y^{n-3} + \cdots + a_{n-1} y + a_n = 0.$$

Note that r is a root of $P(x) = 0$ if and only if $1/r$ is a root of $Q(y) = 0$. Therefore, the roots of $Q(y) = 0$ are $s_1, s_2, \ldots, s_n$, where $s_i = 1/r_i$, $i = 1, \ldots, n$. It follows that

$$\sum_{i=1}^{n} s_i = -1,$$

$$\sum_{i<j} s_i s_j = 1,$$

and therefore,

$$\sum_{i=1}^{n} s_i^2 = \left(\sum_{i=1}^{n} s_i \right)^2 - 2 \sum_{i<j} s_i s_j = 1 - 2 = -1.$$

This equation implies that not all the s_i's are real; equivalently, not all the r_i's are real.

Problems

4.3.11. Let k be a positive integer. Find all polynomials
$$P(x) = a_n x^n + \cdots + a_1 x + a_0,$$
where the a_i are real, which satisfy the equation
$$P(P(x)) = [P(x)]^k.$$

4.3.12.

(a) Prove that $\log x$ cannot be expressed in the form $f(x)/g(x)$ where $f(x)$ and $g(x)$ are polynomials with real coefficients.
(b) Prove that e^x cannot be expressed in the form $f(x)/g(x)$ where $f(x)$ and $g(x)$ are polynomials with real coefficients.

4.3.13. Show that
$$(1 + x)^n - x(1 + x)^n + x^2(1 + x)^n - + \cdots \pm x^k(1 + x)^n$$
$$= (1 + x)^{n-1}(1 - (-x)^{k+1}),$$
and use this identity to prove that
$$\binom{n-1}{k} = \binom{n}{k} - \binom{n}{k-1} + \cdots \pm \binom{n}{0}.$$

4.3.14.

(a) Differentiate each side of the identity
$$(1 + x)^n = \sum_{k=0}^{n} \binom{n}{k} x^k.$$
By comparing the coefficients of x^{k-1} in the resulting identity, show that
$$n\binom{n-1}{k-1} = k\binom{n}{k}$$
(b) Use the result of part (a) to show that
$$\sum_{i=0}^{n-1} (-1)^i \binom{n-1}{i} \frac{1}{i+1} = \frac{1}{n}.$$

4.3.15. Let $x^{(n)} = x(x - 1) \cdots (x - n + 1)$ for n a positive integer, and let $x^{(0)} = 1$. Prove that for all real numbers x and y
$$(x + y)^{(n)} = \sum_{k=0}^{n} \binom{n}{k} x^{(k)} y^{(n-k)}.$$

(Hint: This can be done by induction, but consider a proof similar to 4.3.3 which first establishes the result for positive integers x and y. For this, count in two different ways, the number of one-to-one functions from $\{1, 2, \ldots, n\}$ into $A \cup B,$ where A is a set with x elements and B is a set, disjoint from A, with y elements. Prove the identity for all real numbers by making use of the identity theorem.)

4.3.16. Is $x^4 + 3x^3 + 3x^2 - 5$ reducible over the integers?

4.3.17. Let p be a prime number. Show that

(a) $\binom{p-1}{k} \equiv (-1)^k \pmod{p}$, $0 \leqslant k \leqslant p - 1$,
(b) $\binom{p+1}{k} \equiv 0 \pmod{p}$, $2 \leqslant k \leqslant p - 1$,
(c) $\binom{pa}{pb} \equiv \binom{a}{b} \pmod{p}$, $a \geqslant b \geqslant 0$,
(d) $\binom{2p}{p} \equiv 2 \pmod{p}$.

4.3.18. Let $\omega = \cos(2\pi/n) + i\sin(2\pi/n)$.

(a) Show that $1, \omega, \omega^2, \ldots, \omega^{n-1}$ are the n roots of $x^n - 1 = 0$.
(b) Show that $(1 - \omega)(1 - \omega^2) \cdots (1 - \omega^{n-1}) = n$.
(c) Show that $\omega + \cdots + \omega^{n-1} = -1$.

4.3.19.

(a) Solve the equation $x^3 - 3x^2 + 4 = 0$, given that two of its roots are equal.
(b) Solve the equation $x^3 - 9x^2 + 23x - 15 = 0$, given that its roots are in arithmetical progression.

4.3.20. Given r, s, t are the roots of $x^3 + ax^2 + bx + c = 0$.

(a) Evaluate $1/r^2 + 1/s^2 + 1/t^2$, provided that $c \neq 0$.
(b) Find a polynomial equation whose roots are r^2, s^2, t^2.

4.3.21. Given numbers x, y, z such that

$$x + y + z = 3,$$
$$x^2 + y^2 + z^2 = 5,$$
$$x^3 + y^3 + z^3 = 7,$$

find $x^4 + y^4 + z^4$. (Hint: Use an argument similar to that used in 4.3.9.)

We close this section with three problems which draw attention to some additional results about polynomials that are very useful in certain problems.

4.3.22 (Theorem). If $x_1, x_2, \ldots, x_n$ are distinct numbers, and $y_1, \ldots, y_n$ are any numbers, not all zero, there is a unique polynomial $f(x)$ of degree

not exceeding $n - 1$ with the property that $f(x_1) = y_1$, $f(x_2) = y_2, \ldots,$
$f(x_n) = y_n$.

Outline of Proof:

(a) Let $g(x) = (x - x_1)(x - x_2) \cdots (x - x_n)$. Show that

$$\frac{g(x)}{(x - x_1)g'(x_1)} \left(= \frac{(x - x_2)(x - x_3) \cdots (x - x_n)}{(x_1 - x_2) \cdots (x_1 - x_n)} \right)$$

is a polynomial of degree $n - 1$ with zeros at $x_2, \ldots, x_n$ and which equals 1 at $x = x_1$.

(b) *Lagrange interpolation formula*. Show that

$$f(x) = \frac{g(x)}{(x - x_1)g'(x_1)} y_1 + \frac{g(x)}{(x - x_2)g'(x_2)} y_2 + \cdots$$

$$+ \frac{g(x)}{(x - x_n)g'(x_n)} y_n$$

takes the values $y_1, y_2, \ldots, y_n$ at the points $x_1, \ldots, x_n$ respectively.

(c) *Application*. Suppose that $P(x)$ is a polynomial which when divided by $x - 1, x - 2, x - 3$ gives remainders of $3, 5, 2$ respectively. Determine the remainder when $P(x)$ is divided by $(x - 1)(x - 2)(x - 3)$. (Hint: Write $P(x) = Q(x)(x - 1)(x - 2)(x - 3) + R(x)$, where $R(x)$ is of degree less than 3. Find $R(x)$ by the Lagrange interpolation formula, since $R(1) = 3$, $R(2) = 5$, $R(3) = 2$.)

4.3.23 (Partial Fractions).

(a) Show that if $f(x)$ is a polynomial whose degree is less than n, then the fraction

$$\frac{f(x)}{(x - x_1)(x - x_2) \cdots (x - x_n)},$$

where $x_1, x_2, \ldots, x_n$ are n distinct numbers, can be represented as a sum of n partial fractions

$$\frac{A_1}{x - x_1} + \frac{A_2}{x - x_2} + \cdots + \frac{A_n}{x - x_n},$$

where $A_1, \ldots, A_n$ are constants (independent of x). (Hint: Use Lagrange's interpolation formula: divide each side by $g(x)$, etc.)

(b) *Application*. Let $f(x)$ be a monic polynomial of degree n with distinct zeros $x_1, x_2, \ldots, x_n$. Let $g(x)$ be any monic polynomial of degree $n - 1$. Show that

$$\sum_{j=1}^{n} \frac{g(x_j)}{f'(x_j)} = 1.$$

(Hint: Write $g(x)/f(x)$ as a sum of partial fractions.)

4.3.24. A sequence of numbers $u_0, u_1, u_2, \ldots$ is called a *sequence of kth order* if there is a polynomial of degree k,

$$P(x) = a_k x^k + a_{k-1} x^{k-1} + \cdots + a_1 x + a_0$$

such that $u_i = P(i)$ for $i = 0, 1, 2, \ldots$.

The *first-difference sequence* of the sequence $u_0, u_1, u_2 \ldots$ is the sequence $u_0^{(1)}, u_1^{(1)}, u_2^{(1)}, \ldots$ defined by

$$u_n^{(1)} = u_{n+1} - u_n, \qquad n = 0, 1, 2, 3, \ldots .$$

(a) Prove that if $u_0, u_1, u_2, \ldots$ is a sequence of order k, then the first-difference sequence is a sequence of order $k - 1$. Define the second-difference sequence of $u_0, u_1, u_2, \ldots$ to be the first-difference sequence of the first-difference sequence, that is, the sequence $u_0^{(2)}, u_1^{(2)}, u_2^{(2)}, \ldots$ defined by

$$u_n^{(2)} = u_{n+1}^{(1)} - u_n^{(1)}$$

$$= u_{n+2} - 2u_{n+1} + u_n, \qquad n = 0, 1, 2, \ldots .$$

From part (a) it follows that $u_0^{(2)}, u_1^{(2)}, u_2^{(2)}, \ldots$ is a sequence of order $k - 2$. Similarly, define the third-difference sequence, the fourth-difference sequence, and so forth. Repeated application of part (a) shows that if $u_0, u_1, u_2, \ldots$ is a sequence of order k, the $(k + 1)$st difference sequence will be identically zero. We aim to establish the converse: if the successive difference sequences of an arbitrary sequence $u_0, u_1, u_2, \ldots$ eventually become identically equal to zero, then the terms of original sequence are successive values of a polynomial expression; that is, there is a polynomial $P(x)$ such that $u_n = P(n)$, $n = 0, 1, 2, \ldots$.

(b) Use induction to prove that

$$u_n = \binom{n}{0} u_0 + \binom{n}{1} u_0^{(1)} + \binom{n}{2} u_0^{(2)} + \cdots + \binom{n}{n} u_0^{(n)}.$$

(c) Suppose that the original sequence is described by the function $F(x)$. That is, suppose that $F(n) = u_n$, $n = 0, 1, 2, \ldots$. For $k = 0, 1, 2, 3, \ldots$, let $\Delta^k F(0) = u_0^{(k)}$, and for x a real number and i a positive integer, let $x^{(i)} = x(x - 1)(x - 2) \cdots (x - i + 1)$. Show that the result of part (b) can be written in the form

$$F(n) = \sum_{k=0}^{\infty} \frac{\Delta^k F(0)}{k!} n^{(k)}.$$

Note the similarity to the Taylor expansion of $F(x)$:

$$F(x) = \sum_{k=0}^{\infty} \left(F^{(k)}(0)/k! \right) x^k.$$

(d) Prove that if the $(k + 1)$st difference sequence is identically zero, then the original sequence is given by

$$P(n) = \sum_{i=0}^{k} \frac{\Delta^i F(0)}{i!} n^{(i)}.$$

(e) Use the result of part (d) to find a closed formula for the sum of the series $1^4 + 2^4 + \cdots + n^4$. (Hint: Notice that the first-difference sequence is given by a polynomial of degree 4, and therefore, the sum will be a polynomial of degree 5.)

Additional Examples

4.4.30, 4.4.31, 7.2.10, 8.2.2, 8.2.3, 8.2.10, 8.4.11.

4.4. Abstract Algebra

A *group* is a set G together with a binary operation $*$ on G such that:

(i) *Associative property*. For all elements a, b, c in G

$$(a*b)*c = a*(b*c).$$

(ii) *Identity*. There is a unique element e in G (called the identity of G) such that for every element a in G,

$$a*e = a = e*a.$$

(iii) *Inverse*. For each element a in G, there is a unique element a^{-1} in G (called the inverse of a) such that

$$a^{-1}*a = e = a*a^{-1}.$$

When working with groups, we sometimes think of the operation $*$ as "multiplication," and in this case we often suppress the $*$ in writing products. Thus, $a*b$ is written simply as ab, and $a*(b*c)$ is written as $a(bc)$, or abc, and so forth. Furthermore, when we think of $*$ as a product, we sometimes denote the identity element as "1." In addition, we use exponential notation to simplify expressions; e.g., $a^4 = aaaa$, etc. It is not difficult to show that the usual laws of exponents hold in a group, namely,

$$a^n a^m = a^{n+m}, \qquad (a^n)^m = a^{nm}, \qquad n, m \text{ integers.}$$

The group operation need not be commutative; i.e., it may not be the case that $ab = ba$ for all elements a, b of G. An example of such a group is the set of n-by-n nonsingular matrices over the real numbers.

In any group G, it is the case that

$$(ab)^{-1} = b^{-1}a^{-1}, \qquad a, b \in G.$$

This identity is fundamental and can be proved in the following way. Observe that $(ab)(b^{-1}a^{-1}) = a(b(b^{-1}a^{-1})) = a((bb^{-1})a^{-1}) = a(ea^{-1}) = aa^{-1} = e$, and $(b^{-1}a^{-1})(ab) = b^{-1}(a^{-1}(ab)) = b^{-1}((a^{-1}a)b) = b^{-1}(eb) = b^{-1}b = e$. Therefore $b^{-1}a^{-1}$ is an inverse for ab. But ab has a unique inverse, denoted by $(ab)^{-1}$. It follows that $(ab)^{-1} = b^{-1}a^{-1}$.

If the group G is commutative (i.e., if $ab = ba$ for all $a, b \in G$), it is easy to show that

$$(ab)^n = a^n b^n, \qquad a, b \in G \quad n \text{ an integer.}$$

4.4.1. Suppose that G is a set and $*$ is a binary operation on G such that:

(i) *Associative property*. For all a, b, c in G, $a*(b*c) = (a*b)*c$;
(ii) *Right identity*. There is an element e in G such that for every element a in G, $a*e = a$; and
(iii) *Right inverse*. For each element a in G, there is an element a^{-1} in G such that $a*a^{-1} = e$.

Prove that G is a group.

Solution. We will show that the right identity e is also a left identity, and the right inverse a^{-1} is also a left inverse for a. Then we will show that e and a^{-1} are unique.

Observe that a^{-1} is an element of G, and therefore by (ii), there is an element $(a^{-1})^{-1}$ in G such that $(a^{-1})*(a^{-1})^{-1} = e$. We now compute

$$a^{-1}a = (a^{-1}a)e = (a^{-1}a)\left(a^{-1}(a^{-1})^{-1}\right)$$
$$= a^{-1}\left[a\left(a^{-1}(a^{-1})^{-1}\right)\right]$$
$$= a^{-1}\left[(aa^{-1})(a^{-1})^{-1}\right]$$
$$= a^{-1}\left[e(a^{-1})^{-1}\right] = (a^{-1}e)(a^{-1})^{-1}$$
$$= a^{-1}(a^{-1})^{-1}$$
$$= e.$$

This shows that a^{-1} is an inverse (left inverse and right inverse).

Also, $ea = (aa^{-1})a = a(a^{-1}a) = ae = a$, and therefore e is an identity for G (that is, for each a, $ea = a = ae$).

Suppose e' is also an identity for G. Then $e = e * e'$ (because e' is an identity) $= e'$ (because e is an identity). This shows the identity element of G is unique.

Suppose $(a^{-1})'$ is also an inverse for a. Then $(a^{-1})' = (a^{-1})'e = (a^{-1})'(aa^{-1}) = [(a^{-1})'a]a^{-1} = ea^{-1} = a^{-1}$. This shows the inverse of a is unique.

It follows that G is a group.

4.4.2. Let G be a group.

(a) *Cancellation property.* For all a, b, c in G, show that

$$ab = ac \quad \text{implies} \quad b = c,$$
$$ba = ca \quad \text{implies} \quad b = c.$$

(b) Let a be an element in G, and consider the sequence

$$1, a, a^2, a^3, \ldots .$$

Show that either all the elements in the sequence are different, or there is a smallest integer n such that $a^n = 1$ and $1, a, \ldots, a^{n-1}$ are distinct. In the latter situation, n is called the *order of a*, denoted by ord(a); in the former case we say that a has infinite order.

Solution. (a) This follows immediately by multiplying each side on the left (and right respectively) by a^{-1}.

(b) Suppose that not all elements in the sequence are different, and let n be the smallest integer such that a^n is a repetition of a previous element in the sequence. Then $a^n = 1$, for if $a^n = a^i$, $0 < i < n$, then by the cancellation property, $a^{n-1} = a^{i-1}$, and this contradicts our choice of n.

4.4.3. Let a and b be two elements in a group such that $aba = ba^2b$, $a^3 = e$, and $b^{2n-1} = e$ for some positive integer n. Prove that $b = e$.

Solution. Note that if $ab = ba$, then $aba = ba^2b$ is the same as $a^2b = a^2b^2$, and the cancellation property implies $b = e$. Although the group may not be commutative, we shall prove that this particular set of equations for a and b does imply that $ab = ba$.

Notice that $ab = ba$ is the same as $ab^{2n} = b^{2n}a$, since by assumption $b^{2n} = b$. To show that $ab^{2n} = b^{2n}a$ it suffices to show that $ab^2 = b^2a$, since $ab^{2n} = a(b^2)^n = (b^2)^n a$ (by repeated application of $ab^2 = b^2a$) $= b^{2n}a$.

Thus, the proof is complete after observing that $ab^2 = (aba)(a^{-1}b)$ $= (ba^2b)(a^{-1}b) = (ba^2)(ba^{-1}b) = (ba^2)(ba^2b) = (ba^2)(aba) = ba^3ba = b^2a$ (since $a^3 = e$).

Let G be a group. We say that H is a *subgroup* of G if H is a subset of G which is itself a group (under the operation of G). The *order of H* is defined to be the number elements in H, and this number is denoted by ord(H).

An important class of subgroups are the following. Let $a \in G$, and let

$$\langle a \rangle = \{ a^n : n \text{ is an integer} \}.$$

It is easy to check that $\langle a \rangle$ is a subgroup of G; it is called the *cyclic subgroup generated by a*. Note that ord$(a) = $ ord$(\langle a \rangle)$.

The following theorem constitutes one of the most important results in the theory of finite groups.

Lagrange's Theorem. *If H is a subgroup of a finite group G, then the order of H divides the order of G.*

Here are three important corollaries.

(i) If G is a group of order n and $a \in G$, then $a^n = 1$.
(ii) If G is a group of order p, where p is a prime, then G is a cyclic group (i.e., $G = \langle a \rangle$ for some $a \in G$).
(iii) If G is a group and $a^n = 1$, then the order of a divides n.

We will leave the proof of Lagrange's theorem as a problem (see 4.4.18); however, it is instructive to see the arguments for the corollaries.

Proof of (i). Let $a \in G$ and let $m = \text{ord}(a)$. By Lagrange's theorem m divides n, so suppose $n = mq$ for some integer q. Then $a^n = a^{mq} = (a^m)^q = 1^q = 1$.

Proof of (ii). Let a be an element of G different from the identity. Then $\langle a \rangle$ is a subgroup of G with more than one element (namely, 1 and a). By Lagrange's theorem, the order of $\langle a \rangle$ divides p, but since p is prime, it must be the case that $\langle a \rangle$ is of order p; that is, $\langle a \rangle = G$.

Proof of (iii). Let $m = \text{ord}(a)$. By the division algorithm there are integers q and r such that $n = qm + r$, $0 \leqslant r < m$. Thus $1 = a^n = a^{qm+r} = (a^m)^q a^r = a^r$. Since $1, a, \ldots, a^{m-1}$ are distinct, it must be the case that $r = 0$, and it follows that m divides n (this is a typical application of the division algorithm for integers).

4.4.4. If in the group G we have $a^5 = 1$, $aba^{-1} = b^2$ for some $a, b \in G$, find $\text{ord}(b)$.

Solution. Since $a^5 = 1$, the order of a is either 1 or 5. If $\text{ord}(a) = 1$, then $a = 1$ and it follows that $b = b^2$, or $b = 1$, and so $\text{ord}(b) = 1$.

Suppose $\text{ord}(a) = 5$. We have $(aba^{-1})(aba^{-1}) = (b^2)^2$, or equivalently, $ab^2a^{-1} = b^4$. Substituting aba^{-1} for b^2 on the left side of this equation yields $a^2ba^{-2} = b^4$. Squaring this, we get $(a^2ba^{-2})(a^2ba^{-2}) = (b^4)^2$, or equivalently, $a^2b^2a^{-2} = b^8$. Again, substituting aba^{-1} for b^2 on the left, we get $a^3ba^{-3} = b^8$. Squaring gives $a^3b^2a^{-3} = b^{16}$, and substituting gives $a^4ba^{-4} = b^{16}$. One more time: $a^4b^2a^{-4} = b^{32}$, or equivalently, $a^5ba^{-5} = b^{32}$. But $a^5 = a^{-5} = 1$, so $b = b^{32}$, and on cancellation, we get $b^{31} = 1$. Since 31 is a prime number, the order of b is 1 (if b is the identity) or 31.

4.4.5. If G is a finite group and m is a positive integer relatively prime to the order of G, then for each a in G there is a unique b in G such that $b^m = a$.

Solution. Let $T: G \to G$ be defined by $T(x) = x^m$. We aim to show that T is a one-to-one function. So suppose that $T(x) = T(y)$ for elements x and y

of G. Then $x^m = y^m$. Let $n = \text{ord}(G)$. Since n and m are relatively prime, there are integers s and t such that $sn + tm = 1$. Hence $x = x^{sn+tm}$ $= (x^n)^s (x^m)^t = (x^m)^t$ (since $x^n = 1$) $= (y^m)^t$ (since $x^m = y^m$) $= (y^n)^s$ $(y^m)^t$ (since $y^n = 1$) $= y^{sn+tm} = y$.

Therefore T is a one-to-one function, and since G is a finite set, T is onto G. That is, for $a \in G$, there is a unique b in G such that $T(b) = a$ (equivalently, $b^m = a$).

The first corollary to Lagrange's theorem states that $a^{\text{ord}(G)} = 1$ for each element in the finite group G. This has a number of interesting and important consequences when applied to particular groups. For example, let V_n denote the set of positive integers less than n that are relatively prime to n. The elements of V_n form a group under multiplication modulo n. Let $\varphi(n) = \text{ord}(V_n)$. (The function φ is called the Euler φ-function.) Then Lagrange's theorem implies the following.

Euler's Theorem. *If a is any integer relatively prime to n, then*

$$a^{\varphi(n)} \equiv 1 \pmod{n}.$$

When n is a prime number, say $n = p$, we have $\varphi(p) = p - 1$, so that $a^{p-1} \equiv 1 \pmod{p}$ whenever a is not a multiple of p. If we multiply each side by a, we get $a^p \equiv a \pmod{p}$. This congruence holds even when a is a multiple of p, and thus we have the following result.

Fermat's Little Theorem. *If a is an integer and p is a prime, then*

$$a^p \equiv a \pmod{p}.$$

4.4.6. Prove that each prime divisor of $2^p - 1$, where p is a prime, is greater than p. (It is a corollary that the number of primes is infinite.)

Solution. The result is true for $p = 2$, so henceforth assume that p is odd. Suppose that q is a prime that divides $2^p - 1$. Then q is odd and $2^p \equiv 1$ (mod q). By Fermat's little theorem, $2^{q-1} \equiv 1 \pmod{q}$. If $q = p$ we have $2 = 2 \times 1 \equiv 2 \times 2^{q-1} \equiv 2^q = 2^p \equiv 1 \pmod{q}$, a contradiction. If $q < p$, then $q - 1$ and p are relatively prime, so there are integers s and t such that $sp + t(q - 1) = 1$. It follows that $2 = 2^{sp+t(q-1)} \equiv (2^p)^s (2^{q-1})^t \equiv 1 \pmod{q}$, a contradiction. Thus q must be larger than p.

4.4.7. Show that if n is an integer greater than 1, then n does not divide $2^n - 1$.

Solution. Suppose that n divides $2^n - 1$; that is, $2^n \equiv 1 \pmod{n}$. Clearly, n is an odd number, since $2^n - 1$ is odd. Suppose that p is a prime divisor of n. Then $2^n \equiv 1 \pmod{p}$. Now, regard 2 as an element of the group V_p. We know that $2^{p-1} \equiv 1 \pmod{p}$ (Fermat's little theorem, since $\gcd(2, p) = 1$). By the third corollary to Lagrange's theorem, $p - 1$ divides n. So far there is no contradiction. However, suppose p is chosen as the *smallest* prime which divides n. Then these same conclusions hold, but now, the fact that $\text{ord}(2)$ divides n and $\text{ord}(2)$ divides $p - 1$ produces a contradiction to our choice of p. Therefore, n can never divide $2^n - 1$.

4.4.8. Show that for any positive integer n there exists a power of 2 with a string of more than n successive zeros (in its decimal representation).

Solution. For any positive integer s, there exists a positive integer t such that $2^t \equiv 1 \pmod{5^s}$ (for example, take $t = \varphi(5^s)$). Let $s = 2n$. There exist positive integers q and r such that $2^r - 1 = q \times 5^{2n}$. Multiply each side by 2^{2n}, rewrite as

$$2^{r+2n} = 2^{2n} + q \times 10^{2n},$$

and notice that 2^{r+2n} has at least n consecutive zeros in its decimal representation, since $2^{2n} < 10^n$.

4.4.9. Given positive integers a and b, show that there exists a positive integer c such that infinitely many numbers of the form $an + b$ (n a positive integer) have all their prime factors $\leq c$.

Solution. The result is obviously true when $a = 1$, so suppose $a > 1$. First, consider the case in which $\gcd(a, b) = 1$. We will prove there are an infinite number of terms of the arithmetic sequence $an + b$ among the terms of the sequence $(a + b)^k$, $k = 1, 2, 3, \ldots$.

From Euler's theorem, $b^{\varphi(a)} \equiv 1 \pmod{a}$, since b is relatively prime to a. It follows that for each positive integer s,

$$(a + b)^{s\varphi(a)+1} \equiv b^{s\varphi(a)+1} \equiv (b^{\varphi(a)})^s b \equiv b \pmod{a}.$$

This means that for each positive integer s there is an integer q_s such that

$$(a + b)^{s\varphi(a)+1} = q_s a + b.$$

It follows that each of the terms $q_s a + b$, $s = 1, 2, 3, \ldots$ has only those prime factors that occur in $a + b$.

Now consider the case in which $\gcd(a, b) = d > 1$. Then $\gcd(a/d, b/d) = 1$, so from our preceding argument, there is a c such that infinitely many members of the sequence $(a/d)n + (b/d)$ have all their prime factors $\leq c$. From this it follows that infinitely many members of the form $an + b$ have all their prime factors $\leq cd$. This completes the proof.

A *ring* is a set R with two binary operations, $+$ and $\cdot$, such that

(i) R is a commutative group with respect to the operation $+$;
(ii) For all a, b, c in R, $a(bc) = (ab)c$ ("$\cdot$" suppressed);
(iii) For all a, b, c in R,

$$a(b + c) = ab + ac,$$

$$(b + c)a = ba + ca.$$

R need not have a multiplicative identity: if it does, we say R is a *ring with identity*. The multiplication in R need not be commutative: if it is, we say that R is a *commutative ring*.

4.4.10. Let a and b be elements of a finite ring such that $ab^2 = b$. Prove that $bab = b$.

Solution. Obviously, if the ring were commutative the result would be immediate, but we must show the result holds even when the ring is noncommutative. In addition, we cannot assume the ring has a multiplicative identity.

Suppose $b = b^2$. Then $bab = bab^2 = b^2 = b$, and we are done. Suppose $b = b^m$ for some integer $m > 2$. Then $bab = bab^m = b(ab^2)b^{m-2} = b^2 b^{m-2} = b^m = b$, and we are done. Therefore it is sufficient to show that $b = b^m$ for some integer $m \geqslant 2$.

Suppose the ring has n elements. By the pigeonhole principle, at least two elements in the sequence $b, b^2, \ldots, b^n, b^{n+1}$ are equal. Let i be the *smallest* integer such that b^i equals some subsequent power of b in the preceding sequence; that is, $b^i = b^{i+j}$, for some $j > 0$. Suppose $i > 1$. Then multiply each side of $ab^2 = b$ on the right by b^{i+j-2} to get $ab^{i+j} = b^{i+j-1}$. But since $b^i = b^{i+j}$, we have $ab^i = b^{i+j-1}$. From here there are two cases to consider.

Suppose $i = 2$. Then $b = ab^2 = b^{j+1}$ (from the last equation), and this contradicts our choice of i. So, suppose $i > 2$. Then $b^{i-1} = b \cdot b^{i-2} = (ab^2) \times b^{i-2} = ab^i = b^{i+j-1}$, which again contradicts our choice of i. Therefore, $i = 1$; that is, $b = b^j$ for some j. By the argument in the first paragraph, the proof is complete.

An *integral domain* D is a commutative ring with unity in which for a, b in D, $ab = 0$ implies $a = 0$ or $b = 0$. The cancellation property holds in an integral domain. For, suppose $ab = ac$ and $a \neq 0$. Then $a(b - c) = 0$, so $b - c = 0$, or equivalently, $b = c$. Similarly, $ba = ca$, $a \neq 0$, implies $b = c$.

A *field* is a commutative ring with identity in which every nonzero element has a multiplicative inverse.

4.4.11. Show that a finite integral domain (an integral domain with only a finite number of elements) is a field.

Solution. We must show that every nonzero element of the integral domain has a multiplicative inverse. So, let $D^* = \{a_1, \ldots, a_n\}$ be the nonzero elements of the integral domain, and consider an arbitrary element a of D^*. Define $T: D^* \to D^*$ by $T(a_i) = aa_i$. If $T(a_i) = T(a_j)$ then $aa_i = aa_j$, so by the cancellation property, $a_i = a_j$. Thus we see that T is a one-to-one function. Since D^* is finite, the mapping T is onto D^*. But one of the elements in D^* is the multiplicative identity, denoted by 1. Therefore, $T(a_k) = 1$ for some $a_k \in D^*$; that is, $aa_k = 1$. This shows that a has a multiplicative inverse.

Problems

4.4.12. Let G be a set, and $*$ a binary operation on G which is associative and is such that for all a, b in G, $a^2 b = b = ba^2$ (suppressing the $*$). Show that G is a commutative group.

4.4.13. A is a subset of a finite group G, and A contains more than one-half of the elements of G. Prove that each element of G is the product of two elements of A.

4.4.14. Let H be a subgroup with h elements of a group G. Suppose that G has an element a such that for all x in H, $(xa)^3 = 1$, the identity. In G, let P be the set of all products $x_1 a x_2 a \cdots x_n a$, with n a positive integer and the x_i in H. Show that P has no more than $3h^2$ elements.

4.4.15. If $a^{-1}ba = b^{-1}$ and $b^{-1}ab = a^{-1}$ for elements a, b of a group, prove that $a^4 = b^4 = 1$.

4.4.16. Let a and b be elements of a finite group G.

(a) Prove that $\operatorname{ord}(a) = \operatorname{ord}(a^{-1})$.
(b) Prove that $\operatorname{ord}(ab) = \operatorname{ord}(ba)$.
(c) If $ba = a^4 b^3$, prove that $\operatorname{ord}(a^4 b) = \operatorname{ord}(a^2 b^3)$.

4.4.17. Let a and b be elements of a group. If $b^{-1}ab = a^k$, prove that $b^{-r}a^s b^r = a^{sk^r}$ for all positive integers r and s.

4.4.18 (Outline for the proof of Lagrange's Theorem). Let G be a finite group and H a subgroup with m distinct elements, say $H = \{1, h_2, h_3, \ldots, h_m\}$. For each $a \in G$, let $Ha = \{a, h_2 a, h_3 a, \ldots, h_m a\}$.

(a) Prove that Ha contains m distinct elements.
(b) Prove that $Hh_i = H$.
(c) If $b \notin Ha$, prove that Ha and Hb are disjoint sets.

(d) Prove that there are elements $a_1, a_2, \ldots, a_k$ in G such that $G = Ha_1 \cup Ha_2 \cup \cdots \cup Ha_k$ and $Ha_i \cap Ha_j = \emptyset$ if $i \neq j$.

(e) Use the previous results to formulate a proof of Lagrange's theorem.

4.4.19. Find the smallest integer n such that $2^n - 1$ is divisible by 47.

4.4.20. Prove that if p is a prime, $p > 3$, then $ab^p - ba^p$ is divisible by $6p$.

4.4.21. Let a and b be relatively prime integers. Show that there exist integers m and n such that $a^m + b^n = 1 \pmod{ab}$.

4.4.22. If a, b, c, d are positive integers, show that 30 divides $a^{4b+d} - a^{4c+d}$.

4.4.23. Let $T_n = 2^n + 1$ for all positive integers. Let φ be the Euler φ-function, and let k be any positive integer and $m = n + k\varphi(T_n)$. Show that T_m is divisible by T_n.

4.4.24. Prove that there exists a positive integer k such that $k2^n + 1$ is composite for every positive integer n. (Hint: Consider the congruence class of n modulo 24 and apply the Chinese Remainder Theorem.)

4.4.25. A *Boolean ring* is a ring for which $a^2 = a$ for every element a of the ring. An element a of a ring is *nilpotent* if $a^n = 0$ for some positive integer n. Prove that a ring R is a Boolean ring if and only if R is commutative, R contains no nonzero nilpotent elements, and $ab(a + b) = 0$ for all a, b in R. (Hint: Show that $a^4 - a^5 = 0$, and consider $(x^2 - x^3)^2$.)

4.4.26. Let R be a ring with identity, and let $a \in R$. Suppose there is a unique element a' such that $aa' = 1$. Prove that $a'a = 1$.

4.4.27. Let R be a ring with identity, and a be a nilpotent element of R (see 4.4.25). Prove that $1 - a$ is invertible (that is, prove there exists an element b in R such that $b(1 - a) = 1 = (1 - a)b$).

4.4.28. Let R be a ring, and let $C = \{x \in R : xy = yx \text{ for all } y \text{ in } R\}$. Prove that if $x^2 - x \in C$ for all x in R, then R is commutative. (Hint: Show that $xy + yx \in C$ by considering $x + y$, and then show that $x^2 \in C$.)

4.4.29. Let p be a prime number. Let J be the set of all 2-by-2 matrices $\begin{pmatrix} a & b \\ c & d \end{pmatrix}$ whose entries are chosen from $\{0, 1, 2, \ldots, p - 1\}$ and satisfy the conditions $a + d \equiv 1 \pmod{p}$, $ad - bc \equiv 0 \pmod{p}$. Determine how many members J has.

4.4.30. Let p be a prime number, and let $Z_p = \{0, 1, 2, \ldots, p - 1\}$. Z_p is a field under the operations of addition and multiplication modulo p.

(a) Show that $0, 1, \ldots, p - 1$ are the zeros of $x^p - x$ (considered as a polynomial over Z_p). Conclude that $x^p - x = x(x - 1)(x - 2) \cdots (x - (p - 1)) \pmod{p}$.

(b) *Wilson's theorem.* From part (a), show that

$$(p-1)! \equiv -1 \pmod{p}.$$

(c) Consider the determinant $|a_{ij}|$ of order 100 with $a_{ij} = i \cdot j$. Prove that the absolute value of each of the 100! terms in the expansion of this determinant is congruent to 1 modulo 101.

4.4.31. Let F be a finite field having an odd number m of elements. Let $p(x)$ be an irreducible polynomial over F of the form $x^2 + bx + c$, $b, c \in F$. For how many elements k in F is $p(x) + k$ irreducible over F?

Additional Examples

1.1.5, 1.1.12.

Chapter 5. Summation of Series

In this chapter we turn our attention to some of the most basic summation formulas. The list is quite short (e.g., the binomial theorem, arithmetic- and geometric-series formulas, elementary power-series formulas) but we shall see that a few standard techniques (e.g. telescoping, differentiation, integration) make them extremely versatile and powerful.

5.1. Binomial Coefficients

Here are some basic identities; we are assuming that n and k are integers, $n \geqslant k \geqslant 0$.

Factorial representation:

$$\binom{n}{k} = \frac{n!}{k!\,(n-k)!}. \tag{1}$$

Symmetry condition:

$$\binom{n}{k} = \binom{n}{n-k}. \tag{2}$$

In-and-out formula:

$$\binom{n}{k} = \frac{n}{k}\binom{n-1}{k-1}, \qquad k \neq 0. \tag{3}$$

Addition formula:

$$\binom{n}{k} = \binom{n-1}{k} + \binom{n-1}{k-1}, \qquad k \neq 0. \tag{4}$$

The next formula is obtained by repeated application of the addition formula.

Summation formula:

$$\binom{n}{0} + \binom{n+1}{1} + \cdots + \binom{n+k}{k} = \binom{n+k+1}{k}. \tag{5}$$

Sums of products (see 4.3.2 and 1.3.4):

$$\binom{r}{0}\binom{s}{n} + \binom{r}{1}\binom{s}{n-1} + \cdots + \binom{r}{n}\binom{s}{0} = \binom{r+s}{n}. \tag{6}$$

Binomial theorem (see 2.1.1, 2.1.11, 4.3.3):

$$\sum_{k=0}^{n} \binom{n}{k} x^k y^{n-k} = (x+y)^n. \tag{7}$$

5.1.1. Use the summation formula to show that

(a) $1 + 2 + 3 + \cdots + n = \dfrac{n(n+1)}{2}$;

(b) $1^2 + 2^2 + \cdots + n^2 = \dfrac{n(n+1)(2n+1)}{6}$.

Solution. (a) We have

$$1 + 2 + \cdots + n = \binom{1}{1} + \binom{2}{1} + \binom{3}{1} + \cdots + \binom{n}{1}$$

$$= \binom{1}{0} + \binom{2}{1} + \binom{3}{2} + \cdots + \binom{n}{n-1}$$

$$= \binom{n+1}{n-1} = \binom{n+1}{2} = \frac{n(n+1)}{2}.$$

(b) We first look for constants a and b such that

$$k^2 = a\binom{k}{2} + b\binom{k}{1} = a\frac{k(k-1)}{2} + bk$$

for $k = 1, 2, \ldots, n$. Think of each side as a polynomial in k of degree 2. The identity will hold if and only if the coefficients of like powers of k are equal; that is, if and only if

$$1 = a/2,$$

$$0 = -a/2 + b.$$

This yields $a = 2$ and $b = 1$. It follows that

$$1^2 + 2^2 + \cdots + n^2$$

$$= \left[2\binom{1}{2} + \binom{1}{1}\right] + \left[2\binom{2}{2} + \binom{2}{1}\right] + \cdots + \left[2\binom{n}{2} + \binom{n}{1}\right]$$

$$= 2\left[\binom{2}{2} + \binom{3}{2} + \cdots + \binom{n}{2}\right] + \left[\binom{1}{1} + \binom{2}{1} + \cdots + \binom{n}{1}\right]$$

$$= 2\left[\binom{2}{0} + \binom{3}{1} + \cdots + \binom{n}{n-2}\right] + \left[\binom{1}{0} + \binom{2}{1} + \cdots + \binom{n}{n-1}\right]$$

$$= 2\binom{n+1}{n-2} + \binom{n+1}{n-1} = 2\binom{n+1}{3} + \binom{n+1}{2}$$

$$= 2\frac{(n+1)(n)(n-1)}{6} + \frac{(n+1)n}{2}$$

$$= \frac{n(n+1)(2n+1)}{6}.$$

(Another approach for part (b) is given in 5.3.11.)

The preceding sums occur so often that it is desirable to memorize them or in some way be able to recall them easily. One way to remember the first formula is shown in Figure 5.1 (for $n = 5$).

The diagram also prompts the following argument for the general case. Let S denote the sum of the first n positive integers. Then

$$S = 1 + \quad 2 \quad + \cdots + n,$$
$$S = n + (n-1) + \cdots + 1.$$

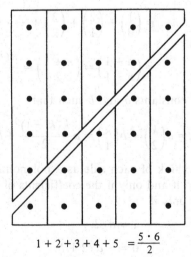

$$1 + 2 + 3 + 4 + 5 \;\; = \frac{5 \cdot 6}{2}$$

Figure 5.1.

Adding, we get

$$2S = (n + 1) + (n + 1) + \cdots + (n + 1)$$
$$= n(n + 1),$$

and it follows that

$$S = \frac{n(n + 1)}{2}.$$

The technique of evaluating a sum by rearranging the terms is a common one. In particular, when the terms are represented as a double summation, it is often advantageous to exchange the order of summation. The next example is an illustration of this idea.

5.1.2. Sum

$$\sum_{j=0}^{n} \sum_{i=j}^{n} \binom{n}{i}\binom{i}{j}.$$

Solution. The terms of this sum are indexed by ordered pairs (i, j), where (i, j) vary over the elements in the following triangular array:

i \ j	0	1	2	3	4	$\cdots$
0	*					
1	*	*				
2	*	*	*			
3	*	*	*	*		
4	*	*	*	*	*	
$\vdots$	$\vdots$	$\vdots$	$\vdots$	$\vdots$	$\vdots$	$\vdots$

In the given sum, the elements are first added columnwise. When we interchange the order of summation, so that the terms are first added rowwise, the sum is expressed in the form

$$\sum_{i=0}^{n} \sum_{j=0}^{i} \binom{n}{i}\binom{i}{j},$$

or, equivalently,

$$\sum_{i=0}^{n} \binom{n}{i} \sum_{j=0}^{i} \binom{i}{j}.$$

This is easily evaluated in the following manner. According to the binomial theorem,

$$(1 + x)^i = \sum_{j=0}^{i} \binom{i}{j} x^j.$$

When $x = 1$, we get

$$\sum_{j=0}^{i} \binom{i}{j} = 2^i.$$

The original sum is therefore

$$\sum_{i=0}^{n} \binom{n}{i} 2^i,$$

which by the binomial theorem is $(1 + 2)^n = 3^n$.

5.1.3. Sum the following:

(a) $\binom{n}{1} + 2\binom{n}{2} + 3\binom{n}{3} + \cdots + n\binom{n}{n}$,

(b) $1 + \frac{1}{2}\binom{n}{1} + \frac{1}{3}\binom{n}{2} + \cdots + \frac{1}{n+1}\binom{n}{n}$.

Solution. The first sum is

$$\sum_{i=1}^{n} i\binom{n}{i}.$$

Our aim, in summations of this type, is to use the in-and-out formula to bring the index of summation "inside" the binomial coefficent. Since

$$\binom{n}{i} = \frac{n}{i}\binom{n-1}{i-1},$$

it follows that

$$i\binom{n}{i} = n\binom{n-1}{i-1},$$

and therefore

$$\sum_{i=1}^{n} i\binom{n}{i} = \sum_{i=1}^{n} n\binom{n-1}{i-1}$$

$$= n\sum_{i=1}^{n} \binom{n-1}{i-1}$$

$$= n\sum_{i=0}^{n-1} \binom{n-1}{i}$$

$$= n \times 2^{n-1}.$$

The second sum is

$$\sum_{i=0}^{n} \frac{1}{i+1} \binom{n}{i}.$$

But

$$\binom{n+1}{i+1} = \frac{n+1}{i+1} \binom{n}{i},$$

and therefore,

$$\sum_{i=0}^{n} \frac{1}{i+1} \binom{n}{i} = \sum_{i=0}^{n} \frac{1}{n+1} \binom{n+1}{i+1}$$

$$= \frac{1}{n+1} \sum_{i=0}^{n} \binom{n+1}{i+1}$$

$$= \frac{1}{n+1} \sum_{i=1}^{n+1} \binom{n+1}{i}$$

$$= \frac{1}{n+1} \left[\left(\sum_{i=0}^{n+1} \binom{n+1}{i} \right) - \binom{n+1}{0} \right]$$

$$= \frac{1}{n+1} [2^{n+1} - 1].$$

There is another instructive way to handle these sums, based on differentiating and integrating each side of

$$\sum_{i=0}^{n} \binom{n}{i} x^i = (1+x)^n.$$

For part (a), we differentiate to get

$$\sum_{i=0}^{n} i \binom{n}{i} x^{i-1} = n(1+x)^{n-1},$$

and, with $x = 1$, we get

$$\sum_{i=1}^{n} i \binom{n}{i} = n \times 2^{n-1}.$$

For part (b), we integrate to get

$$\sum_{i=0}^{n} \binom{n}{i} \frac{x^{i+1}}{i+1} = \frac{(1+x)^{n+1}}{n+1} + C.$$

When $x = 0$, the left side of this equation is 0, and this implies that $C = -1/(n+1)$. Thus, when $x = 1$, we get (as before)

$$\sum_{i=0}^{n} \binom{n}{i} \frac{1}{i+1} = \frac{1}{n+1} [2^{n+1} - 1].$$

5.1.4. Show that

$$\binom{n}{1} - \frac{1}{2}\binom{n}{2} + \frac{1}{3}\binom{n}{3} - \cdots + (-1)^{n+1}\frac{1}{n}\binom{n}{n} = 1 + \frac{1}{2} + \cdots + \frac{1}{n}.$$

Solution. The left side of the identity looks like the definite integral of a binomial series, and this provides the idea for the following argument:

$$(1-x)^n = \binom{n}{0} - \binom{n}{1}x + \binom{n}{2}x^2 - \cdots,$$

$$1 - (1-x)^n = \binom{n}{1}x - \binom{n}{2}x^2 + \binom{n}{3}x^3 - \cdots,$$

$$\frac{1-(1-x)^n}{x} = \binom{n}{1} - \binom{n}{2}x + \binom{n}{3}x^2 - \cdots.$$

We are now set up to integrate each side from 0 to 1, and we get

$$\int_0^1 \frac{1-(1-x)^n}{x}\,dx = \binom{n}{1} - \frac{1}{2}\binom{n}{2} + \frac{1}{3}\binom{n}{3} - \cdots.$$

To finish the problem, we must show the integral on the left is equal to $1 + 1/2 + 1/3 + \cdots + 1/n$. Let $y = 1 - x$. Then

$$\int_0^1 \frac{1-(1-x)^n}{x}\,dx = \int_0^1 \frac{1-y^n}{1-y}\,dy$$

$$= \int_0^1 (1 + y + y^2 + \cdots + y^{n-1})\,dy$$

$$= y + \frac{1}{2}y^2 + \cdots + \frac{1}{n}y^n \Big]_0^1$$

$$= 1 + \frac{1}{2} + \cdots + \frac{1}{n}.$$

The problem can be done without calculus, using the basic identities of this section, but it is technically harder. However, since it is instructive, we will sketch the idea.

First, by repeated use of the addition formula and the in-and-out formula, we have, for $n \geqslant i \geqslant 1$,

$$\frac{1}{i}\binom{n}{i} = \frac{1}{i}\left[\binom{n-1}{i} + \binom{n-1}{i-1}\right] = \frac{1}{i}\binom{n-1}{i} + \frac{1}{n}\binom{n}{i}$$

$$= \frac{1}{i}\left[\binom{n-2}{i} + \binom{n-2}{i-1}\right] + \frac{1}{n}\binom{n}{i}$$

$$= \frac{1}{i}\binom{n-2}{i} + \frac{1}{n-1}\binom{n-1}{i} + \frac{1}{n}\binom{n}{i},$$

and continuing in this way, we get

$$\frac{1}{i}\binom{n}{i} = \frac{1}{n}\binom{n}{i} + \frac{1}{n-1}\binom{n-1}{i} + \cdots + \frac{1}{i}\binom{i}{i}.$$

Therefore,

$$\sum_{i=1}^{n}(-1)^{i+1}\frac{1}{i}\binom{n}{i} = \sum_{i=1}^{n}(-1)^{i+1}\left[\sum_{j=0}^{n-i}\binom{n-j}{i}\frac{1}{n-j}\right],$$

When we interchange the order of summation, we obtain

$$\sum_{j=0}^{n-1}\left[\sum_{i=1}^{n-j}(-1)^{i+1}\binom{n-j}{i}\frac{1}{n-j}\right].$$

Let $k = n - j$, so that the right side is

$$\sum_{k=1}^{n}\sum_{i=1}^{k}(-1)^{i+1}\binom{k}{i}\frac{1}{k} = \sum_{k=1}^{n}\frac{1}{k}\left[\sum_{i=1}^{k}(-1)^{i+1}\binom{k}{i}\right]$$

$$= \sum_{k=1}^{n}\frac{1}{k} \qquad \text{(see 5.1.9(a))}.$$

5.1.5. Sum

$$\sum_{i=0}^{n-1}\sum_{j=i+1}^{n+1}\binom{n+1}{j}\binom{n}{i}.$$

Solution. This can be evaluated using the basic identities of this section; however, we want to illustrate another technique. Although this approach will seem artificial and unlikely, the fact is that the thinking is not as unusual as it might at first appear. The idea is to interpret the sum in probabilistic terms, in the following manner.

Multiply the sum by $1/2^{2n+1}$ and write it in the form

$$\sum_{i=0}^{n-1}\left[\binom{n}{i}\left(\frac{1}{2}\right)^{n}\sum_{j=i+1}^{n+1}\binom{n+1}{j}\left(\frac{1}{2}\right)^{n+1}\right].$$

Now consider the following matching game between players A and B. Player A flips $n + 1$ coins and keeps n of the coins to maximize the number of heads. Player B flips n coins. The player with the maximum number of heads wins, with ties awarded to B.

Observe that the above sum represents the probability that A wins. We will now calculate this probability in another way.

The game is equivalent to the following. Let A and B each flip n coins. The player with the most heads wins. If they each have the same number of heads, but not all heads, A flips the $(n + 1)$st coin, winning if it is heads and losing if it is tails. At this point, A and B have equal chances of winning.

In the remaining case, both A and B have all heads. In this case B wins regardless of A's last toss. Thus, B wins in exactly 2 more cases than A.

That is, out of the 2^{2n+1} total flips, B wins in the 2 cases last described, and B wins in exactly one-half of the other cases ($\frac{1}{2}(2^{2n+1} - 2)$). Thus, the probability that A wins is

$$1 - \Pr(B \text{ wins}) = 1 - \frac{2 + \frac{1}{2}(2^{2n+1} - 2)}{2^{2n+1}} = \frac{2^{2n+1} - 2 - 2^{2n} + 1}{2^{2n+1}}$$

$$= \frac{2^{2n} - 1}{2^{2n+1}}.$$

It follows that the original sum is $2^{2n} - 1$.

Problems

5.1.6.

(a) Sum all the numbers between 0 and 1000 which are multiples of 7 or 11.

(b) Sum all the numbers between 0 and 1000 which are multiples of 7, 11, or 13.

5.1.7.

(a) Prove that for any integer $k > 1$ and any positive integer n, n^k is the sum of n consecutive odd numbers.

(b) Let n be a positive integer and m be any integer with the same parity as n. Prove that the product mn is equal to the sum of n consecutive odd integers.

5.1.8. Use the summation formula (5) to sum (a) $\sum_{k=1}^{n} k^3$; (b) $\sum_{k=1}^{n} k^4$.

5.1.9. Sum each of the following:

(a) $1 - \binom{n}{1} + \binom{n}{2} - \binom{n}{3} + \cdots + (-1)^n \binom{n}{n}$.

(b) $1 \times 2\binom{n}{2} + 2 \times 3\binom{n}{3} + \cdots + (n-1)n\binom{n}{n}$.

(c) $\binom{n}{1} + 2^2\binom{n}{2} + 3^2\binom{n}{3} + \cdots + n^2\binom{n}{n}$.

(d) $\binom{n}{1} - 2^2\binom{n}{2} + 3^2\binom{n}{3} - \cdots + (-1)^{n+1}n^2\binom{n}{n}$.

(e) $\binom{n}{0} - \frac{1}{2}\binom{n}{1} + \frac{1}{3}\binom{n}{2} - \cdots + (-1)^n \frac{1}{n+1}\binom{n}{n}$.

(f) $\sum_{j>1}\left[(-1)^j\binom{n}{j-1}\right] \Big/ \sum_{1 \leqslant k \leqslant j} k$.

5.1.10.

(a) What is the probability of an odd number of sixes turning up in a random toss of n fair dice? (To evaluate the sum, consider $\frac{1}{2}[(x+y)^n - (x-y)^n]$.)

(b) Show that if n is a positive multiple of 6,

$$\binom{n}{1} - 3\binom{n}{3} + 3^2\binom{n}{5} - \cdots = 0,$$

$$\binom{n}{1} - \frac{1}{3}\binom{n}{3} + \frac{1}{3^2}\binom{n}{5} - \cdots = 0.$$

5.1.11. Prove the following identities:

(a) $\dfrac{\binom{n}{1}}{1 \times 2} - \dfrac{\binom{n}{2}}{2 \times 3} + \dfrac{\binom{n}{3}}{3 \times 4} - \cdots + (-1)^{n+1} \dfrac{\binom{n}{n}}{n(n+1)}$

$$= \frac{1}{2} + \frac{1}{3} + \cdots + \frac{1}{n+1},$$

(b) $\dfrac{\binom{n}{0}}{1^2} - \dfrac{\binom{n}{1}}{2^2} + \dfrac{\binom{n}{2}}{3^2} - \cdots + (-1)^n \dfrac{\binom{n}{n}}{(n+1)^2}$

$$= \frac{1}{n+1}\left[1 + \frac{1}{2} + \cdots + \frac{1}{n} \right].$$

5.1.12. Show that

(a) $\binom{r}{0}\binom{s}{n} + \binom{r}{1}\binom{s}{n+1} + \binom{r}{2}\binom{s}{n+2} + \cdots + \binom{r}{n}\binom{s}{n+n} = \binom{r+s}{s-n}$,

(b) $\binom{n}{0}^2 + \binom{n}{1}^2 + \binom{n}{2}^2 + \cdots + \binom{n}{n}^2 = \binom{2n}{n}$.

5.1.13. Use the identities of this section to show that

$$\sum_{k=0}^{n}\left[\frac{n-2k}{n}\binom{n}{k} \right]^2 = \frac{2}{n}\binom{2n-2}{n-1}.$$

5.1.14. Sum

$$\sum_{i=n}^{2n-1}\binom{i-1}{n-1}2^{1-i}.$$

(Hint: For $i = n, n+1, \ldots, 2n-1$, compute $P(E_i)$, the probability that i tosses of a fair coin are required before obtaining n heads or n tails.)

5.1.15. A certain student, having just finished a particularly hairy summation, stared glassy-eyed at an "$x_1 y_2$" which was written on the scratch paper. After some doodling the student wrote:

$$x_1 y_2$$

(1) $x_1 y_2 y_3 x_4$

(2) $x_1 y_2 y_3 x_4 y_5 x_6 x_7 y_8$

(3) $x_1 y_2 y_3 x_4 y_5 x_6 x_7 y_8 y_9 x_{10} x_{11} y_{12} x_{13} y_{14} y_{15} x_{16}$.

On each line, the student copied the line above exactly, and then copied it again, changing x's to y's, y's to x's, and continuing the subscripts in order. The student noticed that the sum of the x-subscripts equals the sum of the y-subscripts in line (1). In line (2) the same equation holds, a similar one for the sums of the squares of the x- and y-subscripts, i.e., $1^2 + 4^2 + 6^2 + 7^2 = 2^2 + 3^2 + 5^2 + 8^2$. The student immediately made the inductive leap that in line (n), the sum of the kth powers of the x-subscripts would equal the sum of the kth powers of the y-subscripts for $k = 1, 2, \ldots, n$. Prove this for all $n > 0$.

Additional Examples

1.3.4, 1.3.15, 1.11.4, 2.1.1, 2.1.2, 4.3.5, 4.3.13, 4.3.14, 4.3.15, 4.3.24, 5.4.8, 6.8.3, 7.2.9, 7.3.8. Applications of the binomial theorem: 1.1.1 (Solution 4), 1.1.2, 1.3.8, 1.6.6(b), 1.12.4, 3.5.8, 3.5.10, 3.5.11, 3.5.12, 3.5.13, 4.2.13, 4.3.5, 4.4.9, 5.1.2, 5.1.15, 5.2.13, 6.8.3, 7.1.5, 7.1.15.

5.2. Geometric Series

The geometric series arises naturally in many problems, and it is therefore imperative to know its sum:

$$\sum_{i=0}^{n} x^i = \frac{1 - x^{n+1}}{1 - x}, \qquad x \neq 1,$$

$$\sum_{i=0}^{\infty} x^i = \lim_{n \to \infty} \sum_{i=0}^{n} x^i = \lim_{n \to \infty} \frac{1 - x^{n+1}}{1 - x} = \frac{1}{1 - x}, \qquad |x| < 1.$$

5.2.1. For a positive integer n, find a formula for $\sigma(n)$, the sum of the divisors of n.

Solution. Clearly $\sigma(1) = 1$. If p is a prime, the only divisors are 1 and p, so $\sigma(p) = p + 1$.

If n is a power of a prime, say $n = p^m$, the divisors are $1, p, p^2, \ldots, p^m$, so $\sigma(p^m) = 1 + p + \cdots + p^m = (1 - p^{m+1})/(1 - p)$.

Suppose $n = ab$, where a and b are relatively prime integers, each larger than one. Suppose the divisors of a are $a_1, a_2, \ldots, a_s$ and the divisors of b are $b_1, b_2, \ldots, b_t$. Then the divisors of n are $a_i b_j$, $i = 1, 2, \ldots, s$, $j = 1, 2, \ldots, t$, and the sum of these is

$$(a_1 b_1 + \cdots + a_1 b_t) + (a_2 b_1 + \cdots + a_2 b_t) + \cdots + (a_s b_1 + \cdots + a_s b_t),$$

or equivalently,

$$(a_1 + a_2 + \cdots + a_s)(b_1 + b_2 + \cdots + b_t).$$

Thus, $\sigma(n) = \sigma(a)\sigma(b)$.

Consider now an arbitrary positive integer n, and suppose its unique factorization is

$$n = p_1^{e_1} p_2^{e_2} \cdots p_k^{e_k}.$$

From the preceding work, we find that

$$\sigma(n) = \left(\frac{1 - p_1^{e_1 + 1}}{1 - p_1} \right) \left(\frac{1 - p_2^{e_2 + 1}}{1 - p_2} \right) \cdots \left(\frac{1 - p_k^{e_k + 1}}{1 - p_k} \right).$$

5.2.2. Let $n = 2m$, where m is an odd integer greater than 1. Let $\theta = e^{2\pi i/n}$. Express $(1 - \theta)^{-1}$ explicitly as a polynomial in θ,

$$a_k \theta^k + a_{k-1} \theta^{k-1} + \cdots + a_1 \theta + a_0,$$

with integer coefficients a_i.

Solution. Notice that θ is an nth root of unity, and that $\theta^m = (e^{2\pi i/2m})^m = e^{\pi i} = -1$. Thus,

$$1 + \theta + \theta^2 + \cdots + \theta^{m-1} = \frac{1 - \theta^m}{1 - \theta} = \frac{2}{1 - \theta}. \tag{1}$$

Also, since m is odd, we have

$$1 - \theta + \theta^2 - \cdots + \theta^{m-1} = \frac{1 - (-\theta)^m}{1 - (-\theta)} = 0. \tag{2}$$

Now, adding equations (1) and (2), we get

$$2 + 2\theta^2 + \cdots + 2\theta^{m-1} = \frac{2}{1 - \theta},$$

or, equivalently,

$$\frac{1}{1 - \theta} = 1 + \theta^2 + \theta^4 + \cdots + \theta^{m-1}.$$

5.2.3. Sum the finite series $\cos \theta + \cos 2\theta + \cdots + \cos n\theta$.

Solution. The series we wish to evaluate is the real part of the geometric series

$$e^{i\theta} + e^{2i\theta} + e^{3i\theta} + \cdots + e^{ni\theta},$$

whose sum is

$$\frac{e^{i(n+1)\theta}-1}{e^{i\theta}-1}-1 = \frac{e^{i(n+1)\theta}-1-(e^{i\theta}-1)}{e^{i\theta}-1} = \frac{e^{i(n+1)\theta}-e^{i\theta}}{e^{i\theta}-1}$$

$$= \frac{e^{i(n+\frac{1}{2})\theta}-e^{i(\frac{1}{2}\theta)}}{e^{i(\frac{1}{2}\theta)}-e^{-i(\frac{1}{2}\theta)}} = \frac{1}{2i}\left[\frac{e^{i(n+\frac{1}{2})\theta}-e^{i(\frac{1}{2}\theta)}}{\sin\frac{1}{2}\theta}\right]$$

$$= \frac{1}{2i}\frac{1}{\sin\frac{1}{2}\theta}\left[\left(\cos(n+\tfrac{1}{2})\theta - \cos\tfrac{1}{2}\theta\right)\right.$$

$$\left. + i\left(\sin(n+\tfrac{1}{2})\theta - \sin\tfrac{1}{2}\theta\right)\right]$$

$$= \frac{1}{2\sin\frac{1}{2}\theta}\left[\left(\sin(n+\tfrac{1}{2})\theta - \sin\tfrac{1}{2}\theta\right)\right.$$

$$\left. + i\left(\cos\tfrac{1}{2}\theta - \cos(n+\tfrac{1}{2})\theta\right)\right].$$

Equating real parts, we have

$$\cos\theta + \cos 2\theta + \cdots + \cos n\theta = \frac{1}{2\sin\frac{1}{2}\theta}\left[\sin(n+\tfrac{1}{2})\theta - \sin\tfrac{1}{2}\theta\right]$$

$$= \frac{\sin(n+\tfrac{1}{2})\theta}{2\sin\frac{1}{2}\theta} - \frac{1}{2}.$$

5.2.4. Prove that the fraction

$$\frac{1\times3\times5\cdots\times(2n-1)}{2\times4\times6\cdots\times2n},$$

when reduced to lowest terms, is of the form $a/2^w$ where a is odd and $w < 2n$.

Solution. We can write the fraction in the form

$$\frac{(2n)!}{2^{2n}n!n!} = \frac{1}{2^{2n}}\binom{2n}{n}.$$

Now, $\binom{2n}{n}$ is an integer, so the only question that remains is to show that $w < 2n$. The highest power of 2 in $(2n)!$ is

$$\left[\frac{2n}{2}\right] + \left[\frac{2n}{4}\right] + \left[\frac{2n}{8}\right] + \cdots + \left[\frac{2n}{2^k}\right] + \cdots$$

(see 3.3.10), and in $n!$ is

$$\left[\frac{n}{2}\right] + \left[\frac{n}{4}\right] + \left[\frac{n}{8}\right] + \cdots + \left[\frac{n}{2^k}\right] + \cdots.$$

It follows that

$$w = 2n + 2 \sum_{k=1}^{\infty} \left[\!\left[\frac{n}{2^k} \right]\!\right] - \sum_{k=1}^{\infty} \left[\!\left[\frac{2n}{2^k} \right]\!\right].$$

But

$$\sum_{k=1}^{\infty} \left[\!\left[\frac{2n}{2^k} \right]\!\right] = \sum_{k=1}^{\infty} \left[\!\left[\frac{n}{2^{k-1}} \right]\!\right] = \sum_{k=0}^{\infty} \left[\!\left[\frac{n}{2^k} \right]\!\right]$$

$$= n + \sum_{k=1}^{\infty} \left[\!\left[\frac{n}{2^k} \right]\!\right],$$

and therefore

$$w = 2n + 2 \sum_{k=1}^{\infty} \left[\!\left[\frac{n}{2^k} \right]\!\right] - n - \sum_{k=1}^{\infty} \left[\!\left[\frac{n}{2^k} \right]\!\right]$$

$$= n + \sum_{k=1}^{\infty} \left[\!\left[\frac{n}{2^k} \right]\!\right]$$

$$< n + \sum_{k=1}^{\infty} n/2^k = 2n.$$

$$\left(\text{Alternatively,} \ \binom{2n}{n} = \frac{2n}{n}\binom{2n-1}{n-1} = 2\binom{2n-1}{n-1}, \text{ so } w < 2n. \right)$$

5.2.5. For $x \geqslant 0$, evaluate in closed form

$$\sum_{n=1}^{\infty} \frac{(-1)^{[\![2^n x]\!]}}{2^n} \ .$$

Solution. Write x in the form

$$x = [\![x]\!] + \sum_{n=1}^{\infty} \frac{a_n}{2^n} \ ,$$

where a_n is 0 or 1, and where, if x has the form $m/2^n$ (m an odd integer), we take $a_k = 0$ for all sufficiently large k.

For each n, $[\![2^n x]\!]$ is even if and only if a_n is 0. It follows that for each n, $(-1)^{[\![2^n x]\!]} = 1 - 2a_n$. Therefore,

$$\sum_{n=1}^{\infty} \frac{(-1)^{[\![2^n x]\!]}}{2^n} = \sum_{n=1}^{\infty} \frac{1 - 2a_n}{2^n}$$

$$= \sum_{n=1}^{\infty} \frac{1}{2^n} - 2 \sum_{n=1}^{\infty} \frac{a_n}{2^n}$$

$$= 1 - 2(x - [\![x]\!]).$$

5.2.6. Evaluate in closed form

$$\sum_{(p,q)=1} \frac{1}{x^{p+q} - 1}, \qquad |x| > 1,$$

where the sum extends over all positive integers p and q such that p and q are relatively prime.

Solution.

$$\sum_{(p,q)=1} \frac{1}{x^{p+q} - 1} = \sum_{(p,q)=1} \frac{1}{x^{p+q}} \left(\frac{1}{1 - 1/x^{p+q}} \right)$$

$$= \sum_{(p,q)=1} \frac{1}{x^{p+q}} \left(\sum_{n=0}^{\infty} \left(\frac{1}{x^{p+q}} \right)^n \right)$$

$$= \sum_{(p,q)=1} \sum_{n=1}^{\infty} \left(\frac{1}{x^{n(p+q)}} \right).$$

As p, q, and n vary over the index set in this sum, the powers of $1/x$ will vary over all possible ordered pairs of positive integers (i, j). Since the series is absolutely convergent ($1/|x| < 1$), we can rearrange the terms of the series into the form

$$\sum_{(p,q)=1} \sum_{n=1}^{\infty} \frac{1}{x^{n(p+q)}} = \sum_{i=1}^{\infty} \sum_{j=1}^{\infty} \frac{1}{x^{i+j}}$$

$$= \sum_{i=1}^{\infty} \frac{1}{x^i} \left(\sum_{j=1}^{\infty} \frac{1}{x^j} \right)$$

$$= \left(\sum_{i=1}^{\infty} \frac{1}{x^i} \right)^2$$

$$= \left(\frac{1}{1 - 1/x} - 1 \right)^2 = \left(\frac{x}{x - 1} - 1 \right)^2$$

$$= \left(\frac{1}{x - 1} \right)^2.$$

Problems

5.2.7. Let $n = 2^{p-1}(2^p - 1)$, and suppose that $2^p - 1$ is a prime number. Show that the sum of all (positive) divisors of n, not including n itself, is exactly n. (A number having this property is called a *perfect* number.)

5.2.8. Sum the series $1 + 22 + 333 + \cdots + n(\underbrace{11 \ldots 1}_{n})$.

5.2.9. Let $E(n)$ denote the largest integer k such that 5^k is an integral divisor of the product $1^1 2^2 3^2 \cdots n^n$. Find a closed-form formula for $E(5^m)$, m a positive integer. What happens to $E(5^m)/5^{2m}$ as $m \to \infty$?

5.2.10. A sequence is defined by $a_1 = 2$ and $a_n = 3a_{n-1} + 1$. Find the sum $a_1 + a_2 + \cdots + a_n$.

5.2.11. Verify the following formulas:

(a) $\sum_{i=1}^{n} \sin (2k - 1)\theta = \dfrac{\sin^2 n\theta}{\sin \theta}$,

(b) $\sum_{i=1}^{n} \sin^2 (2k - 1)\theta = -\frac{1}{2}n - \dfrac{\sin 4n\theta}{4 \sin 2\theta}$.

5.2.12.

(a) If one tosses a fair coin until a head first appears, what is the probability that this event occurs on an even-numbered toss?

(b) The game of craps is played in the following manner: A player tosses a pair of dice. If the number is 2, 3, or 12 he loses immediately; if it is 7 or 11, he wins immediately. If any other number is obtained on the first toss, then that number becomes the players "point" and he must keep tossing the dice until either he "makes his point" (that is, obtains the first number again), in which case he wins, or he obtains 7, in which case he loses. Find the probability of winning.

5.2.13. If a, b and c are the roots of the equation $x^3 - x^2 - x - 1 = 0$,

(a) show that a, b and c are distinct;

(b) show that

$$\frac{a^{1000} - b^{1000}}{a - b} + \frac{b^{1000} - c^{1000}}{b - c} + \frac{c^{1000} - a^{1000}}{c - a}$$

is an integer.

5.2.14.

(a) Prove $\sqrt[n]{n!} < \prod_{p|n} p^{1/(p-1)}$, where the product on the right is over those positive primes p which divide n. (Hint: First prove that

$$\left[\frac{n}{p}\right] + \left[\frac{n}{p^2}\right] + \left[\frac{n}{p^3}\right] + \cdots \leqslant \frac{n}{p-1}.)$$

(b) Use part (a) to prove there are an infinite number of primes. (Hint: First prove that $(n!)^2 \geqslant n^n$.)

5.2.15. Prove that $\prod_{n=0}^{\infty}(1 + x^{2^n}) = \sum_{n=0}^{\infty} x^n = 1/(1 - x)$, $|x| < 1$.

5.2.16. Evaluate in closed form: $\sum_{n=0}^{\infty}(x^{2^n}/(1 - x^{2^{n+1}}))$, $|x| < 1$.

5.2.17. (a) Let $p_1, p_2, \ldots, p_n$ be all the primes less than m, and define

$$\lambda(m) = \prod_{i=1}^{n}\left(1 - \frac{1}{p_i}\right)^{-1}.$$

Show that $\lambda(m) = \sum(p_1^{a_1} p_2^{a_2} \cdots p_n^{a_n})^{-1}$, where the sum is over all n-tuples of nonnegative integers $(a_1, a_2, \ldots, a_n)$. (Hint:

$$\left(1 - \frac{1}{p_i}\right)^{-1} = 1 + \left(\frac{1}{p_i}\right) + \left(\frac{1}{p_i}\right)^2 + \cdots .$$

(b) Show that $1 + 1/2 + 1/3 + \cdots + 1/m < \lambda(m)$, and conclude that there are an infinite number of primes.

Additional Examples

1.12.1, 4.1.4, 4.1.8, 4.1.9, 4.2.5, 4.2.8, 4.2.12, 4.2.18, 4.3.13, 4.3.18(c), 5.1.4, 5.1.11, 5.4.1, 5.4.7, 5.4.9, 7.6.6.

5.3. Telescoping Series

Infinite series and infinite products can sometimes be evaluated by means of "telescoping." The examples are self-explanatory.

5.3.1. Sum the infinite series

$$\sum_{i=1}^{\infty} \frac{1}{(3i - 2)(3i + 1)} .$$

Solution. The trick is to break the summand into a sum of partial fractions, with the result that most of the terms in the partial sum will cancel. We look for numbers A and B such that

$$\frac{1}{(3i - 2)(3i + 1)} = \frac{A}{3i - 2} + \frac{B}{3i + 1} .$$

This leads to

$$1 = A(3i + 1) + B(3i - 2)$$

and equating coefficients, we have

$$3A + 3B = 0,$$
$$A - 2B = 1.$$

It follows that $A = \frac{1}{3}$, $B = -\frac{1}{3}$. Thus

$$S_n = \sum_{i=1}^{n} \left[\frac{\frac{1}{3}}{3i - 2} - \frac{\frac{1}{3}}{3i + 1} \right]$$

$$= \frac{1}{3}\left[\left(1 - \frac{1}{4}\right) + \left(\frac{1}{4} - \frac{1}{7}\right) + \left(\frac{1}{7} - \frac{1}{10}\right) + \cdots + \left(\frac{1}{3n - 2} - \frac{1}{3n + 1}\right) \right].$$

In this sum we have the "telescoping" property: the second term in each pair cancels with the first term in the successive pair, with the result that

$$S_n = \frac{1}{3}\left[1 - \frac{1}{3n+1} \right].$$

It follows that the infinite series adds to $\lim_{n\to\infty} S_n = \frac{1}{3}$.

5.3.2. Sum the infinite series

$$\frac{3}{1\times 2\times 3} + \frac{5}{2\times 3\times 4} + \frac{7}{3\times 4\times 5} + \frac{9}{4\times 5\times 6} + \cdots .$$

Solution. Again, by partial fractions, we look for real numbers A, B, C such that

$$\frac{2n+1}{n(n+1)(n+2)} = \frac{A}{n} + \frac{B}{n+1} + \frac{C}{n+2}.$$

We get

$$2n+1 = A(n+1)(n+2) + Bn(n+2) + Cn(n+1).$$

Setting $n = 0$ yields $A = \frac{1}{2}$; setting $n = -1$ yields $B = 1$; setting $n = -2$ yields $C = -\frac{3}{2}$. Thus, the nth partial sum is

$$S_n = \left[\frac{\frac{1}{2}}{1} + \frac{1}{2} - \frac{\frac{3}{2}}{3} \right] + \left[\frac{\frac{1}{2}}{2} + \frac{1}{3} - \frac{\frac{3}{2}}{4} \right]$$

$$+ \left[\frac{\frac{1}{2}}{3} + \frac{1}{4} - \frac{\frac{3}{2}}{5} \right] + \cdots + \left[\frac{\frac{1}{2}}{n-2} + \frac{1}{n-1} - \frac{\frac{3}{2}}{n} \right]$$

$$+ \left[\frac{\frac{1}{2}}{n-1} + \frac{1}{n} - \frac{\frac{3}{2}}{n+1} \right] + \left[\frac{\frac{1}{2}}{n} + \frac{1}{n+1} - \frac{\frac{3}{2}}{n+2} \right].$$

In this case, the telescoping takes place across groupings: the last term of one triple cancels with the sum of the middle term of the next grouping and the first term of the third grouping after that:

$$-\frac{\frac{3}{2}}{i} + \frac{1}{i} + \frac{\frac{1}{2}}{i} = 0.$$

The resulting sum is therefore equal to

$$S_n = \left[\frac{\frac{1}{2}}{1} + \frac{1}{2} \right] + \left[\frac{\frac{1}{2}}{2} \right] + \left[\frac{-\frac{3}{2}}{n+1} \right] + \left[\frac{1}{n+1} - \frac{\frac{3}{2}}{n+2} \right]$$

$$= \frac{5}{4} - \frac{\frac{1}{2}}{n+1} - \frac{\frac{3}{2}}{n+2}.$$

It follows that the infinite sum adds to $\lim_{n\to\infty} S_n = \frac{5}{4}$.

5.3.3. Express

$$\sum_{n=1}^{\infty} \sum_{m=1}^{\infty} \frac{1}{m^2 n + mn^2 + 2mn}$$

as a rational number.

Solution. Let S be the desired sum. Then, leaving out the details of the partial-fraction decomposition, we have

$$S = \sum_{n=1}^{\infty} \sum_{m=1}^{\infty} \frac{1}{mn(m+n+2)}$$

$$= \sum_{n=1}^{\infty} \frac{1}{n} \sum_{m=1}^{\infty} \frac{1}{m(m+n+2)}$$

$$= \sum_{n=1}^{\infty} \frac{1}{n} \sum_{m=1}^{\infty} \left[\frac{1/(n+2)}{m} - \frac{1/(n+2)}{m+n+2} \right]$$

$$= \sum_{n=1}^{\infty} \frac{1}{n(n+2)} \left[\left(1 - \frac{1}{n+3}\right) + \left(\frac{1}{2} - \frac{1}{n+4}\right) \right.$$

$$\left. + \left(\frac{1}{3} - \frac{1}{n+5}\right) + \cdots \right]$$

$$= \sum_{n=1}^{\infty} \frac{1}{n(n+2)} \left[1 + \frac{1}{2} + \frac{1}{3} + \cdots + \frac{1}{n+2} \right]$$

$$= \sum_{n=1}^{\infty} \left(\frac{\frac{1}{2}}{n} - \frac{\frac{1}{2}}{n+2} \right) \left(1 + \frac{1}{2} + \cdots + \frac{1}{n+2} \right)$$

$$= \frac{1}{2} \left[(1 - \frac{1}{3})(1 + \frac{1}{2} + \frac{1}{3}) + (\frac{1}{2} - \frac{1}{4})(1 + \frac{1}{2} + \frac{1}{3} + \frac{1}{4}) \right.$$

$$\left. + (\frac{1}{3} - \frac{1}{5})(1 + \frac{1}{2} + \frac{1}{3} + \frac{1}{4} + \frac{1}{5}) + \cdots \right]$$

$$= \frac{1}{2} \left[(1 + \frac{1}{2} + \frac{1}{3}) + \frac{1}{2}(1 + \frac{1}{2} + \frac{1}{3} + \frac{1}{4}) + \frac{1}{3}(\frac{1}{4} + \frac{1}{5}) \right.$$

$$\left. + \frac{1}{4}(\frac{1}{5} + \frac{1}{6}) + \frac{1}{5}(\frac{1}{6} + \frac{1}{7}) + \cdots \right]$$

$$= \frac{1}{2} \left[\frac{11}{6} + \frac{1}{2} \cdot \frac{25}{12} + \left(\frac{1}{3 \times 4} + \frac{1}{4 \times 5} + \frac{1}{5 \times 6} + \cdots \right) \right.$$

$$\left. + \left(\frac{1}{3 \times 5} + \frac{1}{4 \times 6} + \frac{1}{5 \times 7} + \cdots \right) \right]$$

$$= \frac{1}{2} \left[\frac{11}{6} + \frac{25}{24} + \left((\frac{1}{3} - \frac{1}{4}) + (\frac{1}{4} - \frac{1}{5}) + \cdots \right) \right.$$

$$\left. + \frac{1}{2} \left((\frac{1}{3} - \frac{1}{5}) + (\frac{1}{4} - \frac{1}{6}) + (\frac{1}{5} - \frac{1}{7}) + \cdots \right) \right]$$

$$= \frac{1}{2} \left[\frac{11}{6} + \frac{25}{24} + \frac{1}{3} + \frac{1}{6} + \frac{1}{8} \right]$$

$$= \frac{1}{2} \left[\frac{44 + 25 + 8 + 7}{24} \right] = \frac{84}{48} = \frac{7}{4}.$$

5.3.4. Sum the series

$$\sum_{n=1}^{\infty} 3^{n-1}\sin^3\left(\frac{x}{3^n}\right).$$

Solution. Using de Moivre's Theorem, we have

$$\sin 3\theta = \text{Im}(e^{3i\theta}) = \text{Im}\big((e^{i\theta})^3\big)$$

$$= \text{Im}\big[\cos\theta + i\sin\theta\big]^3$$

$$= \text{Im}\big[\cos^3\theta + 3\cos^2\theta\, i\sin\theta + 3\cos\theta\, i^2\sin^2\theta + i^3\sin^3\theta\big]$$

$$= 3\cos^2\theta\sin\theta - \sin^3\theta$$

$$= 3\big[(1 - \sin^2\theta)\sin\theta\big] - \sin^3\theta$$

$$= 3\sin\theta - 4\sin^3\theta.$$

It follows that

$$\sin^3\theta = \tfrac{3}{4}\sin\theta - \tfrac{1}{4}\sin 3\theta.$$

Thus,

$$S_k = \sum_{n=1}^{k} 3^{n-1}\sin^3\left(\frac{x}{3^n}\right) = \sum_{n=1}^{k} 3^{n-1}\left[\frac{3}{4}\sin\left(\frac{x}{3^n}\right) - \frac{1}{4}\sin\left(\frac{x}{3^{n-1}}\right)\right]$$

$$= \left(\frac{3}{4}\sin\frac{x}{3} - \frac{1}{4}\sin x\right) + \left(\frac{3^2}{4}\sin\left(\frac{x}{3^2}\right) - \frac{3}{4}\sin\left(\frac{x}{3}\right)\right)$$

$$+ \left(\frac{3^3}{4}\sin\left(\frac{x}{3^3}\right) - \frac{3^2}{4}\sin\left(\frac{x}{3^2}\right)\right) + \cdots$$

$$+ \left(\frac{3^k}{4}\sin\left(\frac{x}{3^k}\right) - \frac{3^{k-1}}{4}\sin\left(\frac{x}{3^{k-1}}\right)\right)$$

$$= \frac{3^k}{4}\sin\left(\frac{x}{3^k}\right) - \frac{1}{4}\sin x.$$

Therefore, the series adds to

$$\lim_{k\to\infty} S_k = \lim_{k\to\infty}\left[\frac{3^k}{4}\sin\left(\frac{x}{3^k}\right) - \frac{1}{4}\sin x\right]$$

$$= \lim_{k\to\infty}\left[\frac{x}{4}\,\frac{\sin(x/3^k)}{(x/3^k)} - \frac{1}{4}\sin x\right] = \frac{x - \sin x}{4}.$$

The telescoping idea is particularly useful in solving recurrence relations. Here is an example; other examples are given in the next section.

5.3.5. A sequence of numbers satisfies the recursion

$$x_0 = 0, \qquad nx_n = (n-2)x_{n-1} + 1 \qquad \text{for} \quad n > 0.$$

Find a closed-form expression for x_n.

Solution. We see that $x_0 = 0$, $x_1 = 1$, $x_2 = \frac{1}{2}$, $x_3 = \frac{1}{2}$, and consequently $x_n = \frac{1}{2}$ for all $n \geqslant 2$. But finding a pattern for a given recursion is not always so simple, and it is instructive to consider the problem in the following manner.

For $n \geqslant 2$, multiply each side of the recursion by $n - 1$, and for each n, set $y_n = n(n-1)x_n$. The recursion in terms of the y_n's is

$$y_n = y_{n-1} + (n-1), \qquad y_1 = 0.$$

It follows that

$$y_2 - y_1 = 1,$$
$$y_3 - y_2 = 2,$$
$$y_4 - y_3 = 3,$$
$$\vdots$$
$$y_n - y_{n-1} = n - 1.$$

Adding (notice the telescope), we get

$$y_n = 1 + 2 + \cdots + (n-1) = \frac{(n-1)n}{2},$$

and therefore,

$$x_n = \tfrac{1}{2}, \qquad n \geqslant 2.$$

Problems

5.3.6. Sum the following:

(a) $\dfrac{1}{2!} + \dfrac{2}{3!} + \dfrac{3}{4!} + \cdots + \dfrac{n-1}{n!}$,

(b) $1 \times 1! + 2 \times 2! + 3 \times 3! + \cdots + n \times n!$,

(c) $\dfrac{2}{1 \times 2 \times 3} + \dfrac{4}{2 \times 3 \times 4} + \dfrac{6}{3 \times 4 \times 5} + \cdots + \dfrac{2n}{n(n+1)(n+2)}$.

5.3.7. Evaluate the following infinite products:

(a) $\prod_{n=1}^{\infty}(1 - 1/n^2)$,

(b) $\prod_{n=1}^{\infty}(n^3 - 1)/(n^3 + 1)$.

(c) Show that an infinite product can be transformed into an infinite series by means of the identity $P = e^{\log P}$. Work part (a) in this way by evaluating the infinite series $\sum_{n=1}^{\infty}\log(1 - 1/n^2)$.

5.3.8. Prove that for each positive integer m,

$$\frac{m}{(m+1)(2m+1)} < \sum_{r=m+1}^{2m} \frac{1}{r^2} < \frac{m}{(m+1)(2m+1)} + \frac{3m+1}{4m(m+1)(2m+1)}$$

(Hint: Notice that $1/r(r+1) < 1/r^2 < 1/(r+1)(r-1)$.)

5.3.9. Let F_1, F_2, ... be the Fibonacci sequence. Use the telescoping property to prove the following identities:

(a) $F_1 + F_2 + \cdots + F_n = F_{n+2} - 1$. (Hint: $F_{n-2} = F_n - F_{n-1}$.)
(b) $F_1 + F_3 + \cdots F_{2n-1} = F_{2n}$.
(c) $F_1^2 + F_2^2 + \cdots + F_n^2 = F_n F_{n+1}$. (Hint: $F_n^2 = F_n(F_{n+1} - F_{n-1}) = F_n F_{n+1} - F_n F_{n-1}$.)
(d) $\sum_{n=2}^{\infty} 1/F_{n-1}F_{n+1} = 1$. (Show that $1/F_{n-1}F_{n+1} = 1/F_{n-1}F_n - 1/F_n F_{n+1}$.)
(e) $\sum_{n=2}^{\infty} F_n/F_{n-1}F_{n+1} = 1$.

5.3.10. Sum the following infinite series:

(a) $\sin^3 x + \frac{1}{3}\sin^3 3x + \frac{1}{3^2}\sin^3 3^2 x + \cdots$,

(b) $\cos^3 x - \frac{1}{3}\cos^3 3x + \frac{1}{3^2}\cos^3 3^2 x + \cdots$.

5.3.11.

(a) Use the identity $(k+1)^3 - k^3 = 3k^2 + 3k + 1$ to evaluate the sum of the first n squares. (Hint: Let k vary from 1 to n in the given identity, and consider the sum, on the left side and the right side, of the resulting n equations.)
(b) Use the telescoping idea, as in part (a), to evaluate the sum of the first n cubes.
(c) Find

$$\lim_{n\to\infty}\left[\frac{1}{n^5}\sum_{h=1}^{n}\sum_{k=1}^{n}(5h^4 - 18h^2k^2 + 5k^4)\right].$$

5.3.12. Show that the reciprocal of every integer greater than 1 is the sum of a finite number of consecutive terms of the infinite series $\sum_{n=1}^{\infty} 1/n(n+1)$.

5.3.13. If $m > 1$ is an integer and x is real, define

$$f(x) = \sum_{k=0}^{\infty} \sum_{i=1}^{m-1} \left[\!\left[\frac{x + im^k}{m^{k+1}}\right]\!\right].$$

Show that

$$f(x) = \begin{cases} [\![x]\!] & \text{if } x \geqslant 0, \\ [\![x+1]\!] & \text{if } x < 0. \end{cases}$$

(Hint: See 1.2.3.)

5.3.14. Solve the following recurrence relations (by the methods of this section).

(a) $x_0 = 1$, $x_n = 2x_{n-1} + 1$ for $n > 0$. [Hint: Divide each side by 2^n.]
(b) $x_0 = 0$, $nx_n = (n + 2)x_{n-1} + 1$ for $n > 0$.
(c) $x_0 = 1$, $x_1 = 1$, $x_2 = 2$, $x_{n+3} = x_n + 3$ for $n \geqslant 0$.

5.3.15. Show that a plane is divided by n straight lines, of which no two are parallel and no three meet in a point, into $\frac{1}{2}(n^2 + n + 2)$ regions.

5.3.16. Let

$$d_n = \frac{1}{n+1} + \frac{1}{n+2} + \cdots + \frac{1}{2n} .$$

Show that

$$d_n = 1 - \frac{1}{2} + \frac{1}{3} - \frac{1}{4} + \cdots + \frac{1}{2n-1} - \frac{1}{2n} .$$

[Hint: Consider the telescoping series $\sum_{i=1}^{n-1}(d_{i+1} - d_i)$.] Conclude that $d_n \to \log 2$ as $n \to \infty$. (For another proof of the first part, consider the difference of each expression from the harmonic sum $1 + 1/2 + \cdots + 1/(2(n-1))$. Also, see Section 6.8.)

Additional Examples

6.6.6, 7.1.8, 7.2.2.

5.4. Power Series

A *power series* is an expression of the type

$$a_0 + a_1x + a_2x^2 + \cdots + a_nx^n + \cdots$$

where $a_0, a_1, a_2, \ldots$ are real numbers.

Given a power series, we can define a function $f(x)$, whose domain is the set of those real numbers x which make the power series into a convergent infinite series, and whose value is given by

$$f(c) = a_0 + a_1c + a_2c^2 + \cdots + a_nc^n + \cdots$$

for any c such that the right side converges.

Given a power series $\sum_{i=0}^{\infty} a_i x^i$, it can be shown that exactly one of the following holds:

(i) The series is convergent for all real numbers x.
(ii) The series is convergent only for $x = 0$.

(iii) There exists a real number r such that the series is convergent for $|x| < r$ and divergent for $|x| > r$.

We define the radius of convergence to be $+\infty$ if (i) holds, 0 if (ii) holds, and r if (iii) holds.

We can ask the following question: Given a function f, can f be represented by a power series? One result along these lines is Taylor's theorem (with remainder): If f can be differentiated as many times as we like on an open interval about 0, then for each x in this open interval,

$$f(x) = f(0) + \frac{f'(0)}{1!} x + \frac{f''(0)}{2!} x^2 + \cdots + \frac{f^{(n)}(0)}{n!} x^n + R_n(x),$$

where $R_n(x) = f^{(n+1)}(c)x^{n+1}/(n+1)!$ for some point c between 0 and x. The important part here is that if $R_n(x)$ is well behaved, so that $R_n(x) \to 0$ as $n \to \infty$, then

$$f(x) = \sum_{n=0}^{\infty} \frac{f^{(n)}(0)}{n!} x^n.$$

This gives us a method of finding a power series for a given function $f(x)$.

Using this idea, one can find power-series expansions for the most common elementary functions. The following series occur so often they should be memorized:

$$e^x = 1 + x + \frac{x^2}{2!} + \frac{x^3}{3!} + \cdots + \frac{x^n}{n!} + \cdots, \tag{i}$$

$$\sin x = x - \frac{x^3}{3!} + \frac{x^5}{5!} - \cdots + (-1)^n \frac{x^{2n-1}}{(2n-1)!} + \cdots, \tag{ii}$$

$$\cos x = 1 - \frac{x^2}{2!} + \frac{x^4}{4!} - \cdots + (-1)^n \frac{x^{2n}}{(2n)!} + \cdots, \tag{iii}$$

$$\frac{1}{1-x} = 1 + x + x^2 + \cdots + x^n + \cdots, \qquad |x| < 1, \tag{iv}$$

$$\log(1 + x) = x - \frac{x^2}{2} + \frac{x^3}{3} - \cdots + (-1)^n \frac{x^n}{n} + \cdots, \qquad |x| < 1, \tag{v}$$

$$(1 + x)^r = 1 + \binom{r}{1} x + \binom{r}{2} x^2 + \cdots + \binom{r}{n} x^n + \cdots, \qquad r \text{ real}, |x| < 1, \tag{vi}$$

where

$$\binom{r}{n} = \frac{r(r-1)(r-2) \cdots (r-n+1)}{n!}.$$

5.4.1. Prove that e is an irrational number.

Solution. Suppose $e = h/k$, where h and k are integers. Using the power-series expansion for e^x, and setting $x = 1$, we have

$$e = 1 + \frac{1}{1!} + \frac{1}{2!} + \cdots + \frac{1}{k!}$$

$$+ \left[\frac{1}{(k+1)!} + \frac{1}{(k+2)!} + \cdots \right].$$

Multiply each side by $k!$, and write in the form

$$k! \left[\frac{h}{k} - 1 - \frac{1}{1!} - \frac{1}{2!} - \cdots - \frac{1}{k!} \right] = \frac{1}{k+1} + \frac{1}{(k+1)(k+2)} + \cdots$$

Notice that the right side of this equation is positive, and the left side is an integer. Thus, the left side is a positive integer. However, on the right side,

$$\frac{1}{k+1} + \frac{1}{(k+1)(k+2)} + \frac{1}{(k+1)(k+2)(k+3)} + \cdots$$

$$= \frac{1}{k+1} \left[1 + \frac{1}{k+2} + \frac{1}{(k+2)(k+3)} + \cdots \right]$$

$$< \frac{1}{k+1} \left[1 + \frac{1}{k+1} + \left(\frac{1}{k+1} \right)^2 + \cdots \right]$$

$$= \frac{1}{k+1} \left[\frac{1}{1 - \left(\frac{1}{k+1} \right)} \right] = \frac{1}{k+1} \left(\frac{k+1}{k} \right) = \frac{1}{k} < 1.$$

Thus, the right side is not a positive integer, and we have a contradiction. It must therefore be the case that e is irrational.

5.4.2. Show that the power-series representation for the infinite series $\sum_{n=0}^{\infty} x^n (x-1)^{2n}/n!$ cannot have three consecutive zero coefficients.

Solution. The series sums to $f(x) = e^{x(x-1)^2}$. To find its power-series representation, we need to compute $f^{(n)}(0)$ for $n = 1, 2, 3, \ldots$. We have

$$f'(x) = e^{x(x-1)^2}(3x^2 - 4x + 1),$$

which is of the form

$$f'(x) = f(x)g(x)$$

where $g(x)$ is a polynomial of degree 2. It follows that

$$f''(x) = f'(x)g(x) + f(x)g'(x),$$

$$f'''(x) = f''(x)g(x) + 2f'(x)g'(x) + f(x)g''(x),$$

$$f^{(iv)}(x) = f'''(x)g(x) + 3f''(x)g'(x) + 3f'(x)g''(x)$$

(note: $g'''(x) = 0$). An induction argument shows that for $n = 3, 4, 5, \ldots$

$$f^{(n+1)}(x) = f^{(n)}(x) g(x) + a_n f^{(n-1)}(x) g'(x) + b_n f^{(n-2)}(x) g''(x)$$

for some integers a_n and b_n.

Suppose that three successive terms in the power series for $f(x)$ were zero, say $f^{(n)}(0) = f^{(n-1)}(0) = f^{(n-2)}(0) = 0$. Then, the recursion of the last paragraph implies that $f^{(k)}(0) = 0$ for all $k > n$, and this means that $f(x)$ is a polynomial, a contradiction. Therefore we are forced to conclude that the power series for $f(x)$ cannot have three successive zero coefficients.

5.4.3. Evaluate $\lim_{x \to \infty} [(e/2)x + x^2[(1 + 1/x)^x - e]]$.

Solution. We will find the first few terms of the Taylor series for $(1 + 1/x)^x$ in powers of $1/x$. We have

$$\left(1 + \frac{1}{x}\right)^x = e^{x \log(1 + 1/x)} = \exp\left\{x\left[(1/x) - \frac{(1/x)^2}{2} + \frac{(1/x)^3}{3} + \cdots\right]\right\}$$

$$= \exp\left[1 - \frac{1}{2}\left(\frac{1}{x}\right) + \frac{1}{3}\left(\frac{1}{x}\right)^2 - \frac{1}{4}\left(\frac{1}{x}\right)^3 + \cdots\right]$$

$$= e \cdot e^{-\frac{1}{2}(1/x)} e^{\frac{1}{3}(1/x)^2} e^{-\frac{1}{4}(1/x)^3} \cdots$$

$$= e\left[1 - \frac{1}{2x} + \frac{1}{2!}\left(\frac{1}{2x}\right)^2 - \cdots\right]$$

$$\times\left[1 + \frac{1}{3}\left(\frac{1}{x}\right)^2 + \frac{1}{2!}\left(\frac{1}{3}\left(\frac{1}{x}\right)^2\right)^2 + \cdots\right] \cdots$$

$$= e\left[1 - \frac{1}{2x} + \frac{11}{24}\left(\frac{1}{x}\right)^2 + \text{higher powers of } \frac{1}{x}\right].$$

It follows that

$$\lim_{x \to \infty}\left[\frac{e}{2}x + x^2\left[\left(1 + \frac{1}{x}\right)^x - e\right]\right]$$

$$= \lim_{x \to \infty}\left[\frac{e}{2}x + x^2\left[e - \frac{e}{2x} + \frac{11e}{24}\left(\frac{1}{x}\right)^2 + \cdots - e\right]\right]$$

$$= \lim_{x \to \infty}\left[\frac{11e}{24} + \text{higher powers of } \frac{1}{x}\right]$$

$$= \frac{11e}{24}.$$

An extremely useful fact about power series is that they can be differentiated and integrated term by term in the interior of the interval of convergence. By this we mean that if $\sum a_n x^n$ has a radius of convergence r

and if $f(x) = \sum a_n x^n$, then

$$f'(x) = \sum n a_n x^{n-1} \quad \text{and} \quad \int_0^x f(x) \, dx = \sum \frac{a_n}{n+1} x^{n+1},$$

and both resulting series have radius of convergence r.

One consequence of the preceding result is that the power-series representation of a function f is unique; that is, if $f(x) = \sum a_n x^n = \sum b_n x^n$, then $a_n = b_n$ for all n. In fact, $a_n = b_n = f^{(n)}(0)/n!$. To see this, simply differentiate each side of $f(x) = \sum a_n x^n$ repeatedly and evaluate each successive derivative at $x = 0$. For example, $f(0) = a_0$; $f'(x) = \sum n a_n x^{n-1}$, so $f'(0) = a_1$; $f''(x) = \sum n(n-1)a_n x^{n-2}$, so $f''(0) = 2!a_2$, or equivalently, $a_2 = f''(0)/2!$, and so forth.

5.4.4. Sum the infinite series

$$\frac{1^2}{0!} + \frac{2^2}{1!} + \frac{3^2}{2!} + \frac{4^2}{3!} + \cdots.$$

Solution. Begin with the series

$$e^x = \sum_{n=0}^{\infty} \frac{x^n}{n!}.$$

Multiply each side by x:

$$xe^x = \sum_{n=0}^{\infty} \frac{x^{n+1}}{n!},$$

and differentiate each side:

$$(1+x)e^x = \sum_{n=0}^{\infty} \frac{(n+1)x^n}{n!}.$$

Multiply each side by x again:

$$(x + x^2)e^x = \sum_{n=0}^{\infty} \frac{(n+1)x^{n+1}}{n!},$$

and differentiate to get

$$(1 + 3x + x^2)e^x = \sum_{n=0}^{\infty} \frac{(n+1)^2 x^n}{n!}.$$

Now set $x = 1$, and find that

$$\sum_{n=0}^{\infty} \frac{(n+1)^2}{n!} = 5e.$$

The following theorem is often useful.

Abel's Limit Theorem. *Let* $r > 0$, *and suppose* $\sum_{n=0}^{\infty} a_n r^n$ *converges. Then* $\sum_{n=0}^{\infty} a_n x^n$ *converges absolutely for* $|x| < r$, *and*

$$\lim_{x \to r^-} \sum_{n=0}^{\infty} a_n x^n = \sum_{n=0}^{\infty} a_n r^n.$$

5.4.5. Sum the infinite series

$$1 - \tfrac{1}{4} + \tfrac{1}{7} - \tfrac{1}{10} + \cdots .$$

Solution. We know that

$$\frac{1}{1 + x^3} = 1 - x^3 + x^6 - x^9 + \cdots, \qquad |x| < 1,$$

and therefore

$$\int_0^x \frac{dx}{1 + x^3} = x - \frac{x^4}{4} + \frac{x^7}{7} - \frac{x^{10}}{10} + \cdots, \qquad |x| < 1.$$

Now, the series on the right side converges for $x = 1$ (by the alternating-series test), and therefore, by Abel's limit theorem,

$$1 - \tfrac{1}{4} + \tfrac{1}{7} - \tfrac{1}{10} + \cdots = \lim_{x \to 1^-} \int_0^x \frac{dx}{1 + x^3}.$$

This integral can be worked by partial fractions (the details are not of interest here), and we get

$$\int_0^x \frac{dx}{1 + x^3} = \frac{1}{3} \left[\log \frac{1 + x}{\sqrt{1 - x + x^2}} \right. $$
$$\left. + \sqrt{3} \left[\arctan \frac{2x - 1}{\sqrt{3}} - \arctan \frac{-1}{\sqrt{3}} \right] \right].$$

Thus, the series sums to

$$\frac{1}{3} \left[\log 2 + \frac{\pi}{\sqrt{3}} \right].$$

5.4.6. Let $S_n = \sum_{i=1}^{n} (-1)^{i+1} / i$ and $S = \lim_{n \to \infty} S_n$. Show that $\sum_{n=1}^{\infty} (S_n - S) = \log 2 - \tfrac{1}{2}$.

Solution. We must evaluate the double series

$$\sum_{n=1}^{\infty} \sum_{i=1}^{\infty} \frac{(-1)^{n+i}}{n + i}.$$

For this purpose, consider the function

$$F(x) = \sum_{j=1}^{\infty} \sum_{i=1}^{\infty} \frac{(-x)^{i+j}}{i+j}, \qquad |x| < 1.$$

Then

$$F'(x) = \sum_{j=1}^{\infty} \sum_{i=1}^{\infty} (-x)^{i+j-1}(-1)$$

$$= \sum_{j=1}^{\infty} (-1)(-x)^{j} \sum_{i=1}^{\infty} (-x)^{i-1}$$

$$= \sum_{j=1}^{\infty} (-1)(-x)^{j} \frac{1}{1+x}$$

$$= \frac{x}{1+x} \sum_{j=0}^{\infty} (-x)^{j}$$

$$= \frac{x}{(1+x)^{2}} = \frac{1}{1+x} - \frac{1}{(1+x)^{2}}.$$

It follows that

$$\int_{0}^{x} F'(x)\,dx = \int_{0}^{x} \frac{dx}{1+x} - \int_{0}^{x} \frac{dx}{(1+x)^{2}},$$

$$F(x) - F(0) = \log(1+x) \Big]_{0}^{x} + \frac{1}{1+x} \Big]_{0}^{x},$$

and we find

$$F(x) = \log(1+x) + \frac{1}{1+x} - 1.$$

The series for $F(x)$ is convergent for $x = 1$, so by Abel's limit theorem,

$$F(1) = \log 2 + \tfrac{1}{2} - 1 = \log 2 - \tfrac{1}{2}.$$

5.4.7. Given the power series $a_0 + a_1 x + a_2 x^2 + \cdots$ with $a_n = (n^2 + 1)3^n$, show that there is a relation of the form

$$a_n + pa_{n+1} + qa_{n+2} + ra_{n+3} = 0,$$

in which p, q, r are independent of n. Find these constants and the sum of the series.

Solution. Substituting the given values of a_n into the recurrence, we find

$$(n^2 + 1)3^n + p(n^2 + 2n + 2)3^{n+1}$$
$$+ q(n^2 + 4n + 5)3^{n+2} + r(n^2 + 6n + 10)3^{n+3} = 0.$$

Now divide each side by 3^n. Then equating coefficients, we find that p, q, and r must satisfy

$$3p + \ \ 9q + \ \ 27r = -1,$$
$$2p + 12q + \ \ 54r = \ \ \ 0,$$
$$6p + 45q + 270r = -1.$$

These equations have a solution: $p = -1$, $q = \frac{1}{3}$, $r = -\frac{1}{27}$.

For the second half, we wish to sum the series

$$\sum_{n=0}^{\infty} (n^2 + 1)3^n x^n,$$

which breaks into two parts,

$$\sum_{n=0}^{\infty} n^2(3x)^n + \sum_{n=0}^{\infty} (3x)^n.$$

Let $S = \sum_{n=0}^{\infty}(3x)^n$. If $|x| < \frac{1}{3}$, $S = 1/(1 - 3x)$. Therefore, from

$$\sum_{n=0}^{\infty} (3x)^n = \frac{1}{1 - 3x}, \qquad |x| < \frac{1}{3},$$

it follows that

$$\sum_{n=0}^{\infty} n(3x)^{n-1} \cdot 3 = \frac{3}{(1 - 3x)^2},$$

$$\sum_{n=0}^{\infty} n(3x)^n = \frac{3x}{(1 - 3x)^2},$$

$$\sum_{n=0}^{\infty} n^2(3x)^{n-1} \cdot 3 = \frac{d}{dx}\left[\frac{3x}{(1 - 3x)^2}\right] = \frac{9x + 3}{(1 - 3x)^3},$$

$$\sum_{n=0}^{\infty} n^2(3x)^n = \frac{3x(3x + 1)}{(1 - 3x)^3}.$$

Combining, we get

$$\sum_{n=0}^{\infty} (n^2 + 1)(3x)^n = \frac{3x(3x + 1)}{(1 - 3x)^3} + \frac{1}{1 - 3x}$$

$$= \frac{18x^2 - 3x + 1}{(1 - 3x)^3}.$$

5.4.8. Evaluate in closed form:

$$S_n = \sum_{k=0}^{n} (-4)^k \binom{n + k}{2k}.$$

Solution. We can compute the first few terms:

$$S_0 = 1,$$

$$S_1 = \binom{1}{0} - 4\binom{2}{2} = 1 - 4 = -3,$$

$$S_2 = \binom{2}{0} - 4\binom{3}{2} + 16\binom{4}{4} = 1 - 12 + 16 = 5,$$

$$S_3 = \binom{3}{0} - 4\binom{4}{2} + 16\binom{5}{4} - 64\binom{6}{6}$$

$$= 1 - 24 + 80 - 64 = -7.$$

From this pattern, we expect that $S_n = (-1)^n(2n+1)$.

If we think of proving this conjecture by mathematical induction, we are led to look for a recurrence relation. This leads to the following reasoning:

$$\binom{n+k}{2k} = \binom{n+k-1}{2k-1} + \binom{n+k-1}{2k}$$

$$= \left[\binom{n+k-2}{2k-2} + \binom{n+k-2}{2k-1}\right] + \binom{n+k-1}{2k}$$

$$= \binom{n+k-2}{2k-2} + \left[\binom{n+k-1}{2k} - \binom{n+k-2}{2k}\right] + \binom{n+k-1}{2k}$$

$$= \binom{n+k-2}{2k-2} + 2\binom{n+k-1}{2k} - \binom{n+k-2}{2k}.$$

Thus,

$$S_n = \sum_{k=0}^{n} (-4)^k \binom{n+k}{2k}$$

$$= \sum_{k=0}^{n} (-4)^k \binom{n+k-2}{2k-2}$$

$$+ 2\sum_{k=0}^{n} (-4)^k \binom{n+k-1}{2k} - \sum_{k=0}^{n} (-4)^k \binom{n+k-2}{2k}$$

$$= \sum_{k=0}^{n} (-4)^k \binom{n+k-2}{2k-2}$$

$$+ 2\sum_{k=0}^{n-1} (-4)^k \binom{n-1+k}{2k} - \sum_{k=0}^{n-2} (-4)^k \binom{n-2+k}{2k}$$

$$= -4\sum_{k=0}^{n} (-4)^{k-1} \binom{n+k-2}{2k-2} + 2S_{n-1} - S_{n-2}$$

$$= -4\sum_{k=0}^{n-1} (-4)^k \binom{n+k-1}{2k} + 2S_{n-1} - S_{n-2}$$

$$= -4S_{n-1} + 2S_{n-1} - S_{n-2}$$

$$= -2S_{n-1} - S_{n-2}.$$

Using this recurrence relation, we can use mathematical induction to establish the claim that $S_n = (-1)^n(2n + 1)$.

Consider the recurrence relation

$$S_n = -2S_{n-1} - S_{n-2}, \qquad S_0 = 1, \quad S_1 = -3,$$

and, for the sake of the illustration, let us suppose that we are not able to discover the formula for S_n by a consideration of the first few cases. We wish to give here a technique for making this discovery. The method is to make use of a *generating function* $F(x)$, in the following way.

Let $F(x)$ be the name of the power series whose coefficients are S_0, S_1, $S_2, \ldots$, namely,

$$F(x) = S_0 + S_1 x + S_2 x^2 + \cdots + S_n x^n + \cdots.$$

We will act as though the series converges at x to the function $F(x)$. Then

$$2xF(x) = 2S_0 x + 2 S_1 x^2 + 2 S_2 x^3 + \cdots + 2 S_{n-1} x^n + \cdots,$$

$$x^2 F(x) = \qquad S_0 x^2 + \quad S_1 x^3 + \cdots + \quad S_{n-2} x^n + \cdots.$$

Adding, and making use of the fact that $S_n + 2S_{n-1} + S_{n-2} = 0$, we find that

$$(1 + 2x + x^2)F(x) = S_0 + (S_1 + 2S_0)x,$$

or, equivalently,

$$F(x) = \frac{1 - x}{(1 + x)^2}.$$

We now express the right side of this equation as a power series. To do this, first differentiate each side of

$$\frac{1}{1 + x} = \sum_{n=0}^{\infty} (-1)^n x^n$$

to get

$$\frac{-1}{(1 + x)^2} = \sum_{n=0}^{\infty} (-1)^n n x^{n-1}.$$

Then multiply each side by $x - 1$ to get

$$F(x) = \sum_{n=0}^{\infty} (-1)^n n (x - 1) x^{n-1}$$

$$= \sum_{n=0}^{\infty} (-1)^n n x^n - \sum_{n=0}^{\infty} (-1)^n n x^{n-1}$$

$$= \sum_{n=0}^{\infty} (-1)^n n x^n + \sum_{n=0}^{\infty} (-1)^n (n + 1) x^n$$

$$= \sum_{n=0}^{\infty} (-1)^n (2n + 1) x^n.$$

Here again, we find that S_n, the coefficient of x^n, is $S_n = (-1)^n(2n + 1)$.

It is true that the generating-function approach shown here has not been justified in a step-by-step manner, since we have completely disregarded convergence considerations. However, the method can nevertheless be used in similar problems to formulate conjectures (about the solution of the recurrence relation), and these conjectures can then be verified by other means (for example, mathematical induction).

5.4.9. Sum the finite series $a_0 + a_1 + \cdots + a_n$, where $a_0 = 2$, $a_1 = 5$, for $n > 1$, $a_n = 5a_{n-1} - 6a_{n-2}$.

Solution. The first few terms of the a_i-sequence are

$$2, 5, 13, 35, 97, 275, 393, \ldots .$$

Here, a general formula for the nth term is not apparent, so we turn to the technique of generating functions. Consider

$$F(x) = a_0 + a_1 x + a_2 x^2 + \cdots + a_n x^n + \cdots .$$

We have

$$-5xF(x) = -5a_0 x - 5a_1 x^2 - \cdots - 5a_{n-1}x^n - \cdots ,$$

$$6x^2 F(x) = \qquad\qquad 6a_0 x^2 + \cdots + 6a_{n-2}x^n + \cdots .$$

Adding and using the recurrence $a_n - 5a_{n-1} + 6a_{n-2} = 0$, we get

$$(1 - 5x + 6x^2)F(x) = a_0 + (a_1 - 5a_0)x,$$

so that

$$F(x) = \frac{2 - 5x}{(1 - 2x)(1 - 3x)} .$$

Write this as a sum of partial fractions, and make use of the geometric series, to find that

$$F(x) = \frac{1}{1 - 2x} + \frac{1}{1 - 3x}$$

$$= \sum_{i=0}^{\infty} (2x)^i + \sum_{i=0}^{\infty} (3x)^i$$

$$= \sum_{i=0}^{\infty} (2^i + 3^i)x^i.$$

Thus, $a_i = 2^i + 3^i$ for $i = 0, 1, 2, 3, \ldots$. [As a check, we can verify this formula by induction. Note that $a_0 = 2^0 + 3^0 = 2$, $a_1 = 2 + 3 = 5$, and, for $i \geqslant 2$, $a_i = 5a_{i-1} - 6a_{i-2} = 5(2^{i-1} + 3^{i-1}) - 6(2^{i-2} + 3^{i-2}) = 5(2^{i-1} + 3^{i-1}) - 3 \times 2^{i-1} - 2 \times 3^{i-1} = 2^i + 3^i$.]

We are now ready to compute the sum:

$$a_0 + a_1 + \cdots + a_n = \sum_{i=0}^{n} (2^i + 3^i) = \sum_{i=0}^{n} 2^i + \sum_{i=0}^{n} 3^i$$

$$= \frac{2^{n+1} - 1}{2 - 1} + \frac{3^{n+1} - 1}{3 - 1}$$

$$= 2^{n+1} - 1 + \frac{3^{n+1} - 1}{2}$$

$$= \frac{2^{n+2} + 3^{n+1} - 3}{2}.$$

5.4.10. Find a closed-form expression for T_n, if $T_0 = 1$ and for $n \geqslant 1$

$$T_n = T_0 T_{n-1} + T_1 T_{n-2} + \cdots + T_{n-1} T_0.$$

Solution. This recurrence relation arose in 2.5.12. To solve it, let

$$f(x) = T_0 + T_1 x + T_2 x^2 + \cdots + T_n x^n + \cdots$$

and set

$$F(x) = xf(x) = T_0 x + T_1 x^2 + T_2 x^3 + \cdots + T_n x^{n+1} + \cdots.$$

The reason for this step is that

$$(F(x))^2 = T_0^2 x^2 + (T_0 T_1 + T_1 T_0)x^3 + \cdots$$

$$+ (T_0 T_{n-1} + T_1 T_{n-2} + \cdots + T_{n-1} T_0)x^{n+1} + \cdots,$$

so in view of the recurrence relation we have

$$(F(x))^2 = T_1 x^2 + T_2 x^3 + \cdots + T_n x^{n+1} + \cdots$$

$$= F(x) - T_0 x.$$

Using the quadratic formula we find that

$$F(x) = \frac{1 - \sqrt{1 - 4x}}{2}.$$

(We choose the negative sign because $F(0) = 0$; the positive sign would yield $F(0) = 1$.)

Now by the power-series expansion,

$$\sqrt{1 - 4x} = 1 + \binom{\frac{1}{2}}{1}(-4x) + \binom{\frac{1}{2}}{2}(-4x)^2 + \cdots$$

$$+ \binom{\frac{1}{2}}{n+1}(-4x)^{n+1} + \cdots.$$

It follows that the coefficient of x^{n+1} in $F(x)$ is

$$
T_n = -\frac{1}{2}\binom{\frac{1}{2}}{n+1}(-4)^{n+1}
$$

$$
= -\frac{1}{2}\frac{\left(\frac{1}{2}\right)\left(-\frac{1}{2}\right)\left(-\frac{3}{2}\right)\cdots\left(-\frac{2n-1}{2}\right)}{(n+1)!}(-1)^{n+1}4^{n+1}
$$

$$
= -\frac{1}{2}\cdot\frac{1\times3\times5\times\cdots\times(2n-1)}{(n+1)!}\cdot\frac{(-1)^n}{2^{n+1}}\cdot(-1)^{n+1}4^{n+1}
$$

$$
= \frac{1}{2}\cdot\frac{1\times2\times3\times4\times\cdots\times2n}{(n+1)!\cdot2^n n!}\cdot\frac{4^{n+1}}{2^{n+1}}
$$

$$
= \frac{1}{n+1}\cdot\frac{(2n)!}{n!n!}
$$

$$
= \frac{1}{n+1}\binom{2n}{n}.
$$

In a manner analogous to the case for real numbers, we can introduce the notion of a complex valued power series

$$
\sum_{n=0}^{\infty} a_n z^n,
$$

where the coefficients may be complex numbers and z a complex variable. The values of z for which this series converges defines a function

$$
f(z) = \sum_{n=0}^{\infty} a_n z^n.
$$

It can be shown that the power series (i)–(vi) given for the elementary functions at the beginning of the section continue to hold when the real variable x is replaced with the complex variable z.

A useful fact regarding complex power series is that if $f(z) = \sum_{n=0}^{\infty} a_n z^n$, then $\operatorname{Re} f(z) = \sum_{n=0}^{\infty}\operatorname{Re}(a_n z^n)$ and $\operatorname{Im} f(z) = \sum_{n=0}^{\infty}\operatorname{Im}(a_n z^n)$.

As an example, we will justify the use of the formula $e^{i\theta} = \cos\theta + i\sin\theta$ introduced in Section 3.5. We have

$$
\operatorname{Re} e^{i\theta} = \operatorname{Re}\sum_{n=0}^{\infty}\frac{(i\theta)^n}{n!} = \sum_{n=0}^{\infty}\operatorname{Re}\frac{(i\theta)^n}{n!}
$$

$$
= \sum_{k=0}^{\infty}\frac{(-1)^k\theta^{2k}}{(2k)!} = \cos\theta
$$

and

$$
\operatorname{Im} e^{i\theta} = \operatorname{Im}\sum_{n=0}^{\infty}\frac{(i\theta)^n}{n!} = \sum_{n=0}^{\infty}\operatorname{Im}\frac{(i\theta)^n}{n!} = \sum_{k=0}^{\infty}\frac{(-1)^k\theta^{2k+1}}{(2k+1)!} = \sin\theta.
$$

It follows that $e^{i\theta} = \operatorname{Re} e^{i\theta} + i\operatorname{Im} e^{i\theta} = \cos\theta + i\sin\theta$.

5.4.11. Sum the infinite series

$$S = r\cos\theta + \frac{r^2}{2}\cos 2\theta + \frac{r^3}{3}\cos 3\theta + \cdots, \qquad 0 < r < 1, \quad 0 < \theta < \pi.$$

Solution. Consider the infinite series

$$-\log(1 - z) = z + \frac{z^2}{2} + \frac{z^3}{3} + \cdots + \frac{z^n}{n} + \cdots, \qquad |z| < 1,$$

and set $z = r(\cos\theta + i\sin\theta)$. Then

$$-\log(1 - r\cos\theta - ir\sin\theta)$$

$$= r(\cos\theta + i\sin\theta) + \frac{r^2(\cos 2\theta + i\sin 2\theta)}{2} + \cdots,$$

and taking the real part of each side gives

$$\mathrm{Re}(-\log(1 - r\cos\theta - ir\sin\theta))$$

$$= r\cos\theta + \frac{r^2}{2}\cos 2\theta + \frac{r^3}{3}\cos 3\theta + \cdots.$$

Now, for a complex number w, $\log w = \log|w| + i\arg w$. It follows that

$$r\cos\theta + \frac{r^2}{2}\cos 2\theta + \cdots = -\log\sqrt{(1 - r\cos\theta)^2 + (r\sin\theta)^2}$$

$$= -\log\sqrt{1 - 2r\cos\theta + r^2}.$$

Problems

5.4.12. Let p and q be real numbers with $1/p - 1/q = 1, 0 < p \leqslant \frac{1}{2}$. Show that

$$p + \tfrac{1}{2}p^2 + \tfrac{1}{3}p^3 + \cdots = q - \tfrac{1}{2}q^2 + \tfrac{1}{3}q^3 - \cdots.$$

5.4.13. Find the power-series expansions for each of the following:

(a) $1/(x^2 + 5x + 6)$.

(b) $\dfrac{1 + x}{(1 + x^2)(1 - x)^2}$.

(c) $\arcsin x$.

(d) $\arctan x$ (use this to find a series of rational numbers which converges to π).

5.4.14. Sum the following infinite series:

(a) $\displaystyle\sum_{n=0}^{\infty} \frac{(-4\pi^2 r^2)^n}{(2n+1)!}$, r a nonzero integer.

(b) $1 + \dfrac{1}{3} + \dfrac{1\times 3}{2!3^2} + \dfrac{1\times 3\times 5}{3!3^3} + \cdots$.

(c) $\dfrac{2}{9} + \dfrac{2}{2!}\left(\dfrac{2}{9}\right)^2 + \dfrac{2\times 5}{3!}\left(\dfrac{2}{9}\right)^3 + \dfrac{2\times 5\times 8}{4!}\left(\dfrac{2}{9}\right)^4 + \cdots$.

(d) $\dfrac{1^2}{1!}x + \dfrac{1^2+2^2}{2!}x^2 + \dfrac{1^2+2^2+3^2}{3!}x^3 + \dfrac{1^2+2^2+3^2+4^2}{4!}x^4 + \cdots$.

5.4.15. Let $f_0(x) = e^x$ and $f_{n+1}(x) = xf_n'(x)$ for $n = 0, 1, 2, , \ldots$. Show that

$$\sum_{n=0}^{\infty} \frac{f_n(1)}{n!} = e^e.$$

(Hint: Consider $g(x) = e^{e^x}$.)

5.4.16. Prove that the value of the nth derivative of $x^3/(x^2 - 1)$ for $x = 0$ is zero if n is even and $-n!$ if n is odd and greater than 1.

5.4.17. Show that the functional equation

$$f\left(\frac{2x}{1+x^2}\right) = (1 + x^2)f(x)$$

is satisfied by

$$f(x) = 1 + \tfrac{1}{3}x^2 + \tfrac{1}{5}x^4 + \tfrac{1}{7}x^6 + \cdots, \qquad |x| < 1.$$

5.4.18. Using power series, prove that $\sin(x + y) = \sin x \cos y + \cos x \sin y$.

5.4.19. Show that

$$\frac{\sin x}{1-x} = x + x^2 + \left(1 - \frac{1}{3!}\right)x^3 + \left(1 - \frac{1}{3!}\right)x^4$$

$$+ \left(1 - \frac{1}{3!} + \frac{1}{5!}\right)x^5 + \left(1 - \frac{1}{3!} + \frac{1}{5!}\right)x^6$$

$$+ \left(1 - \frac{1}{3!} + \frac{1}{5!} - \frac{1}{7!}\right)x^7 + \left(1 - \frac{1}{3!} + \frac{1}{5!} - \frac{1}{7!}\right)x^8 + \cdots$$

5.4.20. Let $B(n)$ be the number of ones in the base 2 expression for the positive integer n. For example, $B(6) = B(110_2) = 2$ and $B(15) = B(1111_2) = 4$. Determine whether or not

$$\exp \sum_{n=1}^{\infty} \frac{B(n)}{n(n+1)}$$

is a rational number.

5.4.21. For which real numbers a does the sequence defined by the initial condition $u_0 = a$ and the recursion $u_{n+1} = 2u_n - n^2$ have $u_n > 0$ for all $n \geqslant 0$?

5.4.22. Prove that

$$\left(\frac{1+x}{1-x}\right)^3 = 1 + 6x + 18x^2 + \cdots + (4n^2 + 2)x^n + \cdots, \qquad |x| < 1.$$

5.4.23. Let $T_n = \sum_{i=1}^n (-1)^{i+1}/(2i - 1)$, $T = \lim_{n \to \infty} T_n$. Show that

$$\sum_{n=1}^{\infty} (T_n - T) = \frac{\pi}{8} - \frac{1}{4}.$$

5.4.24. Solve the recurrence relation $a_0 = 1$, $a_1 = 0$, $a_2 = -5$, and for $n \geqslant 3$

$$a_n = 4a_{n-1} - 5a_{n-2} + 2a_{n-3}.$$

5.4.25. Use the technique of generating functions to show that the nth Fibonacci number F_n is equal to

$$F_n = \frac{\left(\dfrac{1+\sqrt{5}}{2}\right)^n - \left(\dfrac{1-\sqrt{5}}{2}\right)^n}{\sqrt{5}}.$$

5.4.26. Sum the finite series $a_0 + a_1 + \cdots + a_n$, where $a_0 = 2$, $a_1 = 17$, and for $i > 1$, $a_i = 7a_{i-1} - 12a_{i-2}$.

5.4.27. Show that the power series for the function $e^{ax}\cos bx$, $a > 0$, $b > 0$ in powers of x has either no zero coefficients or infinitely many zero coefficients.

5.4.28. Sum the infinite series

$$S = 1 - 2r\cos\theta + 3r^2\cos 2\theta - 4r^3\cos\theta + \cdots, \qquad |r| < 1.$$

5.4.29. Show that $\sum_{n=0}^{\infty} (\sin n\theta)/n! = \sin(\sin\theta)e^{\cos\theta}$.

5.4.30. Use infinite series to evaluate $\lim_{x \to \infty}[\sqrt[3]{x^3 - 5x^2 + 1} - x]$.

Additional Examples

1.12.1, 5.3.16, 6.8.1, 7.6.7(c). Also, see Section 5.2 (Geometric Series) and Section 7.5 (Inequalities by Series).

Chapter 6. Intermediate Real Analysis

In this chapter we will review, by way of problems, the hierarchy of definitions and results concerning continuous, differentiable, and integrable functions. We will build on the reader's understanding of limits to review the most important definitions (continuity in Section 6.1, differentiability in Section 6.3, and integrability in Section 6.8). We will also call attention to the most important properties of these classes of functions. It is useful to know, for example, that if a problem involves a continuous function, then we might be able to apply the intermediate-value theorem or the extreme-value theorem; or again, if the problem involves a differentiable function, we might expect to apply the mean-value theorem. Examples of these applications are included in this chapter, as well as applications of L'Hôpital's rule and the fundamental theorem of calculus.

Throughout this chapter, R will denote the set of real numbers.

6.1. Continuous Functions

A real-valued function f is *continuous* at a if $f(x) \to f(a)$ as $x \to a$, or more precisely, if

(i) $f(a)$ is defined,
(ii) $\lim_{x \to a} f(x)$ exists, and
(iii) $\lim_{x \to a} f(x) = f(a)$.

(If a is a boundary point in the domain of f, it is understood that the x's in (ii) are restricted to the domain of f. We will assume the reader is familiar with these contingencies.)

A function f is continuous in a domain D if it is continuous at each point of D.

It is not difficult to prove that f is continuous at a if and only if for every sequence $\{x_n\}$ converging to a, the sequence $\{f(x_n)\}$ converges to $f(a)$.

The sequential form for the definition of continuity of f is used most often when one wishes to show a function is discontinuous at a point. For example, the function f defined by

$$f(x) = \begin{cases} \sin \dfrac{1}{x} & \text{if } x \neq 0, \\ 0 & \text{if } x = 0 \end{cases}$$

is discontinuous at 0 because, for instance, the sequence $x_n = 2/(4n+1)\pi$ converges to 0, whereas $\{f(x_n)\} = \{\sin(2\pi n + \frac{1}{2}\pi)\}$ converges to 1 (rather than to $f(0) = 0$).

6.1.1. Define $f:[0,1] \rightarrow [0,1]$ in the following manner: $f(1) = 1$, and if $a = .a_1 a_2 a_3 a_4 \ldots$ is the decimal representation of a (written as a terminating decimal if possible; e.g., .099999 .... is replaced by .1), define $f(a) = .0a_1 0a_2 0a_3 \ldots$. Discuss the continuity of f.

Solution. Observe that f is a monotone increasing function. We will show it is discontinuous at each terminating decimal number (i.e., at each point of the form $N/10^n$, N an integer, $1 \leqslant N < 10^n$).

Consider, as an example, the point $a = .413$. By definition, $f(a) = .040103$. Now define a sequence x_n by

$$x_1 = .4129,$$

$$x_2 = .41299,$$

$$x_3 = .412999,$$

$$\vdots$$

$$x_n = .412\,999 \ldots 9.$$
$$\underbrace{}_{n \text{ times}}$$

The sequence $\{x_n\}$ converges to a; however,

$$f(x_n) = .040102 \underbrace{09\,09 \ldots 09}_{n \text{ pairs}},$$

and we see that $\{f(x_n)\}$ does not converge to $f(a)$. Thus f is not continuous at a.

A similar construction can be made to show that f is discontinuous at each terminating decimal number. The argument is based on the fact that the terminating decimal numbers have two decimal representations, namely,

$$a = .a_1a_2a_3 \ldots a_{n-1}a_n, \qquad a_n \neq 0,$$
$$a = .a_1a_2 \ldots a_{n-1}(a_n - 1)999 \ldots .$$

Now suppose that a in $(0, 1)$ is not a terminating decimal number. We will show that f is continuous at a. Write a in its unique decimal form:

$$a = a_1a_2a_3a_4 \ldots .$$

Because the number a is not a terminating decimal number, there are arbitrarily large integers n such that $a_n \neq 0$ and $a_{n+1} \neq 9$. For each such n, define X_n and Y_n by

$$X_n = .a_1a_2 \ldots a_n \left(= \sum_{i=1}^{n} \frac{a_i}{10^i} \right),$$

$$Y_n = .a_1a_2 \ldots a_n(a_{n+1} + 1) = X_n + \frac{a_{n+1} + 1}{10^{n+1}}.$$

Then $a \in (X_n, Y_n)$. Moreover, the first n digits of each of the numbers in (X_n, Y_n) are the same as those of X_n and Y_n. Consequently, all the numbers within (X_n, Y_n) are mapped to the interval $(f(X_n), f(Y_n))$.

It is clear that the sequences $\{X_n\}$ and $\{Y_n\}$ converge to a; furthermore, the sequences $\{f(X_n)\}$ and $\{f(Y_n)\}$ converge to $f(a)$. Since any sequence $\{x_n\}$ which converges to a must eventually become interior to (X_n, Y_n) for any n, it must be the case that $\{f(x_n)\}$ converges to $f(a)$. It follows that f is continuous at a.

The preceding example is difficult to visualize geometrically, and a thorough understanding of the proof requires a clear understanding of continuity. The next example also demands a precise rendering of the definition: a function f is continuous at a if for every $\varepsilon > 0$ there is a number $\delta > 0$ such that $|x - a| < \delta$ implies $|f(x) - f(a)| < \varepsilon$.

6.1.2. Suppose that $f: R \rightarrow R$ is a one-to-one continuous function with a fixed point x_0 (that is, $f(x_0) = x_0$) such that $f(2x - f(x)) = x$ for all x. Prove that $f(x) \equiv x$.

Solution. Let $S = \{x \mid f(x) = x\}$. Because f is continuous, the set S is a closed subset of R (i.e., if $x_n \in S$ and $x_n \rightarrow x$, then $x \in S$; this is because $x = \lim_{n \rightarrow \infty} x_n = \lim_{n \rightarrow \infty} f(x_n) = f(\lim_{n \rightarrow \infty} x_n) = f(x)$.)

Now suppose that $S \neq R$. Let x_0 be a boundary point of S (every neighborhood of x_0 contains points that are not in S; note that $x_0 \in S$ because S is closed).

If y is a point that is not in S, there is a nonzero real number r such that $f(y) = y + r$. The fact that f is one-to-one and satisfies $f(2x - f(x)) = x$

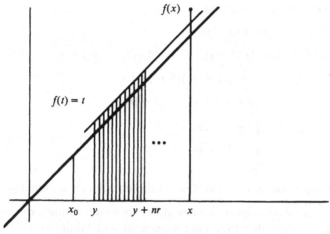

Figure 6.1.

implies that

$$f(y + nr) = (y + nr) + r$$

for every integer n (this is the content of 2.1.12). This identity is crucial to the argument which follows.

Here's the idea: Suppose x is not in S; that is, $f(x) \neq x$. Choose y in $R - S$ so that y is "close" to x_0 and $f(y)$ is "close" to y (this can be done because f is continuous at x_0 and $f(x_0) = x_0$). Then, if r is such that $f(y) = y + r$, and if r is sufficiently small, the fact that $f(y + nr) = (y + nr) + r$ will lead to a contradiction to the continuity of f at x (see Figure 6.1).

A formal proof goes as follows. Suppose, as above, that x_0 is a boundary point of S and x is such that $f(x) \neq x$. Let $\varepsilon = |f(x) - x|$. Because f is continuous at x, there is a $\delta > 0$, and we may assume that $\delta \leqslant \frac{1}{4}\varepsilon$, such that $|z - x| < \delta$ implies $|f(z) - f(x)| < \frac{1}{4}\varepsilon$. Because f is continuous at x_0, there is an $\eta > 0$, $\eta < \delta$, such that $|w - x_0| < \eta$ implies $|f(w) - f(x_0)| < \delta$.

Now choose $y \in (x_0 - \eta, x_0 + \eta)$ such that $f(y) \neq y$ (such a y exists because x_0 is a boundary point of S). Then

$$0 < |f(y) - y| \leqslant |f(y) - f(x_0)| + |f(x_0) - y|$$

$$= |f(y) - f(x_0)| + |x_0 - y|$$

$$< \delta + \eta$$

$$< 2\delta.$$

Let $r = f(y) - y$ (note: r may be negative). Since $0 < |r| < 2\delta$, there is an integer n such that $y + nr \in (x - \delta, x + \delta)$. But we know that $f(y + nr)$

$= (y + nr) + r$. It follows that

$$\varepsilon = |f(x) - x|$$
$$\leqslant |f(x) - f(y + nr)| + |f(y + nr) - x|$$
$$< \tfrac{1}{4}\varepsilon + |(y + nr) + r - x|$$
$$\leqslant \tfrac{1}{4}\varepsilon + |(y + nr) - x| + |r|$$
$$< \tfrac{1}{4}\varepsilon + \delta + 2\delta$$
$$< \tfrac{1}{4}\varepsilon + \tfrac{1}{4}\varepsilon + \tfrac{1}{2}\varepsilon$$
$$= \varepsilon.$$

This contradiction means that $S = R$ and the solution is complete.

Two of the most important facts about continuous functions over closed intervals $[a, b]$ are that they have maximum and minimum values on the interval and take on every value between these two. This is the content of the following two theorems.

Extreme-Value Theorem. *If f is a continuous function on $[a, b]$, then there are numbers c and d in $[a, b]$ such that $f(c) \leqslant f(x) \leqslant f(d)$ for all x in $[a, b]$ (that is to say, $f(d)$ is the maximum value for f over $[a, b]$, and $f(c)$ is the minimum value).*

Intermediate-Value Theorem. *If f is a continuous function on $[a, b]$ and if $f(a) < y < f(b)$ (or, $f(b) < y < f(a)$), then there is a number c in (a, b) such that $f(c) = y$.*

These results can be proved in a variety of ways; we will sketch a proof of the intermediate-value theorem which makes use of a methodology (repeated bisection) that is applicable in other problems (e.g., see 6.3.6).

Suppose that f is a continuous function on the closed interval $[a, b]$, and suppose that $f(a) < f(b)$ (a similar proof can be given if $f(a) > f(b)$). Let $y \in [f(a), f(b)]$. We wish to find an element c in $[a, b]$ such that $f(c) = y$. The procedure goes as follows (a diagram will help). Let $a_0 = a$, $b_0 = b$, and let x_1 denote the midpoint of the interval $[a, b]$ (the first bisection). If $f(x_1) < y$, define $a_1 = x_1$, $b_1 = b$, whereas if $y \leqslant f(x_1)$, define $a_1 = a$, $b_1 = x_1$. In either case we have $f(a_1) \leqslant y \leqslant f(b_1)$, and the length of $[a_1, b_1]$ is one-half the length of $[a, b]$.

Now, let x_2 denote the midpoint of $[a_1, b_1]$ (the second bisection). If $f(x_2) < y$, define $a_2 = x_2$, $b_2 = b_1$, and if $y \leqslant f(x_2)$, define $a_2 = a_1$, $b_2 = x_2$. Again, it follows that $f(a_2) \leqslant y \leqslant f(b_2)$, and $b_2 - a_2 = (b - a)/4$.

Continue in this way. The result will be an infinite nested sequence of closed intervals

$$[a_0, b_0] \supset [a_1, b_1] \supset [a_2, b_2] \supset \cdots \supset \cdots$$

whose lengths converge to zero (in fact, $b_i - a_i = (b - a)/2^i$). These conditions imply that $\{a_i\}$ and $\{b_i\}$ each converge to the same real number in $[a, b]$: call this number c.

By the continuity of f, $\lim_{i \to \infty} f(a_i) = f(c)$ and $\lim_{i \to \infty} f(b_i) = f(c)$. Furthermore, for each i, $f(a_i) \leqslant y \leqslant f(b_i)$, and therefore (by the squeeze principle, which will be treated in Section 7.6),

$$f(c) = \lim_{i \to \infty} f(a_i) \leqslant y \leqslant \lim_{i \to \infty} f(b_i) = f(c).$$

It follows that $f(c) = y$, and the theorem is proved.

The proof of the extreme-value theorem can be carried out in similar manner and is left as a problem (6.1.5).

Problems

6.1.3. Suppose that f is bounded for $a \leqslant x \leqslant b$ and, for every pair of values x_1, x_2 with $a \leqslant x_1 \leqslant x_2 \leqslant b$,

$$f(\tfrac{1}{2}(x_1 + x_2)) \leqslant \tfrac{1}{2}(f(x_1) + f(x_2)).$$

Prove that f is continuous for $a < x < b$. [Hint: Show that $f(x + \delta) - f(x)$ $\leqslant \tfrac{1}{2}[f(x + 2\delta) - f(x)] \leqslant \cdots \leqslant (1/2^n)[f(x + 2^n\delta) - f(x)]$, $a < x + 2^n\delta$ $< b$. Let $\delta \to 0$.]

6.1.4. A real-valued continuous function satisfies for all real x and y the functional equation

$$f\left(\sqrt{x^2 + y^2}\right) = f(x)f(y).$$

Prove that $f(x) = [f(1)]^{x^2}$. [Hint: First prove the theorem for all numbers of the form $2^{n/2}$ where n is an interger. Then prove the theorem for all numbers of the form $m/2^n$, m an integer, n a nonnegative integer.]

6.1.5. Use the method of repeated bisection to prove the extreme-value theorem.

6.1.6. Let $f(0) > 0$, $f(1) < 0$. Prove that $f(x) = 0$ for some x under the assumption that there exists a continuous function g such that $f + g$ is nondecreasing. (Hint: Use repeated bisection—choose the right half of the interval if there is a point x in it such that $f(x) \geqslant 0$, otherwise choose the left half. This yields a nested sequence of intervals $[a_1, b_1] \supseteq [a_2, b_2] \supseteq \cdots$ which converge to a point c. Note that for each n there is a point y_n in $[a_n, c]$ such that $f(y_n) \geqslant 0$. Prove that $f(c) = 0$.)

6.1.7. Let f be defined in the interval $[0, 1]$ by

$$f(x) = \begin{cases} 0 & \text{if } x \text{ is irrational,} \\ 1/q & \text{if } x = p/q \text{ (in lowest terms).} \end{cases}$$

(a) Prove that f is discontinuous at each rational number in $[0, 1]$.
(b) Prove that f is continuous at each irrational number in $[0, 1]$.

6.1.8. If x is an element of the Cantor set K (see 3.4.6), it can be expressed uniquely in the form

$$x = \sum_{n=1}^{\infty} \frac{2b_n}{3^n}$$

where $b_n = 0$ or 1. Define $g : K \rightarrow [0, 1]$ by setting

$$g(x) = \sum_{n=1}^{\infty} \frac{b_n}{2^n}.$$

Now extend g to $[0, 1]$ in the following manner. If $x \in [0, 1]$ is not in the Cantor set, then, using the notation of 3.4.6, there is a unique integer n such that $x \in I_n$, where $I_n = (X_n, Y_n)$, X_n and Y_n in K. Define $g(x) = g(Y_n)$. (Note that for all n, $g(X_n) = g(Y_n)$, and thus we have simply made g constant on the closed interval $[X_n, Y_n]$.) Prove that g is continuous. (Also, see 6.2.13.)

Additional Examples

6.3.1, 6.3.5, 6.3.6, 6.4.3, 6.7.2, 6.7.7, 6.8.9, 6.8.10, 6.9.5. Continuity is an underlying assumption in most of the examples in Chapter 6; in particular, see Section 6.2 (intermediate-value theorem).

6.2. The Intermediate-Value Theorem

The intermediate-value theorem states that if f is a continuous function on the closed interval $[a, b]$ and if d is between $f(a)$ and $f(b)$, then there is a number c between a and b such that $f(c) = d$. The power of the theorem lies in the fact that it provides a way of knowing about the existence of something without requiring that it be explicitly found.

As an example, let us show that $-2x^5 + 4x = 1$ has a solution in the interval $(0, 1)$. Consider $f(x) = -2x^5 + 4x$, and take two "pot-shots": $f(0)$ is too small, and $f(1)$ is too large. Therefore, by the intermediate-value theorem, there is a number in $(0, 1)$ that is just right.

6.2.1. A cross-country runner runs a six-mile course in 30 minutes. Prove that somewhere along the course the runner ran a mile in exactly 5 minutes.

Solution. Let x denote the distance along the course, measured in miles from the starting line. For each x in $[0, 5]$, let $f(x)$ denote the time that elapsed for the mile from the point x to the point $x + 1$. The function f is

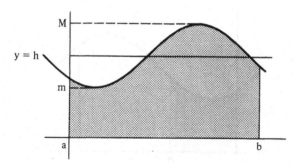

Figure 6.2.

continuous. We are given that $f(0) + f(1) + f(2) + f(3) + f(4) + f(5) = 30$.
It follows that not all of $f(0), \ldots, f(5)$ are smaller than 5, and similarly, not
all of $f(0), \ldots, f(5)$ are larger than 5. Therefore, there are points a and b in
$[0, 5]$ such that $f(a) \leqslant 5 \leqslant f(b)$. Thus, by the intermediate-value theorem,
there is a number c between a and b such that $f(c) = 5$; that is to say, the
mile from c to $c + 1$ was run in exactly 5 minutes.

6.2.2. Suppose that $f : [a, b] \to R$ is a continuous function.

(a) *Mean value theorem for integrals.* Prove that there is a number c in $[a, b]$
such that $\int_a^b f(t) \, dt = f(c)(b - a)$.
(b) Prove there is a number c in $[a, b]$ such that $\int_a^c f(t) \, dt = \int_c^b f(t) \, dt$.
(Note: For this, it is enough to know that f is integrable over $[a, b]$.)

Solution. (a) Let M and m be the maximum and minimum values of f on
$[a, b]$ respectively (guaranteed to exist by the extreme-value theorem), and
let $A = \int_a^b f(t) \, dt$. The intuition for the argument which follows is shown
(for the case of a positive function f) in Figure 6.2. As the line $y = h$ moves
continuously from $y = m$ to $y = M$, the area $A(h)$ in the rectangle bounded
by $y = h$, $y = 0$, $x = a$, $x = b$ moves from being smaller than A (at $A(m)$)
to being greater than A (at $A(M)$). Algebraically (and true independently
of the "area" interpretation), $A(m) = m(b - a) \leqslant \int_a^b f(t) \, dt \leqslant M(b - a)$
$= A(M)$. Since $A(h) \equiv h(b - a)$ is a continuous function of h, it follows
from the intermediate-value theorem (note: $A(h) \equiv h(b - a)$ is a continuous function) that there is a point d such that $A(d) = A$; or equivalently,
$d(b - a) = A$. But d is between m and M, so again by the intermediate-
value theorem, since f is continuous, there is a point c in $[a, b]$ such that
$f(c) = d$. It follows that

$$\int_a^b f(t) \, dt = f(c)(b - a).$$

(b) Again, the intuition is shown in Figure 6.3 (for the case of a positive
function). Let $A = \int_a^b f(t) \, dt$, and let $A(h) = \int_a^h f(t) \, dt$. In the figure, for

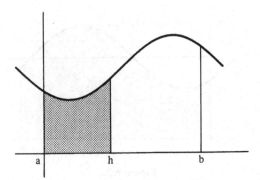

Figure 6.3.

$a < h < b$, $A(h)$ represents the area bounded by $y = f(x)$, $y = 0$, $x = a$, $x = h$ (shaded). The problem asks us to find a point c such that $A(c) = \frac{1}{2}A$. It is clear that as the vertical line $x = h$ moves to the right from $x = a$ to $x = b$, the corresponding integral (area) will move from 0 to A, and must therefore pass through $\frac{1}{2}A$ at some point.

The preceding argument is perfectly valid provided we prove that $A(h)$ is a continuous function of h. To see this, note that

$$A(h + x) - A(h) = \int_h^{h+x} f(t)\, dt.$$

From part (a), we know there is a point c_x between h and $h + x$ such that

$$\int_h^{h+x} f(t)\, dt = f(c_x)x.$$

Therefore,

$$\lim_{x \to 0} \left[A(h + x) - A(h) \right] = \lim_{x \to 0} c_x|x| = 0.$$

(Note: c_x is bounded because f is integrable.) Thus, $A(h + x) \to A(h)$ as $x \to 0$, and this means that $A(h)$ is continuous at h.

6.2.3. Let A be a set of $2n$ points in the plane, no three of which are collinear. Suppose that n of them are colored red and the remaining n blue. Prove or disprove: there are n closed straight line segments, no two with a point in common, such that the endpoints of each segment are points of A having different colors.

Solution. This problem was considered in 1.11.2, but it is instructive to see a proof based on the intermediate-value property.

We shall prove that the result holds by induction on n. Certainly the property holds when $n = 1$. Suppose the result holds when $n = 1, 2, \ldots, k$, and consider a set A with $2(k + 1)$ points, no three of which are collinear, such that $k + 1$ are colored red and $k + 1$ are colored blue.

Suppose that two vertices of the convex hull of A have different colors. Then, there are two consecutive vertices along the perimeter of the convex hull, say P and Q, that have different colors. By the inductive assumption, the set of points $A - \{P, Q\}$ may be connected in the desired way. None of these segments will intersect the line segment PQ because of the way P and Q were chosen, and therefore, the result holds for the set A.

It remains to consider the case in which all the vertices of the convex hull have the same color, say red. If L is any nonhorizontal line in the plane, let $B(L)$ denote the number of points of A to the left of L that are colored blue, let $R(L)$ denote the number of points of A to the left of L that are colored red, and let $D(L) = B(L) - R(L)$. Now choose a nonhorizontal line L that lies to the left of all the points in A and which is not parallel to any of the line segments that can be formed by joining points of A. In this position, $D(L) = 0$. As L moves continuously to the rght, it will encounter points of A one at a time, and in passing such a point, $D(L)$ will change $+1$ if the point is colored blue and -1 if the point is colored red. As L moves to the right, its first nonzero value will be negative (obtained just after passing the first point of A). Since the last-encountered point of A is also red, its last nonzero value will be positive (obtained just before passing the last point of A).

It follows from these observations that $D(L)$ will equal zero somewhere between the first and last encountered points of A (note that $D(L)$ is an integer-valued function). When L is in such a position, the inductive assumption can be applied to the points to the left of L and also to the points to the right of L. Since none of the resulting segments will intersect, the result follows for the set A, and by induction, the proof is complete.

Problems

6.2.4. Suppose $f : [0, 1] \to [0, 1]$ is continuous. Prove that there exists a number c in $[0, 1]$ such that $f(c) = c$.

6.2.5. A rock climber starts to climb a mountain at 7:00 A.M. on Saturday and gets to the top at 5:00 P.M. He camps on top and climbs back down on Sunday, starting at 7:00 A.M. and getting back to his original starting point at 5:00 P.M. Show that at some time of day on Sunday he was at the same elevation as he was at that time on Saturday.

6.2.6. Prove that a continuous function which takes on no value more than twice must take on some value exactly once.

6.2.7. Prove that the trigonometric polynomial

$$a_0 + a_1 \cos x + \cdots + a_n \cos nx,$$

where the coefficients are all real and $|a_0| + |a_1| + \cdots + |a_{n-1}| \leqslant a_n$, has at least $2n$ zeros in the interval $[0, 2\pi)$.

6.2.8. Establish necessary and sufficient conditions on the constant k for the existence of a continuous real-valued function $f(x)$ satisfying $f(f(x)) = kx^9$ for all real x.

6.2.9.

(a) Suppose that $f:[a,b] \to R$ is continuous and $g:[a,b] \to R$ is integrable and such that $g(x) \geqslant 0$ for all $x \in [a,b]$. Prove that there is a number c in $[a,b]$ such that

$$\int_a^b f(x)g(x)\,dx = f(c) \int_a^b g(x)\,dx.$$

(b) Suppose that $f:[a,b] \to R$ is increasing (and therefore integrable), and $g:[a,b] \to R$ is integrable and such that $g(x) \geqslant 0$ for all $x \in [a,b]$. Prove that there is a number c in $[a,b]$ such that

$$\int_a^b f(x)g(x)\,dx = f(a) \int_a^c g(x)\,dx + f(b) \int_c^b g(x)\,dx.$$

6.2.10. Let $f:[0,1] \to R$ be continuous and suppose that $f(0) = f(1)$. Prove that for each positive integer n there is an x in $[0, 1 - 1/n]$ such that $f(x) = f(x + 1/n)$.

6.2.11. A polynomial $P(t)$ of degree at most 3 describes the temperature of a certain body at time t. Show that the average temperature of the body between 9 A.M. and 3 P.M. can always be found by taking the average of the temperature at two fixed times, which are independent of which polynomial occurs. Also, show that these two times are 10:16 A.M. and 1:44 P.M. to the nearest minute. [Hint: Use the mean-value theorem for integrals; see 6.2.2(a).]

6.2.12. For any pair of triangles, prove that there exists a line which bisects them simultaneously.

6.2.13. Give an example of a continuous real-valued function f from $[0, 1]$ to $[0, 1]$ which takes on every value in $[0, 1]$ an infinite number of times. [Hint: One way to do this is to modify the continuous function defined in 6.1.8.]

Additional Examples

6.1.6, 6.5.2, 6.5.3, 6.5.4, 6.5.13, 6.6.4, 6.6.5, 6.6.6, 6.6.9, 7.6.13.

6.3. The Derivative

The derivative of $f: [a, b] \to R$ at a point x in (a, b) is defined by

$$f'(x) = \lim_{h \to 0} \frac{f(x + h) - f(x)}{h},$$

provided this limit exists. We note that if f has a derivative at x, then f is continuous at x, because

$$\lim_{h \to 0} \left[f(x + h) - f(x) \right] = \lim_{h \to 0} \left[\left(\frac{f(x + h) - f(x)}{h} \right) \cdot h \right]$$

$$= \lim_{h \to 0} \left(\frac{f(x + h) - f(x)}{h} \right) \lim_{h \to 0} h$$

$$= f'(x) \lim_{h \to 0} h$$

$$= 0.$$

6.3.1. If the function $xf(x)$ has a derivative at a given point $x_0 \neq 0$, and if f is continuous there, show that f has a derivative there.

Solution. Let

$$L = \lim_{x \to x_0} \frac{xf(x) - x_0 f(x_0)}{x - x_0}.$$

The limit on the right exists, since it represents the derivative of $xf(x)$ at the point x_0 (in the definition of the derivative given above, we have substituted $x - x_0$ for h). For x sufficiently close to, but different from, x_0 (and therefore different from zero),

$$\frac{f(x) - f(x_0)}{x - x_0} = \frac{\dfrac{xf(x)}{x} - \dfrac{x_0 f(x_0)}{x_0}}{x - x_0}$$

$$= \frac{xx_0 f(x) - xx_0 f(x_0)}{xx_0(x - x_0)}$$

$$= \frac{xx_0 f(x) - x^2 f(x) - xx_0 f(x_0) + x^2 f(x)}{xx_0(x - x_0)}$$

$$= \frac{xf(x)(x_0 - x) + x(xf(x) - x_0 f(x_0))}{xx_0(x - x_0)}$$

$$= \frac{1}{x_0} \left(\frac{xf(x) - x_0 f(x_0)}{x - x_0} \right) - \frac{f(x)}{x_0}.$$

It follows that

$$f'(x_0) = \lim_{x \to x_0} \frac{f(x) - f(x_0)}{x - x_0}$$

$$= \lim_{x \to x_0} \left[\frac{1}{x_0} \left(\frac{xf(x) - x_0 f(x_0)}{x - x_0} \right) - \frac{f(x)}{x_0} \right]$$

$$= \frac{1}{x_0} \lim_{x \to x_0} \left(\frac{xf(x) - x_0 f(x_0)}{x - x_0} \right) - \frac{1}{x_0} \lim_{x \to x_0} f(x).$$

$$= \frac{1}{x_0} \left[L - f(x_0) \right].$$

The fact that $\lim_{x \to x_0} f(x) = f(x_0)$ follows from the hypothesis that f is continuous at x_0. However, this assumption is not necessary, because

$$\frac{f(x) - f(x_0)}{x - x_0} = \frac{xx_0 f(x) - xx_0 f(x_0)}{xx_0 (x - x_0)}$$

$$= \frac{xx_0 f(x) - x_0^2 f(x_0) - xx_0 f(x_0) + x_0^2 f(x_0)}{xx_0 (x - x_0)}$$

$$= \frac{1}{x} \left(\frac{xf(x) - x_0 f(x_0)}{x - x_0} \right) - \frac{f(x_0)}{x}.$$

Using this we find that

$$f'(x_0) = \lim_{x \to x_0} \left(\frac{f(x) - f(x_0)}{x - x_0} \right)$$

$$= \lim_{x \to x_0} \left(\frac{1}{x} \left(\frac{xf(x) - x_0 f(x_0)}{x - x_0} \right) - \frac{f(x_0)}{x} \right)$$

$$= \frac{1}{x_0} \left[L - f(x_0) \right].$$

6.3.2. Let $f(x) = a_1 \sin x + a_2 \sin 2x + \cdots + a_n \sin nx$, where $a_1, a_2, \ldots,$ a_n are real numbers and where n is a positive integer. Given that $|f(x)| \leqslant |\sin x|$ for all real x, prove that $|a_1 + 2a_2 + \cdots + na_n| \leqslant 1$.

Solution. We gave an induction proof of this problem in 2.4.4; however, a more natural approach is based on noticing that $f'(x) = a_1 \cos x + \cdots + na_n \cos nx$, from which we see that $f'(0) = a_1 + 2a_2 + \cdots + na_n$ (which is the left side of the inequality we wish to prove). This prompts the following argument:

$$|f'(0)| = \lim_{x \to 0} \left| \frac{f(x) - f(0)}{x - 0} \right|$$

$$= \lim_{x \to 0} \left| \frac{f(x)}{x} \right|$$

$$\leqslant \lim_{x \to 0} \left| \frac{\sin x}{x} \right|$$

$$= 1,$$

and this completes the proof.

6.3.3. Let f be differentiable at $x = a$, and $f(a) \neq 0$. Evaluate

$$\lim_{n \to \infty} \left[\frac{f(a + 1/n)}{f(a)} \right]^n.$$

Solution. It suffices to evaluate

$$\lim_{x \to 0} \left[\frac{f(a + x)}{f(a)} \right]^{1/x}.$$

For x small enough, $f(a + x)$ and $f(a)$ have the same sign, and it follows that

$$\log \left[\lim_{x \to 0} \left(\frac{f(a + x)}{f(a)} \right)^{1/x} \right] = \lim_{x \to 0} \left[\log \left(\frac{|f(a + x)|}{|f(a)|} \right)^{1/x} \right]$$

$$= \lim_{x \to 0} \frac{\log|f(a + x)| - \log|f(a)|}{x}.$$

The last expression on the right is the definition of the derivative of $\log|f(x)|$ at $x = a$, which we know from calculus is $f'(a)/f(a)$. Thus,

$$\lim_{x \to 0} \left[\frac{f(a + x)}{f(a)} \right]^{1/x} = e^{f'(a)/f(a)}.$$

Problems

6.3.4.

(a) Suppose that instead of the usual definition of the derivative, which we will denote by $Df(x)$, we define a new kind of derivative $D^*f(x)$ by the formula

$$D^*f(x) = \lim_{h \to 0} \frac{f^2(x + h) - f^2(x)}{h}.$$

Express $D^*f(x)$ in terms of $Df(x)$.

(b) If f is differentiable at x, compute

$$\lim_{h \to 0} \left(\frac{f(x + ah) - f(x + bh)}{h} \right).$$

(c) Suppose f is differentiable at $x = 0$ and satisfies the functional equation $f(x + y) = f(x) + f(y)$ for all x and y. Prove that f is differentiable at every real number x.

6.3.5. Define f by

$$f(x) = \begin{cases} x^2 \sin \dfrac{1}{x} & \text{if} \quad x \neq 0, \\ 0 & \text{if} \quad x = 0. \end{cases}$$

(a) Show that $f'(x)$ exists for all x but that f' is not continuous at $x = 0$. (The derivative for $x \neq 0$ is $2x \sin(1/x) - \cos(1/x)$; what is the derivative at 0?)

(b) Let $g(x) = x + 2f(x)$. Show that $g'(0) > 0$ but that g is not monotonic in any open interval about 0.

6.3.6. Let $f : [0, 1] \to R$ be a differentiable function. Assume there is no point x in $[0, 1]$ such that $f(x) = 0 = f'(x)$. Show that f has only a finite number of zeros in $[0, 1]$. [Suppose there are an infinite number. Either $[0, \frac{1}{2}]$ or $[\frac{1}{2}, 1]$ contains an infinite number of these zeros (perhaps both will). Choose one that does, and continue by repeated bisection. Along the way, construct a convergent sequence of distinct zeros. Use this to reach a contradiction.]

6.3.7. Prove that if f is differentiable on (a, b) and has an extremum (that is, a maximum or minimum) at a point c in (a, b), then $f'(c) = 0$. [For applications of this result, see 6.4.1, 6.4.2, 6.4.5, 6.4.6, 6.4.7, 6.6.4, 7.4.1.]

Additional Examples

6.6.2, 6.7.2, 6.9.1, 7.6.2.

6.4. The Extreme-Value Theorem

An existence theorem is a theorem which states that something exists (for example, a point within the domain of a function which has some stated property). Quite often this special object occurs at some "extreme" position.

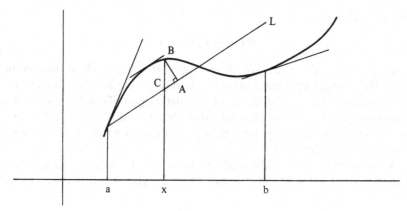

Figure 6.4.

It is in this way that one comes to make use of the extreme-value theorem: If f is a continuous function over a closed interval $[a,b]$, there are points c and d in $[a,b]$ such that $f(c) \leqslant f(x) \leqslant f(d)$ for all x in $[a,b]$.

6.4.1. Suppose that $f:[a,b] \rightarrow R$ is a differentiable function. Show that f' satisfies the conclusion of the intermediate-value theorem (i.e., if d is any number between $f'(a)$ and $f'(b)$, then there is a number c in the interval (a,b) such that $f'(c) = d$).

Solution. If f' were continuous we could get the result by a direct application of the intermediate-value theorem (applied to f'). However, f' may not be continuous (for example, see 6.3.5(a)), so how are we to proceed?

To help generate ideas, consider Figure 6.4. In this figure, a line L of slope d is drawn through the point $(a, f(a))$, where $f'(b) < d < f'(a)$. For each point x in $[a,b]$, let $g(x)$ denote the signed distance from the point $f(x)$ to the line L (the length of AB in the figure). Our intuition is that the point we seek is that point which maximizes the value of g. We shall show that this is indeed the case, but to simplify the computation we look at a slightly different function.

For each x in $[a,b]$, let $h(x)$ denote the signed distance of the vertical segment from the point $(x, f(x))$ to the line L (the length of BC in the figure). We observe that the point which maximizes the value of h on $[a,b]$ is the same as the point which maximizes the value of g on $[a,b]$. (This is because $g(x) = h(x)\cos\alpha$, where α is the inclination of L.) The advantage of considering $h(x)$ is that we can easily get an expression for it in terms of $f(x)$ and the equation of L.

So now return to the problem as stated, and consider the function

$$h(x) = f(x) - \left[f(a) + d(x - a) \right].$$

We see that

$$h'(x) = f'(x) - d.$$

Since $f'(b) < d < f'(a)$, we have $h'(b) < 0 < h'(a)$. These inequalities imply that neither $h(a)$ nor $h(b)$ is a maximum value for h on $[a, b]$ (this is a consequence of the definition of the derivative). Therefore, since h is continuous on $[a, b]$, the extreme-value theorem says that h takes on a maximum value at some point c in (a, b). At this point, by 6.3.7, $h'(c) = 0$, which is to say, $f'(c) = d$.

A similar argument can be made if $f'(a) < d < f'(b)$. In this case, h takes on a minimum value at some point c in (a, b), and at this point, $f'(c) = d$.

6.4.2. P is an interior point of the angle whose sides are the rays OA and OB. Locate X on OA and Y on OB so that the line segment XY contains P and so that the product of distances $(PX)(PY)$ is a minimum.

Solution. The situation is illustrated in Figure 6.5.

The problem is typical of the "max-min" problems encountered in beginning calculus: it does not ask "Is there a minimum value?," but rather, "Where does the minimum value occur?." The technique is to apply the result of 6.3.7: if the minimum is in the interior of an open interval, it will occur at a point where the derivative is zero. Thus, we need to express $(PX)(PY)$ as a function of a single variable, and find where it has a zero derivative.

For each positive number x, there is a unique point X on OA such that $x = |OX|$, and this point in turn determines a unique point Y on OB such that X, P, and Y are collinear. Thus, $(PX)(PY)$ is a function of x. However, an explicit expression for this function is very messy; perhaps there is another way.

Notice that $(PX)(PY)$ is uniquely determined by the angle γ (see Figure 6.5). To obtain an explicit expression for $(PX)(PY)$, first use the Law of

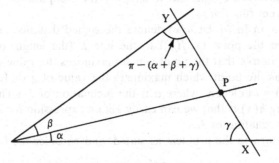

Figure 6.5.

Sines in $\triangle OXP$ and $\triangle OPY$ to get

$$\frac{\sin\alpha}{PX} = \frac{\sin\gamma}{OP} \quad \text{and} \quad \frac{\sin\beta}{PY} = \frac{\sin(\pi - \alpha - \beta - \gamma)}{OP}.$$

Then, it follows that

$$F(\gamma) = (PX)(PY) = \left(\frac{\sin\alpha}{\sin\gamma}\right)\cdot(OP)\cdot\left(\frac{\sin\beta}{\sin(\pi - \alpha - \beta - \gamma)}\right)\cdot(OP)$$

$$= C(\csc\gamma)(\csc(\pi - \alpha - \beta - \gamma)), \qquad 0 < \gamma < \pi$$

where $C = \sin\alpha \sin\beta(OP)^2$ is a constant.

The function F is continuous and differentiable on $(0,\pi)$, and $F(\gamma)\to\infty$ as $\gamma\to 0^+$ and as $\gamma\to\pi^-$, and therefore F will take on a minimum value at a point in $(0,\pi)$. At this point $F'(\gamma) = 0$; that is

$$0 = \csc\gamma \csc(\pi - \alpha - \beta - \gamma)\left[\cot\gamma - \cot(\pi - \alpha - \beta - \gamma)\right].$$

Since neither $\csc\gamma$ nor $\csc(\pi - \alpha - \beta - \gamma)$ equal zero on $(0,\pi)$, the minimum occurs when $\cot\gamma = \cot(\pi - \alpha - \beta - \gamma)$. But this happens for $0 < \gamma < \pi$ and $0 < \pi - \alpha - \beta - \gamma < \pi$ only when $\gamma = \pi - \alpha - \beta - \gamma$. Thus, the minimum occurs when $\triangle OXY$ is an isoceles triangle; that is, when $OX = OY$. (For another proof, see 8.1.3.)

Problems

6.4.3.

(a) Let $f:[a,b]\to R$ be continuous and such that $f(x) > 0$ for all x in $[a,b]$. Show that there is a positive constant c such that $f(x) \geqslant c$ for all x in $[a,b]$.

(b) Show that there is no continuous function f which maps the closed interval $[0,1]$ onto the open interval $(0,1)$.

6.4.4. Let $f:[a,b]\to R$ be differentiable at each point of $[a,b]$, and suppose that $f'(a) = f'(b)$. Prove that there is at least one point c in (a,b) such that

$$f'(c) = \frac{f(c) - f(a)}{c - a}.$$

6.4.5.

(a) *Rolle's theorem.* Suppose $f:[a,b]\to R$ is continuous on $[a,b]$ and differentiable on (a,b). If $f(a) = f(b)$, then there is a number c in (a,b) such that $f'(c) = 0$.

(b) *Mean value theorem.* If $f:[a,b]\to R$ is continuous on $[a,b]$ and differentiable on (a,b), then there is a number c in (a,b) such that

$$\frac{f(b) - f(a)}{b - a} = f'(c).$$

6.4.6. If A, B, and C are the measures of the angles of a triangle, prove that

$$-2 \leqslant \sin 3A + \sin 3B + \sin 3C \leqslant \tfrac{3}{2}\sqrt{3} ,$$

and determine when equality holds.

6.4.7. Given a circle of radius r and a tangent line L to the circle through a point P on the circle. From a variable point R on the circle, a perpendicular RQ is drawn to L with Q on L. Determine the maximum of the area of triangle PQR.

Additional Examples

1.11.5, 6.6.1, 6.6.4, 6.6.5.

6.5. Rolle's Theorem

One of the fundamental properties of differentiable functions is the following existence theorem.

> **Rolle's Theorem.** *Suppose $f:[a,b] \to R$ is continuous on $[a,b]$ and differentiable on (a,b). If $f(a) = f(b)$, then there is a number c in (a,b) such that $f'(c) = 0$.*

This result is a direct consequence of 6.3.7: For let c be a point in (a,b) such that $f(c)$ is an extremum (such a point c exists by the extreme value theorem). Then by 6.3.7, $f'(c) = 0$. Rolle's theorem is important from a theoretical point of view (we shall subsequently show that the mean-value theorem and a host of useful corollaries are easy consequences of Rolle's theorem), but it is also important as a problem-solving method.

6.5.1. Show that $4ax^3 + 3bx^2 + 2cx = a + b + c$ has at least one root between 0 and 1.

Solution. Any attempt to apply the intermediate-value theorem (as in the solution to a similar problem in Section 6.2) leads to complications, because not enough information is given regarding the values for a, b, c. But consider the function $f(x) = ax^4 + bx^3 + cx^2 - (a + b + c)x$. Notice that $f(0) = 0 = f(1)$. Therefore, by Rolle's theorem, there is a point d in $(0, 1)$ such that $f'(d) = 0$; that is to say, d is a root of $4ax^3 + 3bx^2 + 2cx = a + b + c$, and the solution is complete.

6.5.2. Prove that if the differentiable functions f and g satisfy $f'(x)g(x)$ $\neq g'(x)f(x)$ for all x, then between any two roots of $f(x) = 0$ there is a root of $g(x) = 0$.

Solution. Let a and b be two roots of f, $a < b$. The condition implies that neither a nor b are roots of $g(x) = 0$. Suppose that g has no zeros between a and b. Then, as a consequence of the intermediate value theorem, the sign of g on $[a, b]$ is always the same (that is, $g(x) > 0$ for all x in $[a, b]$, or $g(x) < 0$ for all x in $[a, b]$).

Now consider the function $F(x) = f(x)/g(x)$. This function is continuous and differentiable on $[a, b]$ and $F(a) = 0 = F(b)$. Therefore, by Rolle's theorem, there is a point c such that $F'(c) = 0$. But this leads to a contradiction, since

$$F'(c) = \frac{g(c)f'(c) - g'(c)f(c)}{g^2(c)}$$

and, by supposition, $g(c)f'(c) - g'(c)f(c) \neq 0$. This contradiction implies that g must have a zero between a and b, and the proof is complete.

A useful corollary to Rolle's theorem is that if f is a continuous and differentiable function, say on the interval $[a, b]$, and if x_1 and x_2 are zeros of f, $a < x_1 < x_2 < b$, then f' has a zero between x_1 and x_2. More generally, if f has n distinct zeros in $[a, b]$, then f' has at least $n - 1$ zeros (these are interlaced with the zeros of f), f'' has at least $n - 2$ zeros (assuming f' is continuous and differentiable on $[a, b]$), and so forth.

6.5.3. Show that $x^2 = x \sin x + \cos x$ for exactly two real values of x.

Solution. Consider $f(x) = x^2 - x \sin x - \cos x$. Then $f(-\pi/2) > 0$, $f(0)$ < 0, and $f(\pi/2) > 0$, so the intermediate-value theorem implies that f has at least two zeros. If f has three or more zeros, then, by the remarks preceding this example, f' has at least two zeros. However,

$$f'(x) = 2x - \sin x - x \cos x + \sin x$$
$$= x[2 - \cos x],$$

has only one zero. Therefore, f has exactly two zeros and the result follows.

6.5.4. Let $P(x)$ be a polynomial with real coefficients, and form the polynomial

$$Q(x) = (x^2 + 1)P(x)P'(x) + x[(P(x))^2 + (P'(x))^2].$$

Given that the equation $P(x) = 0$ has n distinct real roots exceeding 1, prove or disprove that the equation $Q(x) = 0$ has at least $2n - 1$ distinct real roots.

Solution. Let $a_1, a_2, \ldots, a_n$ be n distinct real roots of $P(x) = 0$, where $1 < a_1 < a_2 < \cdots < a_n$, and write $Q(x)$ in the form

$$Q(x) = (x - 1)^2 P(x) P'(x) + x \left[P(x) + P'(x) \right]^2,$$

Suppose that $P(x)$ has no zeros in the open interval (a_i, a_{i+1}), $i = 1, 2, \ldots, n - 1$. (There is no loss of generality here, for if there are more, say m, $m > n$, relabel the a_i's to include these, and the following proof will show that Q has at least $2m - 1$ distinct real roots.) By Rolle's theorem, there is a point b_i in (a_i, a_{i+1}) such that $P'(b_i) = 0$. Since P is a polynomial, $P'(x) = 0$ has only a finite number of roots in (a_i, a_{i+1}), so for each i, we may assume that b_i is chosen as the largest zero of P' in (a_i, a_{i+1}).

Suppose that $P(x)$ is positive for all x in (a_i, a_{i+1}) (see Figure 6.6), and consider the function $F(x) = P(x) + P'(x)$. Our idea is to find a point c_i in (b_i, a_{i+1}) where $F(c_i) < 0$. Then, since $F(b_i) > 0$, the intermediate-value theorem would imply that there is a point d_i in (b_i, c_i) such that $F(d_i) = 0$, and consequently,

$$Q(b_i) = b_i \big(F(b_i) \big)^2 > 0,$$

$$Q(d_i) = (d_i - 1)^2 P(d_i) P'(d_i) < 0$$

(note that $P'(x) < 0$ for all x in (b_i, a_{i+1})), and

$$Q(a_{i+1}) = a_{i+1} \big(F(a_{i+1}) \big)^2 \geqslant 0.$$

Therefore, by the intermediate-value theorem, there are points x_i in (b_i, d_i) and y_i in $(d_i, a_{i+1}]$ such that $Q(x_i) = 0 = Q(y_i)$.

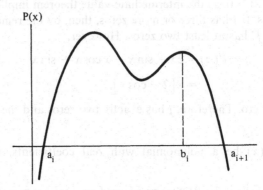

Figure 6.6.

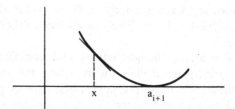

Figure 6.7.

For the preceding argument to work we must show there exists a point c_i in (b_i, a_{i+1}) where $F(c_i) < 0$. If a_{i+1} is a root of multiplicity one, then $F(a_{i+1}) = P'(a_{i+1}) < 0$, and the desired c_i can be found in a sufficiently small neighborhood of a_{i+1}. If a_{i+1} is a root of multiplicity greater than one, then $P(a_{i+1}) = 0 = P'(a_{i+1})$ and there is an interval $(a_{i+1} - \delta, a_{i+1})$ for sufficiently small $\delta > 0$, where $P''(x) > 0$ (see Figure 6.7). For such an x, it is the case that

$$P'(x) < \frac{P(x) - P(a_{i+1})}{x - a_{i+1}} = \frac{P(x)}{x - a_{i+1}}$$

and therefore,

$$F(x) = P(x) + P'(x)$$

$$< P(x)\left[1 + \frac{1}{x - a_{i+1}}\right]$$

$$= P(x)\left[\frac{x - a_{i+1} + 1}{x - a_{i+1}}\right].$$

Therefore, let $c_i = x$, where x is chosen sufficiently close to a_{i+1} so that the numerator of this last expression is positive and the denominator is negative. Then, for such a c_i, $F(c_i) < 0$, $b_i < c_i < a_{i+1}$. This completes the argument: $Q(x) = 0$ has two roots in $(b_i, a_{i+1}]$.

The preceding argument was based on the assumption that $P(x) > 0$ for x in (a_i, a_{i+1}). For the case in which $P(x) < 0$ for all x in (a_i, a_{i+1}), an exactly analogous argument leads to the same conclusion. Thus, we have shown that Q has at least $2n - 2$ zeros (two in each of the intervals (a_i, a_{i+1}), $i = 1, 2, \ldots, n - 1$). The solution will be complete if we can show Q has a zero in $(-\infty, a_1)$. Again there are several cases to consider.

Suppose that $P'(x) = 0$ has a root in the interval $(0, a_1)$. Then, without going through the details again, the same arguments show that Q has a zero in (b_0, a_1), where b_0 is chosen as the largest zero of P' in $(0, a_1)$.

We are left to consider what happens if $P'(x) = 0$ does not have a zero in $(0, a_1)$. If $P(x) > 0$ for all x in $(0, a_1)$, then $P'(x) < 0$ for all x in $(0, a_1)$ and therefore $Q(0) < 0$ and $Q(a_1) > 0$. By the intermediate-value theorem,

$Q(x) = 0$ has a root in $(0, a_1)$. Similarly, if $P(x) < 0$ for all x in $(0, a_1)$, we get $Q(0) > 0$ and $Q(a_1) < 0$, etc. Thus, in all cases $Q(x) = 0$ has at least $2n - 1$ distinct roots.

The preceding analysis, though tedious and complicated, was based entirely on first principles: Rolle's theorem and the intermediate-value theorem. With these two ideas the conceptual aspects of the proof are quite natural and easy to understand. There is another solution which is much easier going, after a clever, but not uncommon, key step (e.g., see 6.5.11 and 6.9.4). Since it is instructive, we will consider it also.

First, notice that Q can be written as a product in the following manner:

$$Q(x) = (x^2 + 1)P(x)P'(x) + x\left[(P(x))^2 + (P'(x))^2\right]$$
$$= \left[P'(x) + xP(x)\right]\left[xP'(x) + P(x)\right].$$

Let $F(x) = P'(x) + xP(x)$ and $G(x) = xP'(x) + P(x)$. The key step, as we shall see, depends on noticing that $F(x) = e^{-x^2/2}[e^{x^2/2}P(x)]'$ and $G(x) = [xP(x)]'$.

Assume that $P(x)$ has exactly m distinct real zeros a_i exceeding 1, with $1 < a_1 < a_2 < \cdots < a_m$ ($m \geqslant n$). Then $e^{x^2/2}P(x)$ also has zeros at a_1, $a_2, \ldots, a_m$, so by Rolle's theorem, $[e^{x^2/2}P(x)]'$, and hence also $F(x)$, has at least $m - 1$ zeros b_i with $a_i < b_i < a_{i+1}$. Similarly, by Rolle's theorem, $G(x)$ has at least m zeros, $c_0, c_1, \ldots, c_{m-1}$, $0 < c_0 < a_1$, $a_i < c_i < a_{i+1}$, $i = 1, 2, \ldots, m - 1$. We will be done if we can show that $b_i \neq c_i$ for $i = 1, \ldots, m - 1$.

So, assume that for some i, $b_i = c_i$, and let r be this common value. From $F(r) = 0$, we find that $P'(r) = -rP(r)$. Substituting this into $G(r) = 0$, we get $r[-rP(r)] + P(r) = 0$, or equivalently, $(r^2 - 1)P(r) = 0$. Since $r > 1$, the last equation implies $P(r) = 0$. But since $a_i < r < a_{i+1}$, we then have a contradiction to our assumption concerning the roots of $P(x) = 0$ (namely, a_i and a_{i+1} were assumed to be consecutive roots of P; i.e., all the roots of P exceeding 1 were included among the a_i's). It follows that the b_i's and the c_i's are different, and therefore $Q(x) = 0$ has at least $2m - 1$ ($\geqslant 2n - 1$) distinct real roots.

Problems

6.5.5.

(a) Show that $5x^4 - 4x + 1$ has a root between 0 and 1.

(b) If $a_0, a_1, \ldots, a_n$ are real numbers satisfying

$$\frac{a_0}{1} + \frac{a_1}{2} + \cdots + \frac{a_n}{n+1} = 0,$$

show that the equation $a_0 + a_1 x + \cdots + a_n x^n = 0$ has at least one real root.

6.5.6.

(a) Suppose that $f: [0, 1] \to R$ is differentiable, $f(0) = 0$, and $f(x) > 0$ for x in $(0, 1)$. Prove there is a number c in $(0, 1)$ such that
$$\frac{2f'(c)}{f(c)} = \frac{f'(1 - c)}{f(1 - c)}.$$
(Hint: Consider $f^2(x)f(1 - x)$.)

(b) Is there a number d in $(0, 1)$ such that
$$\frac{3f'(d)}{f(d)} = \frac{f'(1 - d)}{f(1 - d)}?$$

6.5.7.

(a) *Cauchy mean-value theorem.* If f and g are continuous on $[a, b]$ and differentiable on (a, b), then there is a number c in (a, b) such that
$$[f(b) - f(a)]g'(c) = [g(b) - g(a)]f'(c).$$

(b) Show that the mean-value theorem (6.4.5(b)) is a special case of part (a).

6.5.8.

(a) Show that $x^3 - 3x + b$ cannot have more than one zero in $[-1, 1]$, regardless of the value of b.

(b) Let $f(x) = (x^2 - 1)e^{cx}$. Show that $f'(x) = 0$ for exactly one x in the interval $(-1, 1)$ and that this x has the same sign as the parameter c.

6.5.9. How many zeros does the function $f(x) = 2^x - 1 - x^2$ have on the real line?

6.5.10. Let $f(x) = a_0 + a_1 x + \cdots + a_n x^n$ be a polynomial with real coefficients such that f has $n + 1$ distinct real zeros. Use Rolle's theorem to show that $a_k = 0$ for $0 \leqslant k \leqslant n$.

6.5.11. If $f: R \to R$ is a differentiable function, prove there is a root of $f'(x) - af(x) = 0$ between any two roots of $f(x) = 0$.

6.5.12. Suppose n is a nonnegative integer and
$$f(x) = c_0 e^{r_0 x} + c_1 e^{r_1 x} + \cdots + c_n e^{r_n x},$$
where c_i and r_i are real numbers. Prove that if f has more than n zeros in R, then $f(x) \equiv 0$. (Hint: Induct on n.)

6.5.13. The nth Legendre polynomial is defined by
$$P_n(x) = \frac{1}{2^n n!} D^n \left[(x^2 - 1)^n \right]$$
where D^n denotes the nth derivative with respect to x. Prove that $P_n(x)$ has exactly n distinct real roots and that they lie in the interval $(-1, 1)$. (Hint:

$(x^2 - 1)^n = (x - 1)^n(x + 1)^n$. Show, by an inductive argument, that the kth derivative of $(x - 1)^n(x + 1)^n$ has 1 as a zero of multiplicity $n - k$, -1 as a zero of multiplicity $n - k$, and at least k distinct zeros between -1 and 1.)

6.6. The Mean-Value Theorem

Suppose that $f: [a, b] \to R$ is continuous on $[a, b]$ and differentiable on (a, b). In a manner similar to that used in the solution to 6.4.1, consider the function

$$F(x) = f(x) - L(x),$$

(see Figure 6.8), where $y = L(x)$ is the equation of the line from $(a, f(a))$ to $(b, f(b))$. Geometrically, $F(x)$ represents the signed distance along the vertical line segment from $(x, f(x))$ to the line $y = L(x)$. Since $F(a) = 0 = F(b)$, we know from Rolle's theorem that there is a point c in (a, b) such that $F'(c) = 0$. At that point, $f'(c) - L'(c) = 0$, or equivalently,

$$f'(c) = L'(c) = (\text{slope of } L) = \frac{f(b) - f(a)}{b - a}.$$

We have just proved the following.

Mean-Value Theorem. *If $f: [a, b] \to R$ is continuous on $[a, b]$ and differentiable on (a, b), then there is a number c in (a, b) such that*

$$\frac{f(b) - f(a)}{b - a} = f'(c).$$

If $f(a) = f(b)$, this is just the statement of Rolle's theorem. Otherwise, it says that there is a point between a and b where the slope of the curve is equal to the slope of the line through $(a, f(a))$ and $(b, f(b))$.

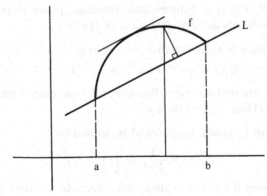

Figure 6.8.

6.6.1. Let $g(x)$ be a function that has a continuous first derivative $g'(x)$ for all values of x. Suppose that the following conditions hold:

(i) $g(0) = 0$,
(ii) $|g'(x)| \leqslant |g(x)|$ for all x.

Prove that $g(x)$ vanishes identically.

Solution. We will give a rather unusual solution, simply to illustrate the use of the mean-value theorem. Begin by considering the interval $[0, 1]$. Let x be an arbitrary point in $(0, 1]$. By the mean-value theorem, there is a point c_1 in $(0, x)$ such that

$$g'(c_1) = \frac{g(x) - g(0)}{x - 0}.$$

It follows that $|g(x)| = |xg'(c_1)| = |x||g'(c_1)| \leqslant |x||g(c_1)|$.

Similarly, there is a point c_2 in $(0, c_1)$ such that $|g(c_1)| \leqslant |c_1||g(c_2)|$, and substituting this into the last inequality, $|g(x)| \leqslant |x||c_1||g(c_2)|$.

Continuing in this way, we are able to find numbers $c_1, c_2, \ldots, c_n$, $0 < c_n < \cdots < c_2 < c_1 < x < 1$, such that $|g(x)| \leqslant |x||c_1| \cdots |c_{n-1}||g(c_n)|$. Since g is continuous on $[0, 1]$, it is bounded (between its minimum and maximum values, which exist by the extreme-value theorem), and therefore, since the right side of this last inequality can be made arbitrarily small by taking sufficiently large n (each of the $|c_i|$'s is less than 1), it must be the case that $g(x) = 0$. Thus, $g(x)$ is identically equal to zero on $[0, 1]$.

The same argument can now be applied to the interval $[1, 2]$ (for x in $(1, 2)$ there is a c_1 in $(1, x)$ such that $|g(x)| \leqslant |x - 1||g(c_1)|$, etc.). As a consequence of this argument, we will get $g(x)$ identically zero on $[1, 2]$.

By an inductive argument, we will get g equal to zero on $[n, n + 1]$ for all integers n. Therefore, g is identically zero. (Notice that we did not use the hypothesis that g' was continuous.)

The mean-value theorem has a number of important corollaries which are useful in practice. Among these are the following.

Suppose f and g are continuous on $[a, b]$ and differentiable on (a, b).

(i) *If $f'(x) = 0$ for all x in (a, b), then f is a constant.*
(ii) *If $f'(x) = g'(x)$ for all x in (a, b), then there is a constant C such that* $f(x) = g(x) + C$.
(iii) *If $f'(x) > 0$ for all x in (a, b), then f is an increasing function. Similarly, if $f'(x) < 0$ ($f'(x) \geqslant 0$, $f'(x) \leqslant 0$) for all x in (a, b) then f is decreasing (nondecreasing, nonincreasing, respectively) on (a, b). [For applications, see Section 7.4.]*

Proof of (i): Let $x \in (a, b)$. By the mean-value theorem, there is a number c in (a, x) such that $[f(x) - f(a)]/[x - a] = f'(c) = 0$. It follows that $f(x) = f(a)$ for all x in (a, b).

Proof of (ii): Apply (i) to the function $h(x) = f(x) - g(x)$.

Proof of (iii): Consider $x, y \in (a, b)$, $x < y$. By the mean-value theorem there is a number c in (x, y) such that $[f(y) - f(x)]/[y - x] = f'(c) > 0$, from which it follows that $f(y) > f(x)$, and f is increasing.

6.6.2. Let $f: R \rightarrow R$ be such that for all x and y in R, $|f(x) - f(y)| \leqslant (x - y)^2$. Prove that f is a constant.

Solution. By the first of the preceding corollaries, it suffices to show that $f'(x) = 0$ for all x. Therefore, we argue as follows:

$$|f'(x)| = \left| \lim_{y \to x} \frac{f(y) - f(x)}{y - x} \right|$$

$$= \lim_{y \to x} \left| \frac{f(y) - f(x)}{y - x} \right|$$

$$= \lim_{y \to x} \frac{|f(y) - f(x)|}{|y - x|}$$

$$\leqslant \lim_{y \to x} \frac{(y - x)^2}{|y - x|}$$

$$= \lim_{y \to x} |y - x|$$

$$= 0.$$

6.6.3. Suppose that $f: R \rightarrow R$ is twice differentiable with $f''(x) \geqslant 0$ for all x. Prove that for all a and b, $a < b$,

$$f\left(\frac{a + b}{2}\right) \leqslant \frac{f(a) + f(b)}{2}.$$

Solution. Figure 6.9 makes the conclusion believable, but it is the mean-value theorem that enables us to translate the local property, $f''(x) \geqslant 0$ (f'' at x is determined by those values of f close to x), into a global property (true for all a and b regardless of their proximity).

By the mean-value theorem there is a number x_1 in $(a, \frac{1}{2}(a + b))$ such that

$$\frac{f(\frac{1}{2}(a + b)) - f(a)}{\frac{1}{2}(a + b) - a} = f'(x_1),$$

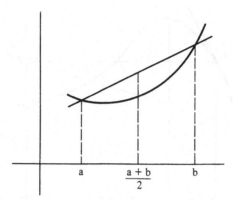

Figure 6.9.

and a number x_2 in $(\frac{1}{2}(a + b), b)$ such that

$$\frac{f(b) - f(\frac{1}{2}(a + b))}{b - \frac{1}{2}(a + b)} = f'(x_2).$$

But $f''(x) \geqslant 0$ for all x in (x_1, x_2), so f' is a nondecreasing function. Thus $f'(x_2) \geqslant f'(x_1)$, or equivalently,

$$\frac{f(b) - f(\frac{1}{2}(a + b))}{\frac{1}{2}(b - a)} \geqslant \frac{f(\frac{1}{2}(a + b)) - f(a)}{\frac{1}{2}(b - a)},$$

$$f(\frac{1}{2}(a + b)) \leqslant \frac{f(a) + f(b)}{2}.$$

In the remainder of the section we will consider problems which make use of all of the major existence theorems considered in this chapter: the intermediate-value theorem, the extreme-value theorem, Rolle's theorem, and the mean-value theorem.

6.6.4. Let f be differentiable with f continuous on $[a, b]$. Show that there is a number c in $(a, b]$ such that $f'(c) = 0$, then we can find a number ξ in (a, b) such that

$$f'(\xi) = \frac{f(\xi) - f(a)}{b - a}.$$

Solution. We begin by getting a geometrical feel for the problem: consider the graph in Figure 6.10, where B is located so that the line CB is horizontal. For a point x between a and b, the right side of the equation, i.e.,

$$\frac{f(x) - f(a)}{b - a},$$

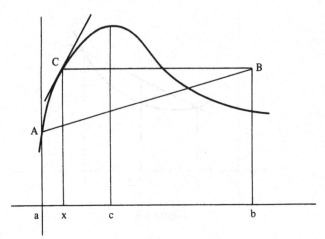

Figure 6.10.

represents the slope of the line AB, whereas the left side, $f'(x)$, represents the slope of the tangent to the curve at C.

Consider, then, the function

$$F(x) = f'(x) - \frac{f(x) - f(a)}{b - a}.$$

This is a continuous function of x (here we use the fact that f' is continuous), so by the intermediate-value theorem, there is a point ξ in (a, b) such that $F(\xi) = 0$ provided we can find points x_1 and x_2 in (a, b) such that $F(x_1) > 0$ and $F(x_2) < 0$.

Observe that $F(x)$ moves from being positive at $x = a$ to being negative at $x = c$. Will this, or something similar, always be the case?

Suppose that $f(c) > f(a)$. Then $f'(c) = 0$, and $[f(c) - f(a)]/[b - a] > 0$, so that

$$F(c) = f'(c) - \frac{f(c) - f(a)}{b - a} < 0.$$

By the mean-value theorem, there is a point d in (a, c) such that $f'(d) = [f(c) - f(a)]/[c - a]$. Therefore,

$$
\begin{aligned}
F(d) &= f'(d) - \frac{f(d) - f(a)}{b - a} \\
&= \frac{f(c) - f(a)}{c - a} - \frac{f(d) - f(a)}{b - a} \\
&\geq \frac{f(c) - f(a)}{b - a} - \frac{f(d) - f(a)}{b - a} \\
&= \frac{f(c) - f(d)}{b - a}.
\end{aligned}
$$

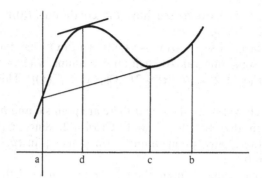

Figure 6.11.

Now, we would be done if it were the case that $f(c) > f(d)$. Unfortunately, this may not be true, as the graph in Figure 6.11 indicates.

To alleviate this difficulty, we can proceed as follows. Consider the function f over the interval $[a, c]$. By the extreme-value theorem, it attains a maximum value on this interval, say at $x = s$ (s may equal c). Since we are assuming that $f(c) > f(a)$, we know that $a < s \leqslant c$. If $s = c$ then $f'(s) = f'(c) = 0$, whereas if $a < s < c$ then $f'(s) = 0$ by 6.3.7. Now proceed as before: There is a point d in (a, s) such that $f'(d) = [f(s) - f(a)]/[s - a]$, and

$$F(d) = f'(d) - \frac{f(d) - f(a)}{b - a}$$

$$= \frac{f(s) - f(a)}{s - a} - \frac{f(d) - f(a)}{b - a}$$

$$\geqslant \frac{f(s) - f(a)}{b - a} - \frac{f(d) - f(a)}{b - a}$$

$$= \frac{f(s) - f(d)}{b - a},$$

and this last expression is nonnegative, since $f(s) \geqslant f(d)$ by our choice of s. This completes the proof for this case. The argument is similar for the cases $f(c) < f(a)$ and $f(c) = f(a)$.

6.6.5. Suppose f is a twice continuously differentiable real-valued function defined for all real numbers such that $|f(x)| \leqslant 1$ for all x and $(f(0))^2 + (f'(0))^2 = 4$. Prove that there exists a real number x_0 such that $f(x_0) + f''(x_0) = 0$.

Solution. There are two natural approaches we might consider. One is to try to apply the intermediate-value theorem: that is, to consider the function $F(x) = f(x) + f''(x)$ and to find a and b for which $F(a) > 0$ and $F(b) < 0$.

Unfortunately, it is hard to see how the condition $(f(0))^2 + (f'(0))^2 = 4$ could be used in this approach.

Another idea is to see if $G(x) = (f(x))^2 + (f'(x))^2$ has an extremum in the interior of some interval. At such an extremum, $G'(x) = 0$. Notice that $G'(x) = 2f(x)f'(x) + 2f'(x)f''(x) = 2f'(x)[f(x) + f''(x)]$. This looks more like it!

Our approach will be to show that there are points a and b, $-2 < a < 0$, $0 < b < 2$, such that $|G(a)| \leqslant 2$ and $|G(b)| \leqslant 2$. Since $G(0) = 4$, it will follow that $G(x)$ attains its maximum at a point x_0 in (a, b), and at this point, $G'(x_0) = 0$.

From the mean-value theorem there is a point a in $(-2, 0)$ and b in $(0, 2)$ such that

$$f'(a) = \frac{f(0) - f(-2)}{2} \quad \text{and} \quad f'(b) = \frac{f(2) - f(0)}{2}.$$

It follows that

$$|f'(a)| = \left| \frac{f(0) - f(-2)}{2} \right| \leqslant \frac{|f(0)| + |f(-2)|}{2} \leqslant \frac{1+1}{2} = 1,$$

$$|f'(b)| = \left| \frac{f(2) - f(0)}{2} \right| \leqslant \frac{|f(2)| + |f(0)|}{2} \leqslant \frac{1+1}{2} = 1.$$

Thus,

$$|G(a)| = |(f(a))^2 + (f'(a))^2| \leqslant |f(a)|^2 + |f'(a)|^2 \leqslant 2,$$

$$|G(b)| = |(f(b))^2 + (f'(b))^2| \leqslant |f(b)|^2 + |f'(b)|^2 \leqslant 2.$$

Let x_0 be the point in (a, b) where $G(x_0)$ is a maximum. Then

$$G'(x_0) = 2f'(x_0)\big[f(x_0) + f''(x_0) \big] = 0.$$

If $f'(x_0) = 0$, then $G(x_0) = (f(x_0))^2 + (f'(x_0))^2 = (f(x_0))^2 \leqslant 1$. But $G(x_0) \geqslant 4$, since $G(0) = 4$. Therefore $f'(x_0) \neq 0$, and it must be the case that $f(x_0) + f''(x_0) = 0$. This completes the proof.

6.6.6. Let $f(x)$ be differentiable on $[0, 1]$ with $f(0) = 0$ and $f(1) = 1$. For each positive integer n, show that there exist distinct points $x_1, x_2, \ldots, x_n$ in $[0, 1]$ such that

$$\sum_{i=1}^{n} \frac{1}{f'(x_i)} = n.$$

Solution. To help generate ideas, consider the case $n = 1$. We wish to find x_1 in $[0, 1]$ such that $1/f'(x_1) = 1$. This is possible by the mean-value theorem, since on the interval $[0, 1]$, there is a point x_1 such that $f'(x_1) = 1$.

Consider the case $n = 2$. Consider the subintervals $[0, x]$ and $[x, 1]$ where x is some number between 0 and 1 yet to be determined. By the mean-value theorem, there is an x_1 in $(0, x)$ and x_2 in $(x, 1)$ such that

$$f'(x_1) = \frac{f(x) - f(0)}{x - 0} \quad \text{and} \quad f'(x_2) = \frac{f(1) - f(x)}{1 - x}.$$

Thus,

$$\frac{1}{f'(x_1)} + \frac{1}{f'(x_2)} = 2$$

if and only if

$$\frac{x}{f(x)} + \frac{1 - x}{1 - f(x)} = 2,$$

$$x(1 - f(x)) + (1 - x)f(x) = 2f(x) - 2(f(x))^2,$$

$$x - xf(x) + f(x) - xf(x) - 2f(x) + 2(f(x))^2 = 0,$$

$$x - 2xf(x) - f(x) + 2(f(x))^2 = 0,$$

$$x(1 - 2f(x)) - f(x)(1 - 2f(x)) = 0,$$

$$[x - f(x)][1 - 2f(x)] = 0.$$

Now, had we chosen x in $(0, 1)$ so that $f(x) = \frac{1}{2}$ (this could be done, by the intermediate-value theorem), the proof would be complete upon reversing the previous steps.

With this background we can consider the case for an arbitrary positive integer n. Let c_i be the smallest number in $[0, 1]$ such that $f(c_i) = i/n$ (the existence of this number is a consequence of the intermediate-value theorem together with the assumption of continuity). Then $0 < c_1 < c_2 < \cdots < c_{n-1} < 1$. Define $c_0 = 0$ and $c_n = 1$, and for each interval (c_{i-1}, c_i), $i = 1, 2, \ldots, n$, choose x_i such that

$$f'(x_i) = \frac{f(c_i) - f(c_{i-1})}{c_i - c_{i-1}}$$

(this can be done, by the mean-value theorem). Then

$$f'(x_i) = \frac{\dfrac{i}{n} - \dfrac{i-1}{n}}{c_i - c_{i-1}} = \frac{1}{n(c_i - c_{i-1})},$$

so that

$$\sum_{i=1}^{n} \frac{1}{f'(x_i)} = \sum_{i=1}^{n} n(c_i - c_{i-1}) = n.$$

Problems

6.6.7.

(a) Show that

$$F(x) = \frac{\sin x + \sin(x + a)}{\cos x - \cos(x + a)}$$

is a constant function by showing that $F'(x) = 0$. (This problem arose in 1.2.1.)

(b) If $P(x)$ is a polynomial of degree three in x, and $y^2 = P(x)$, show that

$$\frac{D(y^3 D^2 y)}{y^2}$$

is a constant, where D denotes the derivative operator. (Hint: First write the above expression in terms of P and its derivatives.)

6.6.8.

(a) If $y = f(x)$ is a solution of the differential equation $y'' + y = 0$, show that $f^2 + (f')^2$ is a constant.

(b) Use part (a) to show that every solution of $y'' + y = 0$ is of the form $y = A \cos x + B \sin x$. (Hint: It is easy to show that all functions $A \cos x + B \sin x$ satisfy the differential equation. Let $f(x)$ be a solution. For $f(x)$ to have the form $f(x) = A \cos x + B \sin x$ it is necessary that $A = f(0)$ and $B = f'(0)$. Now consider $F(x) = f(x) - f(0)\cos x - f'(0)\sin x$. Apply part (a) to $F(x)$, making use of the fact that $F(0) = 0 = F'(0)$.)

(c) Use part (b) to prove the addition formulas

$$\sin(x + y) = \sin x \cos y + \cos x \sin y,$$

$$\cos(x + y) = \cos x \cos y - \sin x \sin y.$$

6.6.9. Let $f(x)$ be differentiable on $[0, 1]$ with $f(0) = 0$ and $f(1) = 1$. For each positive integer n and arbitrary given positive numbers $k_1, k_2, \ldots, k_n$, show that there exist distinct $x_1, x_2, \ldots, x_n$ such that

$$\sum_{i=1}^{n} \frac{k_i}{f'(x_i)} = \sum_{i=1}^{n} k_i.$$

Additional Examples

6.9.6, 6.9.10, Section 7.4.

6.7. L'Hôpital's Rule

We will assume the reader is familiar with the various forms of L'Hôpital's rule.

6.7.1. Evaluate

$$\lim_{x \to \infty} \left(\frac{1}{x} \frac{a^x - 1}{a - 1} \right)^{1/x}, \qquad \text{where} \quad a > 1.$$

Solution. Rewrite the expression in the equivalent form

$$\left(\frac{1}{x} \frac{a^x - 1}{a - 1} \right)^{1/x} = \exp\left(\frac{1}{x} \log\left(\frac{1}{x} \frac{a^x - 1}{a - 1} \right) \right).$$

In this way the problem is transformed to that of evaluating

$$\lim_{x \to \infty} \left(\frac{\log \dfrac{1}{x} \dfrac{a^x - 1}{a - 1}}{x} \right),$$

or equivalently,

$$\lim_{x \to \infty} \left(\frac{\log \dfrac{1}{x}}{x} \right) + \lim_{x \to \infty} \left(\frac{\log(a^x - 1)}{x} \right) - \lim_{x \to \infty} \left(\frac{\log(a - 1)}{x} \right)$$

provided each of these limits exists.

Clearly, $\lim_{x \to \infty} ((\log(a - 1))/x) = 0$, and by L'Hôpital's rule,

$$\lim_{x \to \infty} \frac{\log(1/x)}{x} = \lim_{x \to \infty} \left(\frac{-\log x}{x} \right)$$

$$= \lim_{x \to \infty} \left(\frac{-1/x}{1} \right) = 0.$$

Also, by L'Hôpital's rule,

$$\lim_{x \to \infty} \left(\frac{\log(a^x - 1)}{x} \right) = \lim_{x \to \infty} \left(\frac{a^x \log a}{a^x - 1} \right) = \log a.$$

It follows that

$$\lim_{x \to \infty} \left(\frac{1}{x} \frac{a^x - 1}{a - 1} \right)^{1/x} = \exp \log a = a.$$

6.7.2. Suppose that f is a function with two continuous derivatives and $f(0) = 0$. Prove that the function g defined by $g(0) = f'(0)$, $g(x) = f(x)/x$ for $x \neq 0$ has a continuous derivative.

Solution. For $x \neq 0$,

$$g'(0) = \lim_{x \to 0} \left(\frac{g(x) - g(0)}{x - 0} \right)$$

and since f' is continuous, so also is g' for all $x \neq 0$. It only remains to check that g has a derivative at $x = 0$, and if $g'(0)$ exists, to see if g' is continuous at $x = 0$.

For the existence of $g'(0)$ we must examine the following limit:

$$g'(0) = \lim_{x \to 0} \left(\frac{g(x) - g(0)}{x - 0} \right)$$

$$= \lim_{x \to 0} \left(\frac{f(x)/x - f'(0)}{x} \right)$$

$$= \lim_{x \to 0} \left(\frac{f(x) - xf'(0)}{x^2} \right).$$

Since $f(x) - xf'(0) \to 0$ as $x \to 0$, and since f and f' are differentiable, we may apply L'Hôpital's rule to this limit to get

$$g'(0) = \lim_{x \to 0} \left(\frac{f'(x) - f'(0)}{2x} \right)$$

$$= \frac{1}{2} \lim_{x \to 0} \left(\frac{f'(x) - f'(0)}{x} \right)$$

$$= \tfrac{1}{2} f''(0).$$

(The last step follows from the definition of $f''(0)$.) Thus $g'(0)$ exists.

To check continuity of g' at 0 we have

$$\lim_{x \to 0} g'(x) = \lim_{x \to 0} \left(\frac{xf'(x) - f(x)}{x^2} \right)$$

$$= \lim_{x \to 0} \left(\frac{f'(x) + xf''(x) - f'(x)}{2x} \right)$$

$$= \lim_{x \to 0} \left(\frac{f''(x)}{2} \right) = \tfrac{1}{2} f''(0).$$

The last step follows because we are given that f has a continuous second derivative. Thus $\lim_{x \to 0} g'(x) = g'(0)$, and the proof is complete.

Problems

6.7.3. Evaluate

$$\lim_{n \to \infty} 4^n \left(1 - \cos \frac{\theta}{2^n} \right).$$

6.7.4. Evaluate the following limits:

(a) $\displaystyle \lim_{n\to\infty} \left(1 + \frac{1}{n}\right)^n$

(b) $\displaystyle \lim_{n\to\infty} \left(\frac{n+1}{n+2}\right)^n$

(c) $\displaystyle \lim_{n\to\infty} \left(1 + \frac{1}{n^2}\right)^n$

(d) $\displaystyle \lim_{n\to\infty} \left(1 + \frac{1}{n}\right)^{n^2}$

(e) $\displaystyle \lim_{n\to\infty} \frac{2p_n P_n}{p_n + P_n}$, where $p_n = \left(1 + \frac{1}{n}\right)^n$, and $P_n = \left(1 + \frac{1}{n}\right)^{n+1}$.

6.7.5. Let $0 < a < b$. Evaluate

$$\lim_{t\to 0} \left[\int_0^1 \left[bx + a(1 - x)^t \right] dt \right]^{1/t}.$$

6.7.6. Calculate

$$\lim_{x\to\infty} x \int_0^x e^{t^2 - x^2}\, dt.$$

6.7.7. Prove that the function $y = (x^2)^x$, $y(0) = 1$, is continuous at $x = 0$.

6.8. The Integral

Consider what happens to the sum

$$\frac{1}{n} + \frac{1}{n+1} + \cdots + \frac{1}{2n-1}$$

as $n \to \infty$. One way to think about this is to interpret the sum geometrically: construct rectangles on $[n, 2n]$ as shown in Figure 6.12. From the figure it is clear that

$$\frac{1}{n} + \frac{1}{n+1} + \cdots + \frac{1}{2n-1} > \int_n^{2n} \frac{1}{x}\, dx = \log x \Big|_n^{2n}$$

$$= \log 2n - \log n = \log 2.$$

Similarly, from Figure 6.13 it follows that

$$\frac{1}{n+1} + \frac{1}{n+2} + \cdots + \frac{1}{2n} < \int_n^{2n} \frac{1}{t}\, dt = \log 2.$$

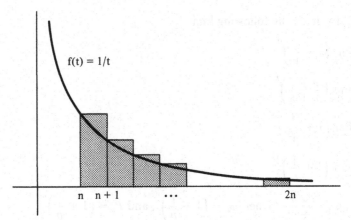

Figure 6.12.

Putting these together, we have

$$\log 2 < \frac{1}{n} + \frac{1}{n+1} + \cdots + \frac{1}{2n-1} < \left(\frac{1}{n} - \frac{1}{2n}\right) + \log 2.$$

Now, as $n \to \infty$ it is apparent that the sum in question approaches $\log 2$.
Another way to see this is to rewrite the sum in the form

$$\sum_{k=0}^{n-1} \frac{1}{n+k} = \sum_{k=0}^{n-1} \left(\frac{1}{1+k/n}\right)\frac{1}{n}$$

and to think of each term,

$$\left(\frac{1}{1+k/n}\right)\frac{1}{n},$$

as the area of the rectangle with base $[k/n, (k+1)/n]$ and height $1/(1 + k/n)$. In this way, the sum represents the area in the shaded rectangles

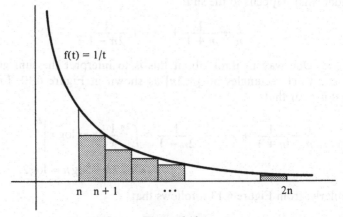

Figure 6.13.

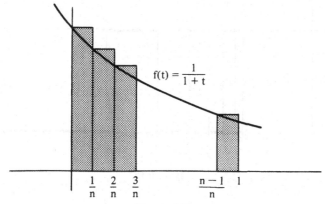

Figure 6.14.

shown in Figure 6.14. As $n \to \infty$, these areas approach the area bounded by $y = 1/(1 + x)$, $y = 0$, $x = 0$, $x = 1$. That is,

$$\lim_{n\to\infty} \sum_{k=0}^{n-1} \frac{1}{n+k} = \lim_{n\to\infty} \sum_{k=0}^{n-1} \left(\frac{1}{1+k/n} \right) \frac{1}{n}$$

$$= \int_0^1 \frac{1}{1+x} \, dx = \log 2.$$

6.8.1. Evaluate

$$\lim_{n\to\infty} \frac{1}{n} \sum_{k=1}^{n} \left(\left[\!\left[\frac{2n}{k} \right]\!\right] - 2 \left[\!\left[\frac{n}{k} \right]\!\right] \right).$$

Solution. The problem asks us to evaluate definite integral

$$\int_0^1 \left(\left[\!\left[\frac{2}{x} \right]\!\right] - 2 \left[\!\left[\frac{1}{x} \right]\!\right] \right) dx.$$

We will do this geometrically by computing the area under the graph of $f(x) = [\![2/x]\!] - 2[\![1/x]\!]$ between $x = 0$ and $x = 1$. The points of discontinuity of $f(x)$ in $(0, 1)$ occur at the points where either $2/x$ or $1/x$ is an integer. In the first case, $2/x = n$ when $x = 2/n$, and in the second case, $1/x = n$ when $x = 1/n$. Thus, we concentrate on the points $1 > 2/3 > 2/4 > 2/5 > 2/6 > \cdots$.

It is easy to check that for each n,

$$f(x) = \begin{cases} 0 & \text{if } x \in \left(\dfrac{2}{2n+1}, \dfrac{2}{2n} \right], \\[2mm] 1 & \text{if } x \in \left(\dfrac{2}{2n+2}, \dfrac{2}{2n+1} \right]. \end{cases}$$

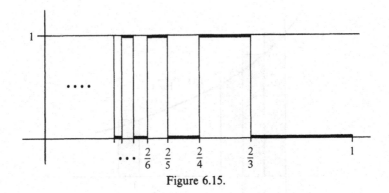

Figure 6.15.

The graph is as shown in Figure 6.15. The integral is therefore equal to

$$\left(\frac{2}{3} - \frac{2}{4}\right) + \left(\frac{2}{5} - \frac{2}{6}\right) + \left(\frac{2}{7} - \frac{2}{8}\right) + \cdots,$$

or

$$2(\tfrac{1}{3} - \tfrac{1}{4} + \tfrac{1}{5} - \tfrac{1}{6} + \tfrac{1}{7} - \tfrac{1}{8} + \cdots).$$

Now recall that

$$\log(1 + x) = x - \frac{x^2}{2} + \frac{x^3}{3} - \frac{x^4}{4} + \cdots, \qquad -1 < x \leqslant 1.$$

This means that

$$2(\tfrac{1}{3} - \tfrac{1}{4} + \tfrac{1}{5} - \tfrac{1}{6} + \cdots) = 2[\log 2 - 1 + \tfrac{1}{2}] = \log 4 - 1,$$

and this completes the solution.

6.8.2. Evaluate

$$\lim_{n \to \infty} \frac{1}{n^4} \prod_{i=1}^{2n} (n^2 + i^2)^{1/n}.$$

Solution. We can change the product into an equivalent form by writing

$$\frac{1}{n^4} \prod_{i=1}^{2n} (n^2 + i^2)^{1/n} = \exp\left[\log \frac{1}{n^4} \prod_{i=1}^{2n} (n^2 + i^2)^{1/n} \right]$$

$$= \exp\left[\sum_{i=1}^{2n} \frac{1}{n} \log(n^2 + i^2) - \log n^4 \right].$$

Therefore, we will examine

$$\lim_{n\to\infty}\left[\sum_{i=1}^{2n}\frac{1}{n}\log(n^2+i^2)-\log n^4\right]$$

$$=\lim_{n\to\infty}\left[\sum_{i=1}^{2n}\frac{1}{n}\log n^2\left(\frac{n^2+i^2}{n^2}\right)-\log n^4\right]$$

$$=\lim_{n\to\infty}\left[\sum_{i=1}^{2n}\frac{1}{n}\left\{\log n^2+\log\left(1+\left(\frac{i}{n}\right)^2\right)\right\}-\log n^4\right]$$

$$=\lim_{n\to\infty}\left[\sum_{i=1}^{2n}\frac{1}{n}\log n^2+\sum_{i=1}^{2n}\frac{1}{n}\log\left(1+\left(\frac{i}{n}\right)^2\right)-\log n^4\right]$$

$$=\lim_{n\to\infty}\left[\frac{2n}{n}\log n^2+\sum_{i=1}^{2n}\frac{1}{n}\log\left(1+\left(\frac{i}{n}\right)^2\right)-\log n^4\right]$$

$$=\lim_{n\to\infty}\left[\sum_{i=1}^{2n}\log\left(1+\left(\frac{i}{n}\right)^2\right)\cdot\frac{1}{n}\right].$$

We recognize this final expression as the definite integral

$$\int_0^2\log(1+x^2)\,dx.$$

Using integration by parts,

$$\int_0^2\log(1+x^2)\,dx=x\log(1+x^2)\Big]_0^2-2\int_0^2\frac{x^2}{1+x^2}\,dx$$

$$=2\log 5-2\int_0^2\left[1-\frac{1}{1+x^2}\right]dx$$

$$=2\log 5-2[x-\arctan x]_0^2$$

$$=2\log 5-2[2-\arctan 2].$$

Thus, the original limit is

$$\exp[2\log 5-4+2\arctan 2],$$

or equivalently,

$$25\exp(2\arctan 2-4).$$

6.8.3. Prove that

$$\sum_{k=0}^{n}(-1)^k\binom{n}{k}\frac{1}{k+m+1}=\sum_{k=0}^{m}(-1)^k\binom{m}{k}\frac{1}{k+n+1}.$$

Solution. The key is to observe that

$$\frac{1}{k+m+1} = \int_0^1 t^{k+m}\, dt.$$

Using this, we find that

$$\sum_{k=0}^{n} (-1)^k \binom{n}{k} \frac{1}{k+m+1} = \sum_{k=0}^{n} (-1)^k \binom{n}{k} \int_0^1 t^{k+m}\, dt$$

$$= \int_0^1 \sum_{k=0}^{n} (-1)^k \binom{n}{k} t^{k+m}\, dt$$

$$= \int_0^1 t^m (1-t)^n\, dt.$$

Now use a change of variable: let $s = 1 - t$. Continuing from the last integral:

$$= \int_0^1 s^n (1-s)^m\, ds$$

$$= \int_0^1 s^n \sum_{k=0}^{m} (-1)^k \binom{m}{k} s^k\, ds$$

$$= \sum_{k=0}^{m} (-1)^k \binom{m}{k} \int_0^1 s^{k+n}\, ds$$

$$= \sum_{k=0}^{m} (-1)^k \binom{m}{k} \frac{1}{k+n+1}.$$

Problems

6.8.4. Evaluate each of the following:

(a) $\displaystyle \lim_{n\to\infty} \left[\frac{1}{2n+1} + \frac{1}{2n+2} + \cdots + \frac{1}{3n} \right].$

(b) $\displaystyle \lim_{n\to\infty} \left[\frac{1^a + 2^a + \cdots + n^a}{n^{1+a}} \right], \qquad a > -1.$

(c) $\displaystyle \lim_{n\to\infty} \left[\frac{n}{1^2 + n^2} + \frac{n}{2^2 + n^2} + \cdots + \frac{n}{n^2 + n^2} \right].$

(d) $\displaystyle \lim_{n\to\infty} \left[\left(1 + \frac{1}{n}\right)\left(1 + \frac{2}{n}\right) \cdots \left(1 + \frac{n}{n}\right) \right]^{1/n}.$

6.8.5. Evaluate each of the following:

(a) $\displaystyle\lim_{n\to\infty} n^{-3/2} \sum_{k=1}^{n} \sqrt{k}$.

(b) $\displaystyle\lim_{n\to\infty} \sum_{k=1}^{n} \frac{1}{\sqrt{k^2 + n^2}}$.

6.8.6. Find the integral part of $\sum_{n=1}^{10^9} n^{-2/3}$. (Hint: Compare the area under $f(x) = x^{-2/3}$ over $[1, 10^9 + 1]$ with the area under $g(x) = (x - 1)^{-2/3}$ over $[2, 10^9 + 1]$.)

6.8.7. Suppose that f and g are continuous functions on $[0, a]$, and suppose that $f(x) = f(a - x)$ and $g(x) + g(a - x) = k$ for all x in $[0, a]$, where k is a fixed number. Prove that

$$\int_0^a f(x)g(x)\, dx = \tfrac{1}{2} k \int_0^a f(x)\, dx.$$

Use this fact to evaluate

$$\int_0^\pi \frac{x \sin x}{1 + \cos^2 x}\, dx.$$

6.8.8.

(a) Let

$$A = \int_0^\pi \frac{\cos x}{(x + 2)^2}\, dx.$$

Compute

$$\int_0^{\pi/2} \frac{\sin x \cos x}{x + 1}\, dx$$

in terms of A.

(b) Let

$$f(x) = \int_1^x \frac{\log t}{1 + t}\, dt \qquad \text{for } x > 0.$$

Compute $f(x) + f(1/x)$.

6.8.9. Find all continuous positive functions $f(x)$, for $0 \leqslant x \leqslant 1$, such that $\int_0^1 f(x)\, dx = 1$, $\int_0^1 x f(x)\, dx = a$, $\int_0^1 x^2 f(x)\, dx = a^2$, where a is a given real number.

6.8.10. Let $f(x, y)$ be a continuous function on the square

$$S = \{(x, y) : 0 \leqslant x \leqslant 1, 0 \leqslant y \leqslant 1\}.$$

For each point (a,b) in the interior of S, let $S_{(a,b)}$ be the largest square that is contained in S, centered at (a,b), and has sides parallel to those of S. If the double integral $\iint f(x,y)\,dx\,dy$ is zero when taken over each square $S_{(a,b)}$, must $f(x,y)$ be identically zero on S?

Additional Examples

1.4.4, 1.6.3, 1.12.3, 1.12.6, 2.5.15, 6.2.2, 6.2.9, 7.6.3.

6.9. The Fundamental Theorem

The fundamental theorem of calculus refers to the inverse relationship that holds between differentiation and integration. The fundamental theorem for integrals of derivatives states that if $F(t)$ has a continuous derivative on an interval $[a,b]$, then

$$\int_a^b F'(t)\,dt = F(b) - F(a).$$

In other words, differentiation followed by integration recovers the function up to a constant, in the sense that

$$F(x) = \int_0^x F'(t)\,dt + C$$

where $C = F(0)$.

For example, the derivative of $F(t) = \sin^2 t$ is $F'(t) = 2\sin t\cos t$. Integration of $F'(t)$ on $[0,x]$ yields

$$\sin^2 x = \int_0^x 2\sin t\cos t\,dt.$$

In this case we have recovered the function exactly because $F(0) = 0$. But also observe that the integration can be carried out in another manner; namely (let $u = \cos t$),

$$\int_0^x 2\sin t\cos t\,dt = -\cos^2 t\Big]_0^x = -\cos^2 x + 1.$$

It follows that $\sin^2 x = -\cos^2 x + 1$, or equivalently, $\sin^2 x + \cos^2 x = 1$ for all x.

6.9.1. Find all the differentiable functions f defined for $x > 0$ which satisfy

$$f(xy) = f(x) + f(y), \qquad x, y > 0.$$

Solution. When $x = y = 1$, we get $f(1) = f(1 \times 1) = f(1) + f(1)$, and it follows that $f(1) = 0$.

If $x \neq 0$, we have $0 = f(1) = f(x \times 1/x) = f(x) + f(1/x)$, and therefore, $f(1/x) = -f(x)$. It follows that $f(x/y) = f(x) + f(1/y) = f(x) - f(y)$.

Now the idea is to look at the derivative of f and then to recover f by integration:

$$f'(x) = \lim_{h \to 0} \left(\frac{f(x+h) - f(x)}{h} \right)$$

$$= \lim_{h \to 0} \left(\frac{f((x+h)/x)}{h} \right)$$

$$= \lim_{t \to 0} \left(\frac{f(1+t)}{tx} \right), \quad \text{where } h/x = t$$

$$= \lim_{t \to 0} \left(\frac{1}{x} \cdot \frac{f(1+t) - f(1)}{t} \right)$$

$$= \frac{1}{x} f'(1).$$

Therefore, by the fundamental theorem,

$$f(x) = f(x) - f(1) = \int_1^x f'(x)\,dx = \int_1^x \frac{f'(1)}{x}\,dx = f'(1)\log x.$$

Thus, the functions we seek are those of the form $f(x) = A \log x$, where A is an arbitrary constant.

6.9.2. Find the sum of the series

$$1 - \frac{1}{5} + \frac{1}{7} - \frac{1}{11} + \cdots + \frac{1}{6n - 5} - \frac{1}{6n - 1} + \cdots .$$

Solution. Consider the function defined by the infinite series

$$f(x) = x - \frac{x^5}{5} + \frac{x^7}{7} - \frac{x^{11}}{11} + \cdots + \frac{x^{6n-5}}{6n - 5} - \frac{x^{6n-1}}{6n - 1} + \cdots$$

for $0 < x \leq 1$. The series is absolutely convergent for $|x| < 1$, and therefore we can rearrange the terms:

$$f(x) = \left(x + \frac{x^7}{7} + \frac{x^{13}}{13} + \cdots + \frac{x^{6n-5}}{6n - 5} + \cdots \right)$$

$$- \left(\frac{x^5}{5} + \frac{x^{11}}{11} + \cdots + \frac{x^{6n-1}}{6n - 1} + \cdots \right).$$

Our idea is to differentiate f, to change its form, and then to recover f by integration by use of the fundamental theorem. We have, for $0 < x < 1$,

$$f'(x) = (1 + x^6 + \cdots + x^{6n-6} + \cdots) - (x^4 + x^{10} + \cdots + x^{6n-2} + \cdots)$$

$$= \frac{1}{1 - x^6} - \frac{x^4}{1 - x^6} = \frac{(1 - x^2)(1 + x^2)}{(1 - x^2)(1 + x^2 + x^4)} = \frac{1 + x^2}{1 + x^2 + x^4}.$$

Integrating (the details are not of interest here), and noting that $f(0) = 0$, we get

$$f(x) = \frac{1}{\sqrt{3}} \left[\arctan\left(\frac{2x - 1}{\sqrt{3}} \right) + \arctan\left(\frac{2x + 1}{\sqrt{3}} \right) \right].$$

Since the series representation of f is convergent for $x = 1$, Abel's theorem (see Section 5.4) implies that the original series converges to

$$f(1) = \frac{1}{\sqrt{3}} \left[\arctan\frac{1}{\sqrt{3}} + \arctan\sqrt{3} \right] = \frac{\pi}{2\sqrt{3}}.$$

The fundamental theorem for derivatives of integrals states that if f is a continuous function in an interval $[a, b]$, then for any x in (a, b)

$$\frac{d}{dx} \int_a^x f(t) \, dt = f(x).$$

In other words, integration followed by differentiation recovers the function exactly.

6.9.3. If $a(x)$, $b(x)$, $c(x)$, and $d(x)$ are polynomials in x, show that

$$\int_1^x a(x)c(x) \, dx \int_1^x b(x)d(x) \, dx - \int_1^x a(x)d(x) \, dx \int_1^x b(x)c(x) \, dx$$

is divisible by $(x - 1)^4$.

Solution. Denote the expression in question by $F(x)$. Notice that $F(x)$ is a polynomial in x. Also, notice that $F(1) = 0$ and therefore $x - 1$ is a factor of $F(x)$.

Because F is a polynomial, we know that $(x - 1)^4$ divides $F(x)$ if and only if $F'''(1) = 0$. We can compute F' by use of the fundamental theorem:

$$F'(x) = ac \int_1^x bd + bd \int_1^x ac - ad \int_1^x bc - bc \int_1^x ad.$$

(Note that $F'(1) = 0$ and hence that $(x - 1)^2$ divides $F(x)$.) The derivatives F'' and F''' are done in a similar manner; it turns out that $F'''(1) = (ac)'bd + (bd)'ac - (ad)'bc - (bc)'ad]_{x=1} = 0$. This completes the proof.

The next three examples combine several ideas from this chapter.

6.9.4. Let $f: (0, \infty) \to R$ be differentiable, and assume that $f(x) + f'(x) \to 0$ when $x \to \infty$. Show that $f(x) \to 0$ as $x \to \infty$.

Solution. First, a digression: If $p(x)$ and $q(x)$ are continuous functions, the equation

$$\frac{dy}{dx} + p(x)y = q(x)$$

can be solved in the following manner. Multiply each side of the equation by $m(x) = e^{\int p(x)\, dx}$, and notice that the resulting equation can be put into the form

$$\frac{d}{dx}(ym(x)) = m(x)q(x).$$

Thus, by the fundamental theorem of calculus, for each constant a, there is a constant C such that

$$ym(x) = \int_a^x m(t)q(t)\, dt + C.$$

From this, we can solve for y.

Now, return to our problem and set $g(x) = f(x) + f'(x)$. According to the reasoning of the last paragraph, we can solve for $f(x)$ (in terms of $g(x)$) by first multiplying each side by e^x. As above, this leads to the equation

$$f(x)e^x = \int_a^x e^t g(t)\, dt + C,$$

or equivalently,

$$f(x) = e^{-x} \int_a^x e^t g(t)\, dt + Ce^{-x}.$$

Let $\varepsilon > 0$. Since $g(x) \to 0$ as $x \to \infty$, choose a so that $|g(x)| < \varepsilon$ for all $x > a$. Then

$$|f(x)| \leqslant e^{-x}\left|\int_a^x e^t g(t)\, dt\right| + |Ce^{-x}|$$

$$\leqslant e^{-x}\int_a^x e^t |g(t)|\, dt + |Ce^{-x}|$$

$$\leqslant \varepsilon e^{-x}\int_a^x e^t\, dt + |Ce^{-x}|$$

$$= \varepsilon e^{-x}(e^x - e^a) + |Ce^{-x}|$$

$$= \varepsilon(1 - e^{a-x}) + |Ce^{-x}|.$$

Now, for sufficiently large x, we will have $|f(x)| < 2\varepsilon$. It follows that $f(x) \to 0$ as $x \to \infty$.

6.9.5. Evaluate

$$\lim_{x \to 0} (1/x) \int_0^x (1 + \sin 2t)^{1/t} \, dt.$$

Solution. Our aim is to apply L'Hôpital's rule, but some preliminary work must be done. First, there is a question concerning the existence of the integral because the integrand is undefined at $t = 0$. However,

$$\lim_{x \to 0} (1 + \sin 2x)^{1/x} = \lim_{x \to 0} \left[\exp \frac{1}{x} \log(1 + \sin 2x) \right]$$

$$= \exp \left[\lim_{x \to 0} \left(\frac{\log(1 + \sin 2x)}{x} \right) \right]$$

which by L'Hôpital's rule is

$$\exp \left[\lim_{x \to 0} \frac{2 \cos 2x}{1 + \sin 2x} \right] = \exp 2 = e^2.$$

Thus, if we define

$$f(x) = \begin{cases} (1 + \sin 2x)^{1/x} & \text{if } x \neq 0, \\ e^2 & \text{if } x = 0, \end{cases}$$

the function f is continuous, and $\int_0^x (1 + \sin 2t)^{1/t} \, dt = \int_0^x f(t) \, dt$.

In order to apply L'Hôpital's rule to this problem, we must show that $\int_0^x (1 + \sin 2t)^{1/t} \, dt \to 0$ as $x \to 0$. To do this, let K be an upper bound for $|f(x)|$ for all x in $(-1, 1)$. Then, for x in $(-1, 1)$,

$$\left| \int_0^x (1 + \sin 2t)^{1/t} \, dt \right| \leqslant \int_0^x |1 + \sin 2t|^{1/t} \, dt \leqslant K|x|.$$

It follows that

$$\int_0^x (1 + \sin 2t)^{1/t} \, dt \to 0 \qquad \text{as } x \to 0.$$

We are now able to apply L'Hôpital's rule to the original problem:

$$\lim_{x \to 0} \frac{\int_0^x (1 + \sin 2t)^{1/t} \, dt}{x} = \lim_{x \to 0} (1 + \sin 2x)^{1/x} = e^2.$$

6.9.6. Suppose that $f : [0, 1] \to R$ has a continuous second derivative, that $f(0) = 0 = f(1)$, and that $f(x) > 0$ for all x in $(0, 1)$. Show that

$$\int_0^1 \left| \frac{f''(x)}{f(x)} \right| \, dx > 4.$$

Solution. Let X denote a point in $(0, 1)$ where $f(x)$ is a maximum, and suppose that $Y = f(X)$. Then

$$\int_0^1 \left| \frac{f''(x)}{f(x)} \right| dx > \frac{1}{|Y|} \int_0^1 |f''(x)| \, dx$$

$$\geqslant \frac{1}{|Y|} \left| \int_0^1 f''(x) \, dx \right| = \left| \frac{f'(1) - f'(0)}{Y} \right|.$$

We appear to be stymied at this point, because it is certainly not necessary that $|f'(1) - f'(0)| \geqslant 4|Y|$. However, by the mean-value theorem, there are points a in $(0, X)$ and b in $(X, 1)$ such that

$$f'(a) = \frac{f(X) - f(0)}{X - 0} = \frac{f(X)}{X} = \frac{Y}{X}$$

and

$$f'(b) = \frac{f(1) - f(X)}{1 - X} = \frac{-Y}{1 - X}.$$

Thus,

$$\int_0^1 \left| \frac{f''(x)}{f(x)} \right| dx \geqslant \int_a^b \left| \frac{f''(x)}{f(x)} \right| dx \geqslant \frac{1}{|Y|} \left| \int_a^b f''(x) \, dx \right|,$$

so applying the fundamental theorem to the last integral, we have

$$\int_0^1 \left| \frac{f''(x)}{f(x)} \right| dx > \frac{1}{|Y|} |f'(b) - f'(a)|$$

$$= \frac{1}{|Y|} \left| \frac{-Y}{1 - X} - \frac{Y}{X} \right|$$

$$= \frac{1}{|Y|} \left| \frac{Y}{1 - X} + \frac{Y}{X} \right| = \left| \frac{1}{X(1 - X)} \right|.$$

But the maximum value of $x(1 - x)$ in $(0, 1)$ is $\frac{1}{4}$ (when $x = \frac{1}{2}$) and therefore

$$\int_0^1 \left| \frac{f''(x)}{f(x)} \right| dx > \frac{1}{|X(1 - X)|} \geqslant 4.$$

Problems

6.9.7. What function is defined by the equation

$$f(x) = \int_0^x f(t) \, dt + 1?$$

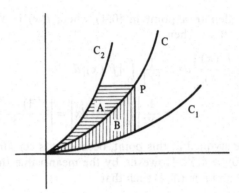

Figure 6.16.

6.9.8. Let $f: [0, 1] \to (0, 1)$ be continuous. Show that the equation

$$2x - \int_0^x f(t)\, dt = 1$$

has one and only one solution in the interval $[0, 1]$.

6.9.9. Suppose that f is a continuous function for all x which satisfies the equation

$$\int_0^x f(t)\, dt = \int_x^1 t^2 f(t)\, dt + \frac{x^{16}}{8} + \frac{x^{18}}{9} + C,$$

where C is a constant. Find an explicit form for $f(x)$ and find the value of the constant C.

6.9.10. Let C_1 and C_2 be curves passing through the origin as shown in Figure 6.16. A curve C is said to *bisect in area* the region between C_1 and C_2 if for each point P of C the two shaded regions A and B shown in the figure have equal areas. Determine the upper curve C_2 given that the bisecting curve C has the equation $y = x^2$ and the lower curve C_1 has the equation $y = \frac{1}{2}x^2$.

6.9.11. Sum the series $1 + \frac{1}{3} - \frac{1}{5} - \frac{1}{7} + \frac{1}{9} + \frac{1}{11} - \frac{1}{13} - \cdots$.

6.9.12. Suppose that f is differentiable, and that $f'(x)$ is strictly increasing for $x \geqslant 0$. If $f(0) = 0$, prove that $f(x)/x$ is strictly increasing for $x > 0$.

Additional Examples

1.5.1, 5.1.3, 5.1.9, 5.1.11, 5.4.6, 7.6.5.

Chapter 7. Inequalities

Inequalities are useful in virtually all areas of mathematics, and inequality problems are among the most beautiful. Among all the possible inequalities that we might consider, we shall concentrate on just two: the arithmetic-mean–geometric-mean inequality in Section 7.2 and the Cauchy–Schwarz inequality in Section 7.3. In addition, we shall consider various algebraic and geometric techniques in Section 7.1, and analytic techniques in Sections 7.4 and 7.5. In the final section, Section 7.6, we shall see how inequalities can be used to evaluate limits.

7.1. Basic Inequality Properties

The most immediate approach for establishing an inequality is to appeal to an algebraic manipulation or a geometric interpretation. For example, the arithmetic-mean–geometric-mean inequality,

$$\frac{a+b}{2} \geqslant \sqrt{ab}, \qquad 0 < a \leqslant b,$$

can be established algebraically by writing it in the equivalent form

$$\left(\sqrt{a} - \sqrt{b}\right)^2 \geqslant 0,$$

or geometrically by considering the semicircle in Figure 7.1. (The semicircle is constructed with diameter AB of length $a + b$, and C is a point chosen so that $AC = a$ and $CB = b$. A perpendicular to AB from C meets the circle at D. Triangles ACD and CDB are similar, and therefore $a/CD = CD/b$.

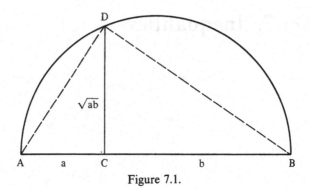

Figure 7.1.

It follows that $CD = \sqrt{ab}$. Clearly, $\sqrt{ab} \leqslant$ radius of circle $= (a + b)/2$.) Both derivations make it clear that equality holds if and only if $a = b$.

In this section we will consider examples of inequalities that can be verified by using only algebraic and geometric ideas.

7.1.1. Show that for positive numbers a, b, c,

$$a^2 + b^2 + c^2 \geqslant ab + bc + ca.$$

Solution. Working backwards,

$$a^2 + b^2 + c^2 \geqslant ab + bc + ca,$$
$$2a^2 + 2b^2 + 2c^2 \geqslant 2ab + 2bc + 2ca,$$
$$(a^2 - 2ab + b^2) + (b^2 - 2bc + c^2) + (c^2 - 2ca + a^2) \geqslant 0,$$
$$(a - b)^2 + (b - c)^2 + (c - a)^2 \geqslant 0.$$

This last inequality is obviously true, and since the steps are reversible, the solution is complete. (The proof also makes it clear that equality holds if and only if $a = b = c$.)

This example illustrates a common theme: manipulate the expression into a form to take advantage of the fact that a squared number is nonnegative.

7.1.2. Prove that for $0 < x < \frac{1}{2}\pi$, $\cos^2 x + x \sin x < 2$.

Solution. Consider the function

$$f(x) = 2 - \cos^2 x - x \sin x,$$

and perform the following manipulations:

$$f(x) = 1 + (1 - \cos^2 x) - x \sin x$$

$$= 1 + \sin^2 x - x \sin x$$

$$= (1 - 2 \sin x + \sin^2 x) - x \sin x + 2 \sin x$$

$$= (1 - \sin x)^2 + (2 - x) \sin x.$$

In this form we see that the desired inequality holds for $0 < x < 2$.

7.1.3. If $0 \leqslant a, b, c \leqslant 1$, show that

$$\frac{a}{b + c + 1} + \frac{b}{c + a + 1} + \frac{c}{a + b + 1} + (1 - a)(1 - b)(1 - c) \leqslant 1.$$

Solution. Here, straightforward algebraic expansion leads to horrendous and unenlightening complications. One simplification is to assume, without loss of generality, that $0 \leqslant a \leqslant b \leqslant c \leqslant 1$. Then, for example, we have

$$\frac{a}{b + c + 1} + \frac{b}{c + a + 1} + \frac{c}{a + b + 1} \leqslant \frac{a + b + c}{a + b + 1},$$

and we might try to prove that

$$\frac{a + b + c}{a + b + 1} + (1 - a)(1 - b)(1 - c) \leqslant 1.$$

This problem is easier algebraically, but still messy, and of course, we may have given away too much (that is to say, this inequality may not even be true). However, we have the following:

$$\frac{a + b + c}{a + b + 1} + (1 - a)(1 - b)(1 - c)$$

$$= \frac{a + b + 1}{a + b + 1} + \frac{c - 1}{a + b + 1} + (1 - a)(1 - b)(1 - c)$$

$$= 1 - \left(\frac{1 - c}{a + b + 1}\right)\left[1 - (1 + a + b)(1 - a)(1 - b)\right].$$

The desired inequality follows from this expression after noting that

$$(1 + a + b)(1 - a)(1 - b) \leqslant (1 + a + b + ab)(1 - a)(1 - b)$$

$$= (1 + a)(1 + b)(1 - a)(1 - b)$$

$$= (1 - a^2)(1 - b^2)$$

$$\leqslant 1.$$

7.1.4. Let n be a positive integer, and $a_i \geqslant 1$, for $i = 1, 2, \ldots, n$. Show that

$$(1 + a_1)(1 + a_2) \cdots (1 + a_n) \geqslant \frac{2^n}{n+1}(1 + a_1 + \cdots + a_n).$$

Solution. Induction is a natural strategy here, and it is not difficult to carry out in this manner. But the following "give a little" argument is more fun:

$$(1 + a_1)(1 + a_2) \cdots (1 + a_n)$$

$$= 2^n \left(\frac{1}{2} + \frac{a_1}{2} \right)\left(\frac{1}{2} + \frac{a_2}{2} \right) \cdots \left(\frac{1}{2} + \frac{a_n}{2} \right)$$

$$= 2^n \left(1 + \frac{a_1 - 1}{2} \right)\left(1 + \frac{a_2 - 1}{2} \right) \cdots \left(1 + \frac{a_n - 1}{2} \right)$$

$$\geqslant 2^n \left(1 + \frac{a_1 - 1}{2} + \frac{a_2 - 1}{2} + \cdots + \frac{a_n - 1}{2} \right)$$

$$\geqslant 2^n \left(1 + \frac{a_1 - 1}{n+1} + \frac{a_2 - 1}{n+1} + \cdots + \frac{a_n - 1}{n+1} \right)$$

$$= \frac{2^n}{n+1}(n + 1 + a_1 - 1 + a_2 - 1 + \cdots + a_n - 1)$$

$$= \frac{2^n}{n+1}(1 + a_1 + a_2 + \cdots + a_n).$$

7.1.5. For each positive integer n, prove that

$$\left(1 + \frac{1}{n} \right)^n < \left(1 + \frac{1}{n+1} \right)^{n+1}.$$

Solution. This is an important inequality that can be proved in a number of ways (see 7.1.11, 7.2.8, 7.4.18). Here we will give a proof based on comparing corresponding terms in the binomial expansions of each side. On the left side,

$$\left(1 + \frac{1}{n} \right)^n = \sum_{k=0}^{n} \binom{n}{k}\left(\frac{1}{n} \right)^k$$

$$= \sum_{k=0}^{n} \frac{n(n-1)(n-2) \cdots (n-k+1)}{n \cdot n \cdot n \cdots n} \frac{1}{k!}$$

$$= \sum_{k=0}^{n} \frac{1}{k!}\left(1 - \frac{1}{n} \right)\left(1 - \frac{2}{n} \right) \cdots \left(1 - \frac{k-1}{n} \right).$$

In a similar manner,

$$\left(1 + \frac{1}{n+1}\right)^{n+1} = \sum_{k=0}^{n+1} \frac{1}{k!}\left(1 - \frac{1}{n+1}\right)\left(1 - \frac{2}{n+1}\right) \cdots \left(1 - \frac{k-1}{n+1}\right)$$

$$= \left(\frac{1}{n+1}\right)^{n+1}$$

$$+ \sum_{k=0}^{n} \frac{1}{k!}\left(1 - \frac{1}{n+1}\right)\left(1 - \frac{2}{n+1}\right) \cdots \left(1 - \frac{k-1}{n+1}\right).$$

The inequality is now obvious, since comparing the coefficients of $1/k!$ in these expressions, we see that for each k, $k = 0, 1, 2, \ldots, n$,

$$\left(1 - \frac{1}{n}\right)\left(1 - \frac{2}{n}\right) \cdots \left(1 - \frac{k-1}{n}\right)$$

$$< \left(1 - \frac{1}{n+1}\right)\left(1 - \frac{2}{n+1}\right) \cdots \left(1 - \frac{k-1}{n+1}\right).$$

It is worth noting that

$$\left(1 + \frac{1}{n}\right)^{n} = \sum_{k=0}^{n} \frac{1}{k!}\left(1 - \frac{1}{n}\right)\left(1 - \frac{2}{n}\right) \cdots \left(1 - \frac{k-1}{n}\right)$$

$$< \sum_{k=0}^{n} \frac{1}{k!} = 1 + \sum_{k=1}^{n} \frac{1}{k!}$$

$$< 1 + \sum_{k=1}^{n} \frac{1}{2^{k-1}} = 1 + \sum_{k=0}^{n-1} \frac{1}{2^{k}}$$

$$< 1 + \sum_{k=0}^{\infty} \frac{1}{2^{k}} = 3.$$

Thus, the sequence $(1 + 1/n)^{n}$ is increasing and bounded above by 3. (It can be shown that the sequence converges to the number e.)

The next result is important theoretically and is very useful (e.g., see 7.4.9 and 7.4.20).

7.1.6. Suppose that $f : R \to R$ satisfies

$$f\left(\frac{x+y}{2}\right) < \frac{f(x) + f(y)}{2}$$

for all x and y in an interval (a, b), $x \neq y$. Show that

$$f\left(\frac{x_1 + x_2 + \cdots + x_n}{n}\right) < \frac{f(x_1) + f(x_2) + \cdots + f(x_n)}{n}$$

whenever the x_i's are in (a, b), with $x_i \neq x_j$ for at least one pair (i, j).

Solution. Assume the result holds for $n = m$; we will show it holds for $n = 2m$. By rearranging notation if necessary, we may assume that $x_1 \neq x_2$ and we have

$$f\left(\frac{x_1 + \cdots + x_{2m}}{2m}\right)$$

$$= f\left(\frac{1}{2}\left(\frac{x_1 + \cdots + x_m}{m} + \frac{x_{m+1} + \cdots + x_{2m}}{m}\right)\right)$$

$$\leq \frac{1}{2}\left[f\left(\frac{x_1 + \cdots + x_m}{m}\right) + f\left(\frac{x_{m+1} + \cdots + x_{2m}}{m}\right)\right]$$

$$< \frac{1}{2}\left(\frac{f(x_1) + \cdots + f(x_m)}{m} + \frac{f(x_{m+1}) + \cdots + f(x_{2m})}{m}\right)$$

$$= \frac{f(x_1) + f(x_2) + \cdots + f(x_{2m})}{2m}.$$

Thus, by induction, the result holds for all positive powers of 2.

Now suppose that $n > 2$ and n is not a power of 2; that is, suppose that $2^{m-1} < n < 2^m$ for some integer m. Let $k = 2^m - n$, and set $y_i = (x_1 + \cdots + x_n)/n$ for $i = 1, 2, \ldots, k$. Then $x_1, x_2, \ldots, x_n, y_1, \ldots, y_k$ are 2^m numbers in the interval (a, b), and therefore our preceding argument implies that

$$f\left(\frac{x_1 + \cdots + x_n + y_1 + \cdots + y_k}{2^m}\right) < \frac{f(x_1) + \cdots + f(y_k)}{2^m}.$$

But note that

$$f\left(\frac{x_1 + \cdots + x_n + y_1 + \cdots + y_k}{2^m}\right)$$

$$= f\left(\frac{x_1 + \cdots + x_n + k(x_1 + \cdots + x_n)/n}{2^m}\right)$$

$$= f\left(\frac{n(x_1 + \cdots + x_n) + (2^m - n)(x_1 + \cdots + x_n)}{n \times 2^m}\right)$$

$$= f\left(\frac{x_1 + \cdots + x_n}{n}\right).$$

Making this substitution into the last inequality,

$$f\left(\frac{x_1 + \cdots + x_n}{n}\right) < \frac{f(x_1) + \cdots + f(x_n) + f(y_1) + \cdots + f(y_k)}{2^m}$$

$$= \frac{f(x_1) + \cdots + f(x_n) + kf((x_1 + \cdots + x_n)/n)}{2^m}.$$

Multiplying each side by 2^m yields

$$2^m f\left(\frac{x_1 + \cdots + x_n}{n}\right) < f(x_1) + \cdots + f(x_n)$$

$$+ (2^m - n) f\left(\frac{x_1 + \cdots + x_n}{n}\right)$$

and from this we get the desired inequality for n:

$$f\left(\frac{x_1 + \cdots + x_n}{n}\right) < \frac{f(x_1) + \cdots + f(x_n)}{n}.$$

Problems

7.1.7. Suppose that a, b, c are positive numbers. Prove that:

(a) $(a + b)(b + c)(c + a) \geqslant 8abc$.
(b) $a^2 b^2 + b^2 c^2 + c^2 a^2 \geqslant abc(a + b + c)$.
(c) If $a + b + c = 1$, then $ab + bc + ca \leqslant \frac{1}{3}$.

7.1.8. Prove that

$$\frac{1}{2} \cdot \frac{3}{4} \cdot \frac{5}{6} \cdots \frac{999999}{1000000} < \frac{1}{1000}.$$

(Hint: Square each side and "give a little" to create a "telescoping" product (see Section 5.3).)

7.1.9.

(a) If a and b are nonzero real numbers, prove that at least one of the following inequalities holds:

$$\left|\frac{a + \sqrt{a^2 + 2b^2}}{2b}\right| < 1, \qquad \left|\frac{a - \sqrt{a^2 + 2b^2}}{2b}\right| < 1.$$

(b) If the n numbers $x_1, x_2, \ldots, x_n$ lie in the interval $(0, 1)$, prove that at least one of the following inequalities holds:

$$x_1 x_2 \cdots x_n \leqslant 2^{-n}, \qquad (1 - x_1)(1 - x_2) \cdots (1 - x_n) \leqslant 2^{-n}.$$

7.1.10.

(a) Let $a_1/b_1, a_2/b_2, \ldots, a_n/b_n$ be n fractions with $b_i > 0$ for $i = 1, 2, \ldots, n$. Show that the fraction

$$\frac{a_1 + a_2 + \cdots + a_n}{b_1 + b_2 + \cdots + b_n}$$

is a number between the largest and the smallest of these fractions.

(Note the special case in which all the fractions a_i/b_i are equal.)

(b) If

$$\frac{a+b}{b+c} = \frac{c+d}{d+a},$$

prove that either $a = c$ or $a + b + c + d = 0$.

7.1.11.

(a) For $0 < a < b$, show that

$$(n+1)(b-a)a^n < b^{n+1} - a^{n+1} < (n+1)(b-a)b^n.$$

(b) Apply this inequality to the special case $a = 1 + 1/(n+1)$ and $b = 1 + 1/n$ to show that $(1 + 1/n)^n < (1 + 1/(n+1))^{n+1}$.

7.1.12. Prove that for all n,

$$\left(\frac{n}{e}\right)^n < n! < e\left(\frac{n}{2}\right)^n.$$

7.1.13 (Cauchy–Schwarz inequality). By mathematical induction on n, prove that for all real numbers $a_1, \ldots, a_n, b_1, \ldots, b_n$,

$$\left(\sum_{k=1}^{n} a_k b_k\right)^2 \leqslant \left(\sum_{k=1}^{n} a_k^2\right)\left(\sum_{k=1}^{n} b_k^2\right).$$

7.1.14. In a convex quadrilateral (the two diagonals are interior to the quadrilateral) prove that the sum lengths of the diagonals is less than the perimeter but greater than one-half the perimeter.

7.1.15. Prove that for any positive integer n, $\sqrt[n]{n} < 1 + \sqrt{2/n}$.

Additional Examples

1.3.3, 1.7.4, 1.7.5, 1.8.2, 1.8.5, 1.8.6, 1.12.7, 2.1.5, 2.1.6, 2.2.4, 2.2.6, 2.4.1, 2.4.4, 2.4.6, 5.3.8, 6.1.3, 7.3.1, 7.4.8, 7.4.9, 7.4.20, 7.4.21, 7.4.22, 7.4.23.

7.2. Arithmetic-Mean–Geometric-Mean Inequality

Let $x_i > 0$ for $i = 1, 2, \ldots, n$. The *arithmetic mean* of $x_1, x_2, \ldots, x_n$ is the number

$$\frac{x_1 + x_2 + \cdots + x_n}{n},$$

and the *geometric mean* of $x_1, x_2, \ldots, x_n$ is the number

$$(x_1 x_2 \cdots x_n)^{1/n}.$$

The arithmetic-mean–geometric-mean inequality states that

$$(x_1 x_2 \cdots x_n)^{1/n} \leqslant \frac{x_1 + x_2 + \cdots + x_n}{n} \,,$$

with equality if and only if all the x_i's are equal.

The special case $n = 2$ was verified both algebraically and geometrically in the beginning paragraphs of Section 7.1. A proof for larger values of n can be handled by mathematical induction (e.g., see 7.2.5 or 2.5.7), or by considering the concavity of the function $f(t) = \log t$ (see 7.4.20). However, a more enlightening heuristic (however, not a proof) can be made as follows.

Consider the geometric mean $(x_1 x_2 \cdots x_n)^{1/n}$ and the arithmetic mean $(x_1 + \cdots + x_n)/n$. If not all the x_i's are equal, replace the largest and the smallest of them, say x_M and x_m respectively, by $\frac{1}{2}(x_M + x_m)$. Then, since $\frac{1}{2}(x_M + x_m) + \frac{1}{2}(x_M + x_m) = x_M + x_m$, and $[\frac{1}{2}(x_M + x_m)]^2 > x_M x_m$, the result of this replacement is that the geometric mean has increased while the arithmetic mean has remained unchanged. If the new set of n numbers are not all equal, we can repeat the process as before. By repeating this process sufficiently often, we can make the quantities as nearly equal as we please (this step needs additional justification, but we won't worry about it here). At each stage of the process, the geometric mean is increased and the arithmetic mean is unchanged. If it should happen that all the numbers become equal (this may never happen, however; e.g., take $x_1 = 1$, $x_2 = 3$, $x_3 = 4$), the two means will coincide. It must be the case, therefore, that the geometric mean is less than or equal to the arithmetic mean, with equality when and only when all the numbers are equal.

As an example of this process, consider the case $x_1 = 2, x_2 = 4$, $x_3 = 8$, $x_4 = 12$. The algorithm described yields the following sequences of sets:

$$\{2, 4, 8, 12\} \rightarrow \{7, 4, 8, 7\} \rightarrow \{7, 6, 6, 7\} \rightarrow \{\tfrac{13}{2}, \tfrac{13}{2}, \tfrac{13}{2}, \tfrac{13}{2}\}.$$

The geometric means of the corresponding sets increase to $\frac{13}{2}$; the arithmetic means remain fixed at $\frac{13}{2}$.

7.2.1. Prove that the cube is the rectangular parallelepiped with maximum volume for a given surface area, and of minimum surface area for a given volume.

Solution. Let the lengths of the three adjacent sides be a, b, and c. Let A and V denote the surface area and volume respectively of the parallelepiped. Then

$$A = 2(ab + bc + ca) \quad \text{and} \quad V = abc.$$

By the arithmetic-mean–geometric-mean inequality,

$$V^2 = a^2b^2c^2 = (ab)(bc)(ca)$$

$$\leqslant \left(\frac{ab + bc + ca}{3}\right)^3 = \left(\frac{2(ab + bc + ca)}{6}\right)^3 = \left(\frac{A}{6}\right)^3.$$

Thus, for all a, b, c,

$$6V^{2/3} \leqslant A.$$

Furthermore, $6V^{2/3} < A$ in all cases except for when $ab = bc = ca$ (or equivalently, when $a = b = c$), and in this case $6V^{2/3} = A$. Thus, if A is fixed, we get the greatest volume (namely $V = (A/6)^{3/2}$) when $a = b = c$ (a cube), and when V is fixed, we get the least surface area (namely $A = 6V^{2/3}$) when $a = b = c$ (a cube).

7.2.2. Prove the following inequality:

$$n\left[(n + 1)^{1/n} - 1\right] < 1 + \frac{1}{2} + \frac{1}{3} + \cdots + \frac{1}{n} < n - (n - 1)n^{-1/(n-1)}.$$

Solution. Let $s_n = 1 + \frac{1}{2} + \cdots + 1/n$. The leftmost inequality is equivalent to proving

$$\frac{n + s_n}{n} > (n + 1)^{1/n},$$

which has vaguely the look of an arithmetic-mean–geometric-mean inequality. We can make the idea work in the following way:

$$\frac{n + s_n}{n} = \frac{n + (1 + 1/2 + \cdots + 1/n)}{n}$$

$$= \frac{(1 + 1) + (1 + 1/2) + \cdots + (1 + 1/n)}{n}$$

$$= \frac{2 + 3/2 + 4/3 + \cdots + (n + 1)/n}{n}$$

$$> \left(2 \cdot \frac{3}{2} \cdot \frac{4}{3} \cdots \frac{n + 1}{n}\right)^{1/n}$$

$$= (n + 1)^{1/n}.$$

For the rightmost inequality, we need to show that

$$\frac{n - s_n}{n - 1} > n^{-1/(n-1)}.$$

Again, using the arithmetic-mean–geometric-mean inequality, we have

$$
\frac{n - s_n}{n - 1} = \frac{n - (1 + 1/2 + 1/3 + \cdots + 1/n)}{n - 1}
$$

$$
= \frac{(1 - 1) + (1 - 1/2) + \cdots + (1 - 1/n)}{n - 1}
$$

$$
= \frac{1/2 + 2/3 + \cdots + (n - 1)/n}{n - 1}
$$

$$
> \left(\frac{1}{2} \cdot \frac{2}{3} \cdot \frac{3}{4} \cdots \frac{n - 1}{n} \right)^{1/(n-1)}
$$

$$
= \left(\frac{1}{n} \right)^{1/(n-1)} = n^{-1/(n-1)}.
$$

7.2.3. If a, b, c are positive numbers such that $(1 + a)(1 + b)(1 + c) = 8$, prove that $abc \leqslant 1$.

Solution. We are given that

$$
1 + (a + b + c) + (ab + bc + ca) + abc = 8.
$$

By the arithmetic-mean–geometric-mean inequality,

$$
a + b + c \geqslant 3(abc)^{1/3} \quad \text{and} \quad ab + bc + ca \geqslant 3(abc)^{2/3},
$$

each with equality if and only if $a = b = c$. Thus,

$$
8 \geqslant 1 + 3(abc)^{1/3} + 3(abc)^{2/3} + abc
$$

$$
= \left[1 + (abc)^{1/3} \right]^3.
$$

It follows that

$$
(abc)^{1/3} \leqslant (2 - 1) = 1,
$$

or equivalently,

$$
abc \leqslant 1
$$

with equality if and only if $a = b = c = 1$.

7.2.4. Suppose that $x_i > 0$, $i = 1, 2, \ldots, n$ and let $x_{n+1} = x_1$. Show that

$$
\sum_{i=1}^{n} \left(\frac{x_{i+1}}{x_i} \right) \leqslant \sum_{i=1}^{n} \left(\frac{x_i}{x_{i+1}} \right)^n.
$$

Solution. Consider the case $n = 3$. By the arithmetic-mean–geometric-mean inequality, we have

$$\frac{x_2}{x_1} = \frac{x_2}{x_3} \cdot \frac{x_3}{x_1} \cdot 1 \;\leqslant\; \frac{1}{3}\left(\frac{x_2}{x_3}\right)^3 + \frac{1}{3}\left(\frac{x_3}{x_1}\right)^3 + \frac{1}{3},$$

$$\frac{x_3}{x_2} = \frac{x_1}{x_2} \cdot \frac{x_3}{x_1} \cdot 1 \;\leqslant\; \frac{1}{3}\left(\frac{x_1}{x_2}\right)^3 + \;\;\;\;\;\; + \frac{1}{3}\left(\frac{x_3}{x_1}\right)^3 + \frac{1}{3},$$

$$\frac{x_1}{x_3} = \frac{x_1}{x_2} \cdot \frac{x_2}{x_3} \cdot 1 \;\leqslant\; \frac{1}{3}\left(\frac{x_1}{x_2}\right)^3 + \frac{1}{3}\left(\frac{x_2}{x_3}\right)^3 + \;\;\;\;\;\; + \frac{1}{3}.$$

Also,

$$1 = \frac{x_1}{x_2} \cdot \frac{x_2}{x_3} \cdot \frac{x_3}{x_1} \;\leqslant\; \frac{1}{3}\left(\frac{x_1}{x_2}\right)^3 + \frac{1}{3}\left(\frac{x_2}{x_3}\right)^3 + \frac{1}{3}\left(\frac{x_3}{x_1}\right)^3.$$

Adding these inequalities gives the desired result. The case for an arbitrary positive integer n is similar.

Problems

7.2.5. Fill in the steps of the following inductive proof of the arithmetic-mean–geometric-mean inequality: For each k, let $A_k = (x_1 + x_2 + \cdots + x_k)/k$, and $G_k = (x_1 x_2 \cdots x_k)^{1/k}$. Assume that we have shown $A_k \geqslant G_k$. Let

$$A = \frac{x_{k+1} + (k-1)A_{k+1}}{k} \quad \text{and} \quad G = (x_{k+1}A_{k+1}^{k-1})^{1/k}.$$

Then, using the inductive assumption, we have $A \geqslant G$, and it follows that $A_{k+1} = \frac{1}{2}(A_k + A) \geqslant (A_k A)^{1/2} \geqslant (G_k G)^{1/2} = (G_{k+1}^{k+1}A_{k+1}^{k-1})^{1/(2k)}$. From this it follows that $A_{k+1} \geqslant G_{k+1}$. On the basis of this argument it is easy to prove the equality holds if and only if all the x_i's are equal.

7.2.6. If a, b, c are positive numbers, prove that

$$(a^2b + b^2c + c^2a)(a^2c + b^2a + c^2b) \geqslant 9a^2b^2c^2.$$

7.2.7. Suppose that $a_1, \ldots, a_n$ are positive numbers and $b_1, \ldots, b_n$ is a rearrangement of $a_1, \ldots, a_n$. Show that

$$\frac{a_1}{b_1} + \frac{a_2}{b_2} + \cdots + \frac{a_n}{b_n} \geqslant n.$$

7.2.8.

(a) For positive numbers a and b, $a \neq b$, prove that

$$(ab^n)^{1/(n+1)} < \frac{a + nb}{n + 1}.$$

(b) In part (a), consider the case $a = 1$ and $b = 1 + 1/n$ and show that

$$\left(1 + \frac{1}{n}\right)^n < \left(1 + \frac{1}{n+1}\right)^{n+1}.$$

(c) In part (a), replace n by $n + 1$, let $a = 1$ and $b = n/(n + 1)$, and show that

$$\left(1 + \frac{1}{n}\right)^{n+1} > \left(1 + \frac{1}{n+1}\right)^{n+2}.$$

7.2.9. For each integer $n > 2$, prove that

(a) $\displaystyle\prod_{k=0}^{n} \binom{n}{k} < \left(\frac{2^n - 2}{n - 1}\right)^{n-1}$,

(b) $n! < \left(\frac{n+1}{2}\right)^n$,

(c) $1 \times 3 \times 5 \times \cdots \times (2n - 1) < n^n$.

7.2.10. Given that all roots of $x^6 - 6x^5 + ax^4 + bx^3 + cx^2 + dx + 1 = 0$ are positive, find a, b, c, d.

7.2.11.

(a) Let $x_i > 0$ for $i = 1, 2, \ldots, n$, and let $p_1, p_2, \ldots, p_n$ be positive integers. Prove that

$$\left(x_1^{p_1} x_2^{p_2} \cdots x_n^{p_n}\right)^{1/(p_1 + \cdots + p_n)} \leqslant \frac{p_1 x_1 + \cdots + p_n x_n}{p_1 + \cdots + p_n}.$$

(b) Prove the same result as in part (a) holds even when the p_i's are positive rational numbers.

7.2.12. Use the arithmetic-mean–geometric-mean inequality for each of the following:

(a) A tank with a rectangular base and rectangular sides is to be open at the top. It is to be constructed so that its width is 4 meters and its volume is 36 cubic meters. If building the tank costs $10 per square meter for the base and $5 per square meter for the sides, what is the cost of the least expensive tank?

(b) A farmer with a field adjacent to a straight river wishes to fence a rectangular region for grazing. If no fence is needed along the river, and he has 1000 feet of fencing, what should be the dimensions of the field so that it has a maximum area? (Hint: It is equivalent to maximize twice the area.)

(c) A farmer with 1000 feet of fencing wishes to construct a rectangular pen and to divide it into two smaller rectangular plots by adding a common fence down the middle. What should the overall dimensions of the pen be in order to maximize the total area?

(d) Prove that the square is the rectangle of maximum area for a given perimeter, and of minimum perimeter for a given area.

(e) Prove that the equilateral triangle is the triangle of maximum area for a given perimeter, and of minimum perimeter for a given area. (Hint: The area of a triangle is related to the perimeter of the triangle by the formula $A = (s(s - a)(s - b)(s - c))^{1/2}$, where a, b, c are the lengths of the sides of the triangle and $s = \frac{1}{2} P$, P the perimeter of the triangle.)

Additional Examples

Introduction to Section 7.6; 7.3.1, 8.1.4.

7.3. Cauchy–Schwarz Inequality

Let $a_i > 0$ and $b_i > 0$ for $i = 1, 2, \ldots, n$. The Cauchy–Schwarz inequality states that

$$\sum_{i=1}^{n} a_i b_i \leq \left(\sum_{i=1}^{n} a_i^2 \right)^{1/2} \left(\sum_{i=1}^{n} b_i^2 \right)^{1/2},$$

with equality if and only if $a_1/b_1 = a_2/b_2 = \cdots = a_n/b_n$.

A proof can be given using mathematical induction (see 7.1.13). But an easier approach is to consider the quadratic polynomial $P(x) = \sum_{i=1}^{n} (a_i x - b_i)^2$. Observe that $P(x) \geq 0$ for all x; in fact, $P(x) = 0$ only under the conditions in which $a_1/b_1 = a_2/b_2 = \cdots = a_n/b_n$ and $x = b_i/a_i$. Now

$$P(x) = \sum_{i=1}^{n} (a_i^2 x^2 - 2a_i b_i x + b_i^2)$$

$$= \left(\sum_{i=1}^{n} a_i^2 \right) x^2 - 2 \left(\sum_{i=1}^{n} a_i b_i \right) x + \sum_{i=1}^{n} b_i^2,$$

and since $P(x) \geq 0$, the discriminant of P cannot be positive, and in fact will equal zero only when $P(x) = 0$. Thus,

$$\left(-2 \sum_{i=1}^{n} a_i b_i \right)^2 - 4 \left(\sum_{i=1}^{n} a_i^2 \right) \left(\sum_{i=1}^{n} b_i^2 \right) \leq 0,$$

or equivalently,

$$\sum_{i=1}^{n} a_i b_i \leq \left(\sum_{i=1}^{n} a_i^2 \right)^{1/2} \left(\sum_{i=1}^{n} b_i^2 \right)^{1/2},$$

with equality if and only if $a_1/b_1 = \cdots = a_n/b_n$.

In this inequality, note that the requirement that the a_i and b_i be positive is redundant, since for all a_i, b_i

$$\sum_{i=1}^{n} a_i b_i \leqslant \sum_{i=1}^{n} |a_i| |b_i| \leqslant \left(\sum_{i=1}^{n} a_i^2 \right)^{1/2} \left(\sum_{i=1}^{n} b_i^2 \right)^{1/2}.$$

7.3.1. If $a, b, c > 0$, is it true that $a\cos^2\theta + b\sin^2\theta < c$ implies $\sqrt{a}\cos^2\theta + \sqrt{b}\sin^2\theta < \sqrt{c}$?

Solution. By the Cauchy–Schwarz inequality

$$\sqrt{a}\cos^2\theta + \sqrt{b}\sin^2\theta$$
$$\leqslant [(\sqrt{a}\cos\theta)^2 + (\sqrt{b}\sin\theta)^2]^{1/2}[(\cos\theta)^2 + (\sin\theta)^2]^{1/2}$$
$$= (a\cos^2\theta + b\sin^2\theta)^{1/2}$$
$$< \sqrt{c}.$$

There is also a nice solution based on the arithmetic-mean–geometric-mean inequality:

$$\left(\sqrt{a}\cos^2\theta + \sqrt{b}\sin^2\theta\right)^2 = a\cos^4\theta + 2\sqrt{a}\sqrt{b}\cos^2\theta\sin^2\theta + b\sin^4\theta$$
$$\leqslant a\cos^4\theta + (a+b)\cos^2\theta\sin^2\theta + b\sin^4\theta$$
$$= (a\cos^2\theta + b\sin^2\theta)(\cos^2\theta + \sin^2\theta)$$
$$< c.$$

Another solution, more geometric in nature, is given in 7.4.19.

7.3.2. Let P be a point in the interior of triangle ABC, and let r_1, r_2, r_3 denote the distances from P to the sides a_1, a_2, a_3 of the triangle respectively. Let R denote the circumradius of ABC. Show that

$$\sqrt{r_1} + \sqrt{r_2} + \sqrt{r_3} \leqslant \frac{1}{\sqrt{2R}} (a_1^2 + a_2^2 + a_3^2)^{1/2}$$

with equality if and only if ABC is equilateral and P is the incenter.

Solution. By the Cauchy–Schwarz inequality

$$\sqrt{r_1} + \sqrt{r_2} + \sqrt{r_3} = \sqrt{a_1 r_1}\sqrt{1/a_1} + \sqrt{a_2 r_2}\sqrt{1/a_2} + \sqrt{a_3 r_3}\sqrt{1/a_3}$$
$$\leqslant (a_1 r_1 + a_2 r_2 + a_3 r_3)^{1/2}\left(\frac{1}{a_1} + \frac{1}{a_2} + \frac{1}{a_3}\right)^{1/2}$$

with equality if and only if

$$\frac{\sqrt{a_1 r_1}}{\sqrt{1/a_1}} = \frac{\sqrt{a_2 r_2}}{\sqrt{1/a_2}} = \frac{\sqrt{a_3 r_3}}{\sqrt{1/a_3}},$$

or equivalently, if and only if

$$a_1^2 r_1 = a_2^2 r_2 = a_3^2 r_3.$$

In the preceding inequality, we recognize that $a_1 r_1 + a_2 r_2 + a_3 r_3 = 2A$, where A is the area of the triangle. Also, we know that the area of a triangle, in terms of the circumradius R, is given by $A = a_1 a_2 a_3 / 4R$ (see 8.1.12). Therefore, $a_1 r_1 + a_2 r_2 + a_3 r_3 = a_1 a_2 a_3 / 2R$, and we have

$$\sqrt{r_1} + \sqrt{r_2} + \sqrt{r_3} \leqslant \left(\frac{a_1 a_2 a_3}{2R}\right)^{1/2} \left(\frac{1}{a_1} + \frac{1}{a_2} + \frac{1}{a_3}\right)^{1/2}$$

$$= \left(\frac{a_1 a_2 a_3}{2R}\right)^{1/2} \left(\frac{a_2 a_3 + a_3 a_1 + a_1 a_2}{a_1 a_2 a_3}\right)^{1/2}$$

$$= \frac{1}{\sqrt{2R}} (a_2 a_3 + a_3 a_1 + a_1 a_2)^{1/2}.$$

Now, again by the Cauchy–Schwarz inequality,

$$a_2 a_3 + a_3 a_1 + a_1 a_2 \leqslant \left(a_2^2 + a_3^2 + a_1^2\right)^{1/2} \left(a_3^2 + a_1^2 + a_2^2\right)^{1/2}$$

$$= \left(a_1^2 + a_2^2 + a_3^2\right)$$

with equality if and only if $a_2/a_3 = a_3/a_1 = a_1/a_2$ $(= (a_2 + a_3 + a_1)/(a_3 + a_1 + a_2) = 1$; see 7.1.10), or equivalently, if and only if

$$a_1 = a_2 = a_3.$$

Thus, we have

$$\sqrt{r_1} + \sqrt{r_2} + \sqrt{r_3} \leqslant \frac{1}{\sqrt{2R}} \left(a_1^2 + a_2^2 + a_3^2\right)^{1/2}$$

with equality if and only if $a_1^2 r_1 = a_2^2 r_2 = a_3^2 r_3$ and $a_1 = a_2 = a_3$; that is, if and only if $a_1 = a_2 = a_3$ and $r_1 = r_2 = r_3$. This completes the proof.

7.3.3. Given that a, b, c, d, e are real numbers such that

$$a + b + c + d + e = 8,$$
$$a^2 + b^2 + c^2 + d^2 + e^2 = 16,$$

determine the maximum value of e.

Solution. The given equations can be put into the form

$$8 - e = a + b + c + d,$$
$$16 - e^2 = a^2 + b^2 + c^2 + d^2.$$

We wish to find an inequality involving only e; the Cauchy–Schwarz inequality provides a way, since

$$(a + b + c + d) \leqslant (1 + 1 + 1 + 1)^{1/2}(a^2 + b^2 + c^2 + d^2)^{1/2}.$$

Making the substitutions given above, and squaring, we have

$$(8 - e)^2 \leqslant 4(16 - e^2),$$
$$64 - 16e + e^2 \leqslant 64 - 4e^2,$$
$$5e^2 - 16e \leqslant 0,$$
$$e(5e - 16) \leqslant 0.$$

It follows that $0 \leqslant e \leqslant \frac{16}{5}$. The upper bound, $\frac{16}{5}$, is attained when $a = b = c = d = \frac{6}{5}$.

7.3.4. Suppose that $a_1, a_2, \ldots, a_n$ are real ($n > 1$) and

$$A + \sum_{i=1}^{n} a_i^2 < \frac{1}{n-1}\left(\sum_{i=1}^{n} a_i\right)^2.$$

Prove that $A < 2a_i a_j$ for $1 \leqslant i < j \leqslant n$.

Solution. By the Cauchy–Schwarz inequality

$$\left(\sum_{i=1}^{n} a_i\right)^2 = \left[(a_1 + a_2) + a_3 + \cdots + a_n\right]^2$$

$$\leqslant (1 + \cdots + 1)\left((a_1 + a_2)^2 + a_3^2 + \cdots + a_n^2\right)$$

$$= (n - 1)\left[\sum_{i=1}^{n} a_i^2 + 2a_1 a_2\right].$$

This, together with the given inequality, implies that

$$A < -\left(\sum_{i=1}^{n} a_i^2\right) + \frac{1}{n-1}\left(\sum_{i=1}^{n} a_i\right)^2$$

$$< -\left(\sum_{i=1}^{n} a_i^2\right) + \frac{1}{n-1}\left[(n-1)\left[\sum_{i=1}^{n} a_i^2 + 2a_1 a_2\right]\right]$$

$$= 2a_1 a_2.$$

In a similar manner, $A < 2a_i a_j$ for $1 \leqslant i < j \leqslant n$.

7.3.5. Let $x_i > 0$ for $i = 1, 2, \ldots, n$. For each nonnegative integer k, prove that

$$\frac{x_1^k + \cdots + x_n^k}{n} \leqslant \frac{x_1^{k+1} + \cdots + x_n^{k+1}}{x_1 + \cdots + x_n}.$$

Solution. We may assume without loss of generality that $x_1 + \cdots + x_n$ $= 1$, for if not, we can replace x_i by $X_i = x_i/(x_1 + \cdots + x_n)$.

The result holds when $k = 0$. Assume the result holds for all nonnegative integers less than k. By the Cauchy–Schwarz inequality,

$$\sum_{i=1}^{n} \frac{x_i^k}{n} = \sum_{i=1}^{n} x_i^{(k+1)/2} \frac{x_i^{(k-1)/2}}{n}$$

$$\leqslant \left(\sum_{i=1}^{n} x_i^{k+1} \right)^{1/2} \left(\sum_{i=1}^{n} \frac{x_i^{k-1}}{n^2} \right)^{1/2}.$$

By the inductive assumption, $\sum_{i=1}^{n} x_i^{k-1}/n \leqslant \sum_{i=1}^{n} x_i^k$, and therefore, continuing from the last inequality, we have

$$\left(\sum_{i=1}^{n} x_i^{k+1} \right)^{1/2} \left(\sum_{i=1}^{n} \frac{x_i^{k-1}}{n^2} \right)^{1/2} \leqslant \left(\sum_{i=1}^{n} x_i^{k+1} \right)^{1/2} \left(\sum_{i=1}^{n} \frac{x_i^k}{n} \right)^{1/2}.$$

Thus,

$$\sum_{i=1}^{n} \frac{x_i^k}{n} \leqslant \left(\sum_{i=1}^{n} x_i^{k+1} \right)^{1/2} \left(\sum_{i=1}^{n} \frac{x_i^k}{n} \right)^{1/2},$$

$$\left(\sum_{i=1}^{n} \frac{x_i^k}{n} \right)^{1/2} \leqslant \left(\sum_{i=1}^{n} x_i^{k+1} \right)^{1/2},$$

$$\sum_{i=1}^{n} \frac{x_i^k}{n} \leqslant \sum_{i=1}^{n} x_i^{k+1}.$$

By induction, the proof is complete.

Problems

7.3.6. Use the Cauchy–Schwarz inequality to prove that if $a_1, \ldots, a_n$ are real numbers such that $a_1 + \cdots + a_n = 1$, then $a_1^2 + \cdots + a_n^2 \geqslant 1/n$.

7.3.7. Use the Cauchy–Schwarz inequality to prove the following:

(a) If $p_1, \ldots, p_n, x_1, \ldots, x_n$ are $2n$ positive numbers,

$$(p_1 x_1 + \cdots + p_n x_n)^2 \leqslant (p_1 + \cdots + p_n)(p_1 x_1^2 + \cdots + p_n x_n^2).$$

(b) If a, b, c are positive numbers,

$$(a^2 b + b^2 c + c^2 a)(ab^2 + bc^2 + ca^2) \geqslant 9a^2 b^2 c^2.$$

(c) If $x_k, y_k, k = 1, 2, \ldots, n$, are positive numbers,

$$\sum_{k=1}^{n} x_k y_k \leqslant \left(\sum_{k=1}^{n} k x_k^2 \right)^{1/2} \left(\sum_{k=1}^{n} y_k^2/k \right)^{1/2}.$$

(d) If a_k, b_k, c_k, $k = 1, 2, \ldots, n$, are positive numbers,

$$\left(\sum_{k=1}^{n} a_k b_k c_k \right)^4 \leqslant \left(\sum_{k=1}^{n} a_k^4 \right) \left(\sum_{k=1}^{n} b_k^4 \right) \left(\sum_{k=1}^{n} c_k^2 \right)^2 .$$

(e) If $C_k = \binom{n}{k}$ for $n > 2$, $1 \leqslant k \leqslant n$,

$$\sum_{k=1}^{n} \sqrt{C_k} \leqslant \sqrt{n(2^n - 1)} \ .$$

7.3.8. For n a positive integer, let $(a_1, a_2, \ldots, a_n)$ and $(b_1, b_2, \ldots, b_n)$ be two (not necessarily distinct) permutations of $(1, 2, \ldots, n)$. Find sharp lower and upper bounds for $a_1 b_1 + \cdots + a_n b_n$.

7.3.9. If a, b, c, d are positive numbers such that $c^2 + d^2 = (a^2 + b^2)^3$, prove that

$$\frac{a^3}{c} + \frac{b^3}{d} \geqslant 1,$$

with equality if and only if $ad = bc$. (Hint: Show that $(a^3/c + b^3/d)(ac + bd) \geqslant (a^2 + b^2)^2 \geqslant ac + bd$.)

7.3.10. Let P be a point in the interior of triangle ABC, and let r_1, r_2, r_3 denote the distances from P to the sides a_1, a_2, a_3 of the triangle respectively. Use the Cauchy–Schwarz inequality to show that the minimum value of

$$\frac{a_1}{r_1} + \frac{a_2}{r_2} + \frac{a_3}{r_3}$$

occurs when P is at the incenter of triangle ABC. (Hint: $a_i = \sqrt{a_i r_i} \sqrt{a_i / r_i}$.)

Additional Example

7.6.14.

7.4. Functional Considerations

In this section we will give examples to show how the techniques of analysis, particularly differentiation, can be used effectively on a wide variety of inequality problems.

7.4.1. Given positive numbers p, q, and r, such that $2p = q + r$, $q \neq r$, show that

$$\frac{p^{q+r}}{q^q r^r} < 1.$$

Solution. Suppose that q and r are positive integers, and consider the q numbers $1/q, \ldots, 1/q$ and the r numbers $1/r, \ldots, 1/r$. By the arithmetic-mean–geometric-mean inequality,

$$\left(\frac{1}{q^q} \cdot \frac{1}{r^r}\right)^{1/(q+r)} < \frac{q(1/q) + r(1/r)}{q+r} = \frac{1}{p},$$

which is equivalent to the desired inequality.

Of course, this method breaks down if either q or r is not an integer, so how shall we proceed? One idea is to rewrite the inequality in the following manner:

$$p^{q+r} < q^q r^r,$$

$$\left(\frac{q+r}{2}\right)^{q+r} < q^q r^r,$$

$$\left(\frac{1}{2}\right)^{q+r} < \left(\frac{q}{q+r}\right)^q \left(\frac{r}{q+r}\right)^r,$$

$$\frac{1}{2} < \left(\frac{q}{q+r}\right)^{q/(q+r)} \left(\frac{r}{q+r}\right)^{r/(q+r)}.$$

Set $x = q/(q+r)$ and $y = r/(q+r)$. Observe that $x + y = 1$ and $0 < x, y < 1$. Then the problem is equivalent to proving that

$$F(x) \equiv x^x (1 - x)^{1-x} > \tfrac{1}{2}, \qquad 0 < x < 1, \quad x \neq \tfrac{1}{2}.$$

By introducing the function in this way, we are able to use the methods of analysis. The idea is to find the minimum value of F on $(0, 1)$. To simplify the differentiation, we will consider the function $G(x) = \log F(x)$. To find the critical points, we differentiate:

$$G'(x) = \frac{d}{dx} \left[x \log x + (1 - x) \log(1 - x) \right]$$

$$= (\log x + 1) - 1 - \log(1 - x)$$

$$= \log \frac{x}{1 - x}.$$

We see that $G'(x) = 0$ if and only if $x = \tfrac{1}{2}$. Furthermore, $G'(x) < 0$ on the interval $(0, \tfrac{1}{2})$, and $G'(x) > 0$ on the interval $(\tfrac{1}{2}, 1)$. Therefore $G(x)$ takes its minimum value on $(0, 1)$ at $x = \tfrac{1}{2}$. Thus, the minimum value of $F(x)$ on $(0, 1)$ is $F(\tfrac{1}{2}) = (\tfrac{1}{2})^{1/2}(\tfrac{1}{2})^{1/2} = \tfrac{1}{2}$. It follows that $F(x) > \tfrac{1}{2}$ for all x in $(0, 1)$, $x \neq \tfrac{1}{2}$, and the proof is complete.

7.4.2. Let p and q be positive numbers with $p + q = 1$. Show that for all x,

$$p e^{x/p} + q e^{-x/q} \leq e^{x^2/8p^2q^2}.$$

Solution. Consider the function

$$F(x) = \frac{pe^{x/p} + qe^{-x/q}}{e^{x^2/8p^2q^2}}.$$

Our problem is to prove that $F(x) \leqslant 1$ for all x. Because of the symmetry in the problem, it suffices to prove that $F(x) \leqslant 1$ for all $x \geqslant 0$.

We note that $F(0) = 1$. By Corollary (iii) of the mean-value theorem (see the discussion preceding 6.6.2), it suffices to prove that $F'(x) \leqslant 0$ for all x. To simplify the computation, consider the function $G(x) = \log F(x)$. Routine differentiation and algebraic simplification yields

$$G'(x) = \frac{F'(x)}{F(x)} = \frac{e^{x/p} - e^{-x/q}}{pe^{x/p} + qe^{-x/q}} - \frac{x}{4p^2q^2}$$

$$= \frac{e^{x/pq} - 1}{pe^{x/pq} + q} - \frac{x}{4p^2q^2}.$$

Since $F(x) > 0$ for all $x \geqslant 0$, $F'(x) \leqslant 0$ if and only if $G'(x) \leqslant 0$. Unfortunately, the preceding expression for $G'(x)$ makes it difficult to determine whether or not $G'(x) \leqslant 0$. Therefore, we will carry the analysis through another step. Namely, $G'(0) = 0$, and (again leaving out the details)

$$G''(x) = -\frac{(pe^{x/pq} - q)^2}{4p^2q^2(pe^{x/pq} + q)^2}.$$

Here it is clear that $G''(x) \leqslant 0$ for all $x \geqslant 0$. This, together with $G'(0) = 0$, implies that $G'(x) \leqslant 0$ for all $x \geqslant 0$, and this in turn implies $F'(x) \leqslant 0$ for all $x \geqslant 0$. Therefore, since $F(0) = 1$, it must be the case that $F(x) \leqslant 1$ for all $x \geqslant 0$, and the proof is complete.

The procedure used in the preceding problem is very common. To recapitulate, it goes like this: To prove an inequality of the form

$$f(x) \geqslant g(x), \qquad x \geqslant a,$$

it is equivalent to prove either that

$$Q(x) \equiv \frac{f(x)}{g(x)} \geqslant 1, \qquad x \geqslant a,$$

or that

$$D(x) \equiv f(x) - g(x) \geqslant 0, \qquad x \geqslant a.$$

Each can be done by showing the inequality for $x = a$ and then by showing that $Q'(x) \geqslant 0$ (or $D'(x) \geqslant 0$ respectively) for all $x \geqslant a$.

In the previous example, if we had considered instead the function

$$D(x) = e^{x^2/8p^2q^2} - pe^{x/p} - qe^{-x/q},$$

this analysis wouldn't have been conclusive: even though $D(0) = 0$, it is not necessarily the case that $D'(x) \geqslant 0$ (for example, when $p = \frac{1}{3}, q = \frac{2}{3}, x = \frac{1}{2}$).

7.4.3. Prove that for all real numbers a and b,
$$|a + b|^p \leqslant |a|^p + |b|^p, \qquad 0 \leqslant p \leqslant 1.$$

Solution. The inequality is trivial in several special cases. For example, the result holds if $a = 0$, or if a and b have opposite signs. Also, if $p = 0$ or $p = 1$, the result is true. Therefore, it suffices to show the result is true when a and b are positive and $0 < p < 1$.

For such a and b and p, let $x = b/a$. Then, the problem is to show that
$$(1 + x)^p \leqslant 1 + x^p, \qquad x > 0, \quad 0 < p < 1.$$

For this, let $D(x) = 1 + x^p - (1 + x)^p$. We have $D(0) = 0$ and $D'(x) = px^{p-1} - p(1 + x)^{p-1} > 0$, so by our earlier remarks, the proof is complete. (Note that if $p > 1$, the inequalities would be reversed.)

7.4.4. On $[0, 1]$, let f have a continuous derivative satisfying $0 < f'(t) \leqslant 1$. Also, suppose that $f(0) = 0$. Prove that
$$\left[\int_0^1 f(t)\, dt \right]^2 \geqslant \int_0^1 [f(t)]^3\, dt.$$

Solution. Here, as in the last example, it is not clear how to make use of differentiation. The idea is to introduce a variable and prove a more general result. For $0 \leqslant x \leqslant 1$, let
$$F(x) \equiv \left[\int_0^x f(t)\, dt \right]^2 - \int_0^x (f(t))^3\, dt.$$
Then $F(0) = 0$, and
$$F'(x) = 2\left[\int_0^x f(t)\, dt \right] f(x) - [f(x)]^3$$
$$= f(x)\left[2\int_0^x f(t)\, dt - [f(x)]^2 \right].$$
We do know that $f(x) \geqslant 0$ for $0 < x < 1$ (since we are given $f(0) = 0$ and $f'(x) > 0$); however, it is not clear that the second factor in the last expression for F' is nonnegative. Therefore, let
$$G(x) = 2\int_0^x f(t)\, dt - [f(x)]^2, \qquad 0 \leqslant x \leqslant 1.$$
Then $G(0) = 0$, and
$$G'(x) = 2f(x) - 2f(x)f'(x)$$
$$= 2f(x)[1 - f'(x)] \geqslant 0$$
(the last inequality holds because $f(x) \geqslant 0$ and, by hypothesis, $1 - f'(x) \geqslant 0$).

It follows from these arguments that $F(x) \geqslant 0$ for all x, $0 \leqslant x \leqslant 1$; in particular, $F(1) \geqslant 0$ and the proof is complete.

7.4.5. Show that if x is positive, then $\log(1 + 1/x) > 1/(1 + x)$.

Solution. Let $f(x) = \log(1 + 1/x) - 1/(1 + x)$ $(= \log(1 + x) - \log x - 1/(1 + x))$. Then

$$f'(x) = \frac{1}{1 + x} - \frac{1}{x} + \frac{1}{(1 + x)^2}$$

$$= \frac{x(1 + x) - (1 + x)^2 + x}{x(1 + x)^2}$$

$$= \frac{-1}{x(1 + x)^2} < 0 \qquad \text{for} \quad x > 0.$$

Furthermore, $\lim_{x \to \infty} f(x) = 0$, and this, together with $f'(x) < 0$ for $x > 0$, implies that $f(x) > 0$ for $x > 0$.

7.4.6. Find all positive integers n such that

$$3^n + 4^n + \cdots + (n + 2)^n = (n + 3)^n.$$

Solution. A direct calculation shows that we get equality when $n = 2$ and when $n = 3$. A parity argument shows that it can't hold when either $n = 4$ or $n = 5$. Based on this beginning, we might expect that the key insight should involve modular arithmetic in some way. However, these attempts aren't fruitful, and we look for another approach. We will show that

$$3^n + 4^n + \cdots + (n + 2)^n < (n + 3)^n$$

for $n \geqslant 6$, and thus equality holds only when $n = 2$ or $n = 3$.

The inequality we wish to prove can be written in the following form:

$$\left(\frac{3}{n + 3}\right)^n + \left(\frac{4}{n + 3}\right)^n + \cdots + \left(\frac{n + 2}{n + 3}\right)^n < 1,$$

$$\left(1 - \frac{n}{n + 3}\right)^n + \left(1 - \frac{n - 1}{n + 3}\right)^n + \cdots + \left(1 - \frac{1}{n + 3}\right)^n < 1,$$

or, reversing the order for convenience,

$$\left(1 - \frac{1}{n + 3}\right)^n + \left(1 - \frac{2}{n + 3}\right)^n + \cdots + \left(1 - \frac{n}{n + 3}\right)^n < 1.$$

To prove this inequality it suffices to show that

$$\left(1 - \frac{k}{n + 3}\right)^n < \left(\frac{1}{2}\right)^k, \qquad k = 1, 2, \ldots, n.$$

For then,

$$\left(1 - \frac{1}{n+3}\right)^n + \left(1 - \frac{2}{n+3}\right)^n + \cdots + \left(1 - \frac{n}{n+3}\right)^n$$

$$< \frac{1}{2} + \left(\frac{1}{2}\right)^2 + \cdots + \left(\frac{1}{2}\right)^n < 1.$$

It remains, then, to prove that

$$\left(1 - \frac{k}{n+3}\right)^n < \left(\frac{1}{2}\right)^k, \qquad k = 1, 2, \ldots, n.$$

By Bernoulli's inequality (a very useful inequality; see 7.4.10),

$$\left(1 - \frac{1}{n+3}\right)^k \ge \left(1 - \frac{k}{n+3}\right),$$

and therefore,

$$\left(1 - \frac{k}{n+3}\right)^n \le \left(1 - \frac{1}{n+3}\right)^{kn} = \left[\left(1 - \frac{1}{n+3}\right)^n\right]^k.$$

The final step is to show that

$$\left(1 - \frac{1}{n+3}\right)^n \le \frac{1}{2} \quad \text{when } n \ge 6.$$

For this, consider the function

$$F(x) = \left(1 - \frac{1}{x+3}\right)^x.$$

It is straightforward to show that $F'(x) < 0$ for $x \ge 6$, and that $F(6) < \frac{1}{2}$. Thus, the proof is complete.

7.4.7. Prove that for $0 \le a < b < \frac{1}{2}\pi$,

$$\frac{b - a}{\cos^2 a} < \tan b - \tan a < \frac{b - a}{\cos^2 b}.$$

Solution. Consider the function $f(x) = \tan x$ on $[a, b]$. According to the mean-value theorem there is a point c in (a, b) such that

$$\frac{f(b) - f(a)}{b - a} = f'(c).$$

In this case, this means that

$$\frac{\tan b - \tan a}{b - a} = \sec^2 c$$

for some c in (a, b). The desired inequality follows from the fact that $\sec^2 a < \sec^2 c < \sec^2 b$ for $0 \le a < b < \pi/2$.

Many inequalities can be established by considering an appropriate convex (or concave) function. The idea is based on the result of 6.6.3: if

$f: R \rightarrow R$ is such that $f''(x) \geqslant 0$, then

$$f\left(\frac{x+y}{2}\right) \leqslant \frac{f(x)+f(y)}{2},$$

and if $f''(x) \leqslant 0$, then

$$f\left(\frac{x+y}{2}\right) \geqslant \frac{f(x)+f(y)}{2}.$$

For example, for real numbers x and y,

$$\left(\frac{x+y}{2}\right)^2 \leqslant \frac{x^2+y^2}{2}$$

because $f(x) = x^2$ is a convex function. As another example, if $0 < x, y < \pi$,

$$\sin\left(\frac{x+y}{2}\right) \geqslant \frac{\sin x + \sin y}{2}$$

because $f(x) = \sin x$ is a concave function on $(0, \pi)$.

7.4.8. Prove that if a and b are positive numbers such that $a + b = 1$, then

$$\left(a + \frac{1}{a}\right)^2 + \left(b + \frac{1}{b}\right)^2 \geqslant \frac{25}{2}.$$

Solution. We have seen that

$$\frac{x^2+y^2}{2} \geqslant \left(\frac{x+y}{2}\right)^2.$$

Take $x = a + 1/a$ and $y = b + 1/b$. Then

$$\frac{1}{2}\left[\left(a+\frac{1}{a}\right)^2 + \left(b+\frac{1}{b}\right)^2\right] \geqslant \left\{\frac{1}{2}\left[\left(a+\frac{1}{a}\right) + \left(b+\frac{1}{b}\right)\right]\right\}^2$$

$$= \left[\frac{1}{2}\left(1+\frac{1}{a}+\frac{1}{b}\right)\right]^2.$$

But by the Cauchy–Schwarz inequality $(1/a + 1/b)(a + b) \geqslant (1 + 1)^2 = 4$, so that

$$\left[\frac{1}{2}\left(1+\frac{1}{a}+\frac{1}{b}\right)\right]^2 \geqslant \left[\frac{1}{2}\left(1+\frac{4}{a+b}\right)\right]^2 = \left(\frac{1+4}{2}\right)^2 = \frac{25}{4}.$$

The result follows after putting together the two preceding inequalities and multiplying each side by 2.

7.4.9. Let $0 < x_i < \pi$, $i = 1, \ldots, n$, and set $x = (x_1 + x_2 + \cdots + x_n)/n$. Prove that

$$\prod_{i=1}^{n}\left(\frac{\sin x_i}{x_i}\right) \leqslant \left(\frac{\sin x}{x}\right)^n.$$

Solution. The problem is equivalent to proving that

$$\sum_{i=1}^{n} \log \frac{\sin x_i}{x_i} \leqslant n \log \frac{\sin x}{x}.$$

Consider the function

$$f(t) = \log \frac{\sin t}{t}.$$

It is a straightforward matter to show that f is concave ($f''(t) < 0$) on the interval $(0, \pi)$. Therefore

$$f\left(\frac{x_1 + x_2}{2}\right) \geqslant \frac{f(x_1) + f(x_2)}{2}.$$

In a manner completely analogous to the proof of 7.1.6, it follows that

$$f\left(\frac{x_1 + \cdots + x_n}{n}\right) \geqslant \frac{f(x_1) + \cdots + f(x_n)}{n}.$$

Direct substitition into this inequality completes the proof:

$$\log\left(\frac{\sin x}{x}\right) \geqslant \frac{1}{n}\left(\log \frac{\sin x_1}{x_1} + \cdots + \log \frac{\sin x_n}{x_n}\right).$$

Problems

7.4.10 (Bernoulli's inequality). Prove that for $0 < a < 1$,

$$(1 + x)^a \leqslant 1 + ax, \qquad x \geqslant -1.$$

How should the inequality go when $a < 0$, or when $a > 1$?

7.4.11. Prove that

$$\frac{x}{1 + x} < \log(1 + x) < \frac{x(x + 2)}{2(x + 1)}, \qquad x > 0.$$

7.4.12 (Huygens's inequality). Prove that

$$2 \sin x + \tan x \geqslant 3x, \qquad 0 < x < \pi/2.$$

7.4.13. For all $x > 0$, $(2 + \cos x)x > 3 \sin x$.

(a) Prove this inequality by considering the function $F(x) = x - (3 \sin x)/(2 + \cos x)$.

(b) Prove this inequality by considering the function $F(x) = (2 + \cos x)x - 3 \sin x$.

7.4.14. Prove that

$$0 \leqslant \frac{x \log x}{x^2 - 1} \leqslant \frac{1}{2}, \qquad x > 0, \ x \neq 1.$$

7.4.15. Prove that

$$\log\left(1 - \frac{1}{x+3}\right) + \frac{x}{(x+2)(x+3)} < 0, \qquad x > -2.$$

7.4.16. Prove that

$$\left(\frac{a+1}{b+1}\right)^{b+1} > \left(\frac{a}{b}\right)^{b}, \qquad a, b > 0, \quad a \neq b.$$

7.4.17. Prove that

$$\frac{\sin a}{\sin b} < \frac{a}{b} < \frac{\tan a}{\tan b}, \qquad 0 < b < a < \tfrac{1}{2}\pi.$$

7.4.18. Use the methods of this section to prove that for each positive integer n,

$$\left(1 + \frac{1}{n}\right)^n < \left(1 + \frac{1}{n+1}\right)^{n+1}.$$

(That is, show that $f(x) = (1 + 1/x)^x$ is an increasing function.)

7.4.19. Use the concavity of $f(x) = \sqrt{x}$ to prove that if a, b, c are positive, then $a \cos^2\theta + b \sin^2\theta < c$ implies $\sqrt{a} \cos^2\theta + \sqrt{b} \sin^2\theta < \sqrt{c}$. (Hint: Sketch the graph of $f(x) = \sqrt{x}$. In the domain, where is the point $a \cos^2\theta + b \sin^2\theta$, and in the range, where is $\sqrt{a} \cos^2\theta + \sqrt{b} \sin^2\theta$?)

7.4.20. Let $x_i > 0$ for $i = 1, 2, \ldots, n$. Consider the function $f(t) = \log t$, and in a manner similar to that used in 7.4.9, prove that

$$(x_1 x_2 \cdots x_n)^{1/n} \leqslant \frac{x_1 + x_2 + \cdots + x_n}{n}$$

with equality if an only if all the x_i are equal.

7.4.21.

(a) Let $x_i > 0$ for $i = 1, 2, \ldots, n$. Use the result of 7.4.20 to show that

$$\frac{n}{\dfrac{1}{x_1} + \dfrac{1}{x_2} + \cdots + \dfrac{1}{x_n}} \leqslant (x_1 x_2 \cdots x_n)^{1/n}.$$

(b) For positive numbers a, b, c such that $1/a + 1/b + 1/c = 1$, show that $(a - 1)(b - 1)(c - 1) \geqslant 8$.

7.4.22. Show that if a, b, c are positive numbers with $a + b + c = 1$, then

$$\left(a + \frac{1}{a}\right)^2 + \left(b + \frac{1}{b}\right)^2 + \left(c + \frac{1}{c}\right)^2 \geqslant \frac{100}{3}.$$

7.4.23. Let a, b, c denote the lengths of the sides of a triangle. Show that

$$\frac{3}{2} \leqslant \frac{a}{b+c} + \frac{b}{c+a} + \frac{c}{a+b} \leqslant 2.$$

Additional Examples

6.4.6, 6.4.7.

7.5. Inequalities by Series

Another way to prove an inequality of the form

$$f(x) \leqslant g(x), \qquad 0 < x < c$$

(see the discussion preceding 7.4.3) is to expand f and g in power series, say $f(x) = \sum_{n=0}^{\infty} a_n x^n$ and $g(x) = \sum_{n=0}^{\infty} b_n x^n$, for x in the interval $(-d, d)$. If it should happen that $a_n \leqslant b_n$ for all n, then it is obvious that $f(x) \leqslant g(x)$ for all x in the interval $(0, d)$.

7.5.1. For which real numbers c is $\frac{1}{2}(e^x + e^{-x}) \leqslant e^{cx^2}$ for all real x?

Solution. If the inequality holds for all x then

$$0 \leqslant e^{cx^2} - \tfrac{1}{2}(e^x + e^{-x})$$

$$= \sum_{n=0}^{\infty} \frac{c^n x^{2n}}{n!} - \sum_{n=0}^{\infty} \frac{x^{2n}}{2^n n!}$$

$$= \sum_{n=0}^{\infty} \left(c^n - \frac{1}{2^n} \right) \frac{x^{2n}}{n!}$$

$$= \sum_{n=1}^{\infty} \left(c^n - \frac{1}{2^n} \right) \frac{x^{2n}}{n!} .$$

To see that $c \geqslant \frac{1}{2}$, divide each side by x^2 and set $x = 0$.
    On the other hand, if $c \geqslant \frac{1}{2}$,

$$\tfrac{1}{2}(e^x + e^{-x}) = \sum_{n=0}^{\infty} \frac{x^{2n}}{(2n)!}$$

$$\leqslant \sum_{n=0}^{\infty} \frac{x^{2n}}{2^n n!}$$

$$= e^{x^2/2}$$

$$\leqslant e^{cx^2}.$$

It follows that the stated inequality holds for all x if and only if $c \geqslant \frac{1}{2}$.

Another important series technique with application to inequality problems concerns alternating series. Recall that if $a_0, a_1, a_2, \ldots$ is a sequence of positive numbers, then the series $\sum_{n=0}^{\infty}(-1)^n a_n$ converges provided the terms steadily decrease to zero (i.e., $a_{n+1} < a_n$ and $a_n \to 0$ as $n \to \infty$). More importantly for our purposes here, the sum of the series lies between any two successive partial sums. (If S denotes the sum of the series and S_n denotes the nth partial sum, then $\{S_{2n+1}\}$ is an increasing sequence, $\{S_{2n}\}$ is a decreasing sequence, and for all n, $S_{2n+1} < S < S_{2n}$.)

7.5.2. Show that for all x,

$$1 + x + \frac{x^2}{2!} + \cdots + \frac{x^{2n}}{(2n)!} > 0.$$

Solution. The claim is true when x is positive or zero. Suppose that x is negative and that k is the nonnegative integer for which $1 \leq |x| \leq |x^2/2| \leq \cdots \leq |x^k/k!|$ and $|x^k/k!| \geq |x^{k+1}/(k+1)!| \geq \cdots$. The result is certainly true if $2n \leq k$, whereas if $2n > k$, the reasoning preceding the problem implies that

$$1 + x + \frac{x^2}{2!} + \cdots + \frac{x^{2n}}{(2n)!}$$

$$> 1 + x + \frac{x^2}{2!} + \cdots + \frac{x^{2n}}{(2n)!} + \cdots = e^x > 0.$$

7.5.3. Prove that $(2 + \cos x)x > 3 \sin x$, $x > 0$.

Solution. This is the same problem as 7.4.13, but here we will give a solution based on series considerations.

On the left side of the desired inequality, we know that for $x > 0$,

$$(2 + \cos x)x > \left(2 + 1 - \frac{x^2}{2!} + \frac{x^4}{4!} - \frac{x^6}{6!}\right)x,$$

and on the right side

$$3 \sin x < 3\left(x - \frac{x^3}{3!} + \frac{x^5}{5!}\right).$$

Therefore, it is sufficient to prove that

$$3x - \frac{x^3}{2!} + \frac{x^5}{4!} - \frac{x^7}{6!} > 3\left(x - \frac{x^3}{3!} + \frac{x^5}{5!}\right).$$

This is true for $x > 0$ if and only if

$$\frac{x^5}{4!} - \frac{x^7}{6!} > \frac{3x^5}{5!},$$

$$\left(\frac{1}{4!} - \frac{3}{5!}\right) > \frac{1}{6!} x^2,$$

$$x^2 < 6!\left(\frac{2}{5!}\right) = 12.$$

This proves the desired inequality for the case in which $0 < x < \sqrt{12}$. But the inequality is obvious for $x \geqslant \sqrt{12}$, and therefore, it is true for all $x > 0$, and the proof is complete.

In the preceding proof, one might ask why these many terms from the infinite series were chosen. Why not more or less? To keep the inequalities going in the right direction, we need to underestimate $\cos x$ and overestimate $\sin x$, thus dictating the signs of the final terms in the series approximations. The crudest estimate would be to replace $\cos x$ by $1 - x^2/2$ and to replace $\sin x$ by x. This leads us to investigate

$$\left(3 - \frac{x^2}{2}\right)x > 3x,$$

which is equivalent to

$$-\frac{x^3}{2} > 0,$$

and this is not true for any positive value of x.

As the number of terms in the series increases, the approximations improve, so the next try might be to replace $\cos x$ by $1 - x^2/2 + x^4/4! - x^6/6!$ and $\sin x$ by $x - x^3/3! + x^5/5!$. This leads to the solution as it was presented.

7.5.4. Prove that

$$\left(\frac{\sin x}{x}\right)^3 \geqslant \cos x, \qquad 0 < x \leqslant \tfrac{1}{2}\pi.$$

Solution. For $x > 0$,

$$\left(\frac{\sin x}{x}\right)^3 > \left(1 - \frac{x^2}{3!}\right)^3 = 1 - \frac{x^2}{2} + \frac{x^4}{12} - \frac{x^6}{216},$$

and

$$\cos x < 1 - \frac{x^2}{2} + \frac{x^4}{4!} - \frac{x^6}{6!} + \frac{x^8}{8!}.$$

Therefore, it suffices to show that

$$1 - \frac{x^2}{2} + \frac{x^4}{12} - \frac{x^6}{216} > 1 - \frac{x^2}{2} + \frac{x^4}{4!} - \frac{x^6}{6!} + \frac{x^8}{8!}$$

or equivalently,

$$\frac{1}{4!} + \left(-\frac{1}{216} + \frac{1}{720}\right)x^2 - \frac{1}{8!}x^4 > 0.$$

The left side is decreasing on the interval $(0, \frac{1}{2}\pi]$, and therefore takes its minimum when $x = \frac{1}{2}\pi$. In particular, for $0 < x \leqslant \pi/2$,

$$\frac{1}{4!} + \left(-\frac{1}{216} + \frac{1}{720}\right)x^2 - \frac{1}{8!}x^4 \geqslant \frac{1}{4!} + \left(-\frac{1}{216} + \frac{1}{720}\right)\left(\frac{\pi}{2}\right)^2 - \frac{1}{8!}\left(\frac{\pi}{2}\right)^4$$

$$> \frac{1}{4!} + \left(-\frac{1}{216}\right)(2)^2 - \frac{1}{8!}(2)^4 > 0.$$

This completes the proof.

Problems

7.5.5. Use infinite series to prove the following inequalities:

(a) $e^x > 1 + (1 + x)\log(1 + x)$, $x > 0$.
(b) $(1 + x)/(1 - x) > e^{2x}$, $0 < x < 1$.
(c) $\arcsin x < x/(1 - x^2)$, $0 < x < 1$.

7.5.6. Prove that $\sqrt[3]{1 + x} - 1 - \frac{1}{3}x + \frac{1}{9}x^2 < \frac{5}{81}x^3$, $x > 0$.

7.5.7. Prove that

$$x < \frac{1}{3}(2\sin x + \tan x), \qquad x > 0.$$

[Hint: Show the equivalent inequality,

$$\sin x(2\cos x + 1) > 3x\cos x, \qquad x > 0.]$$

7.5.8. Show that $\sin^2 x < \sin x^2$ for $0 < x < \sqrt{\pi/2}$.

7.6. The Squeeze Principle

In this section we will see how inequality considerations can play an important role in evaluating limits. The key idea (which has many variations) is expressed in the following result.

The Squeeze Principle. If $\{a_n\}, \{b_n\}, \{c_n\}$ are infinite sequences such that $a_n \leqslant b_n \leqslant c_n$ for all sufficiently large n, and if $\{a_n\}$ and $\{c_n\}$ converge to the same number L, then $\{b_n\}$ also converges to L.

As innocuous as this principle appears (obviously, there is no alternative for $\{b_n\}$; it is "squeezed" between $\{a_n\}$ and $\{c_n\}$, both of which are converging to the same limit), it is surprising that it can be useful in problem solving. Nevertheless, it is applicable in the following situation. Suppose we wish to evaluate the limit of a sequence $\{b_n\}$, and suppose the b_n's are hopelessly complicated, so that they cannot be handled directly. The squeeze principle suggests that we try to "squeeze" $\{b_n\}$ with two simpler sequences $\{a_n\}$ and $\{c_n\}$.

For example, consider the sequence $\{n^{1/n}\}$. We could evaluate this limit by L'Hôpital's rule; however, consider the following argument. By the arithmetic-mean–geometric-mean inequality,

$$1 \leqslant n^{1/n} = \left(\underbrace{1 \times 1 \times \cdots \times 1}_{n-2} \times \sqrt{n} \times \sqrt{n} \right)^{1/n}$$

$$\leqslant \frac{(n-2) + 2\sqrt{n}}{n} = 1 + 2\left(\frac{1}{\sqrt{n}} - \frac{1}{n} \right).$$

Now, by the squeeze principle (with $a_n = 1$, and $c_n = 1 + 2(1/\sqrt{n} - 1/n)$) we see that $n^{1/n}$ is forced to converge to 1.

7.6.1. Prove or disprove that the set of all positive rational numbers can be arranged in an infinite sequence $\{b_n\}$ such that $\{(b_n)^{1/n}\}$ is convergent.

Solution. We begin by ordering the rational numbers by following the usual serpentine path through the square array of rationals shown in Figure 7.2, where we omit all fractions not reduced to lowest terms. The sequence thus begins $1, \frac{1}{2}, 2, 3, \frac{1}{3}, \frac{1}{4}, \frac{2}{3}, \frac{3}{2}, 4, 5, \frac{1}{5}, \frac{1}{6}, \ldots$. If b_n denotes the nth term of this sequence, we would like to prove that $\{b_n^{1/n}\}$ converges to 1.

In Figure 7.2, observe that every element in the nth row is less than or equal to n, and every element in the nth column is greater than or equal to

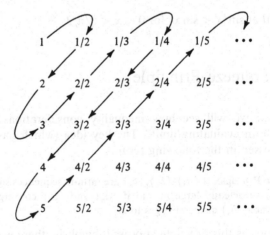

Figure 7.2.

$1/n$. Also, if b_n occurs in row i and column j, then $i \leqslant n$ and $j \leqslant n$. Therefore

$$\frac{1}{n} \leqslant \frac{1}{j} \leqslant b_n \leqslant i \leqslant n \qquad \text{for all } n,$$

and consequently,

$$a_n \equiv \frac{1}{n^{1/n}} = \left(\frac{1}{n}\right)^{1/n} \leqslant b_n^{1/n} \leqslant n^{1/n} \equiv c_n.$$

Now, by the squeeze principle, $\{b_n^{1/n}\}$ converges to 1.

7.6.2. Let $f(x)$ be a real-valued function, defined for $-1 < x < 1$, such that $f'(0)$ exists. Let $\{a_n\}$, $\{b_n\}$ be two sequences such that

$$-1 < a_n < 0 < b_n < 1, \qquad \lim_{n \to \infty} (a_n) = 0 = \lim_{n \to \infty} (b_n).$$

Prove that

$$\lim_{n \to \infty} \frac{f(b_n) - f(a_n)}{b_n - a_n} = f'(0).$$

Solution. The quotient

$$\frac{f(b_n) - f(a_n)}{b_n - a_n}$$

can be interpreted geometrically as the slope of the line segment $P_n(a_n, f(a_n))$, $Q_n(b_n, f(b_n))$ (see Figure 7.3.)

Let R be the point $(0, f(0))$. Either the y-intercept of the line segment $P_n Q_n$ is less than or equal to $f(0)$ (case 1), or it is greater than $f(0)$ (case 2).

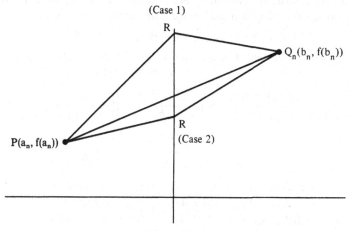

(Case 1)

R

$Q_n(b_n, f(b_n))$

R

(Case 2)

$P(a_n, f(a_n))$

Figure 7.3.

In the first case,

$$\text{Slope } RQ_n \leqslant \text{Slope } P_nQ_n \leqslant \text{Slope } P_nR,$$

or equivalently,

$$\frac{f(b_n) - f(0)}{b_n - 0} \leqslant \frac{f(b_n) - f(a_n)}{b_n - a_n} \leqslant \frac{f(a_n) - f(0)}{a_n - 0}.$$

In the second case,

$$\text{Slope } P_nR \leqslant \text{Slope } P_nQ_n \leqslant \text{Slope } RQ_n,$$

or equivalently,

$$\frac{f(a_n) - f(0)}{a_n - 0} \leqslant \frac{f(b_n) - f(a_n)}{b_n - a_n} \leqslant \frac{f(b_n) - f(0)}{b_n - 0}.$$

In case 2, the inequalities are just reversed from those in case 1. To correct this, we define two new sequences which reverse the roles of a_n and b_n in case 2. Thus, let $\{c_n\}$ and $\{d_n\}$ be defined by

$$c_n = b_n \quad \text{and} \quad d_n = a_n \qquad \text{if case 1 holds for } a_n \text{ and } b_n,$$

$$c_n = a_n \quad \text{and} \quad d_n = b_n \qquad \text{if case 2 holds for } a_n \text{ and } b_n.$$

Then, for all n,

$$\frac{f(c_n) - f(0)}{c_n - 0} \leqslant \frac{f(b_n) - f(a_n)}{b_n - a_n} \leqslant \frac{f(d_n) - f(0)}{d_n - 0}.$$

Since $f'(0)$ exists, and since $\lim_{n\to\infty} c_n = 0 = \lim_{n\to\infty} d_n$,

$$\lim_{n\to\infty} \frac{f(c_n) - f(0)}{c_n - 0} = f'(0) \quad \text{and} \quad \lim_{n\to\infty} \frac{f(d_n) - f(0)}{d_n - 0} = f'(0).$$

The result now follows from the squeeze principle.

Another instructive solution, also based on the squeeze principle, is based on the fact that if a and b are real numbers, $a < b$, then

$$a \leqslant ra + sb \leqslant b$$

for all positive numbers r and s that add to 1 (see 1.2.11). In this problem, write

$$\frac{f(b_n) - f(a_n)}{b_n - a_n} = \left(\frac{f(b_n) - f(0)}{b_n} \right) \left(\frac{b_n}{b_n - a_n} \right)$$

$$+ \left(\frac{f(a_n) - f(0)}{a_n} \right) \left(\frac{-a_n}{b_n - a_n} \right),$$

and set $r = b_n/(b_n - a_n)$ and $s = -a_n/(b_n - a_n)$. Then $r \geqslant 0$, $s \geqslant 0$, and $r + s = 1$. Therefore $[f(b_n) - f(a_n)]/[b_n - a_n]$ lies between $[f(b_n) - f(0)]/b_n$ and $[f(a_n) - f(0)]/a_n$. Since these latter quotients converge to $f'(0)$, so also must $[f(b_n) - f(a_n)]/[b_n - a_n]$ by the squeeze principle.

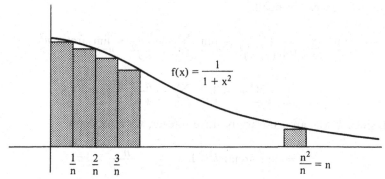

Figure 7.4.

7.6.3. Evaluate

$$\lim_{n \to \infty} \sum_{j=1}^{n^2} \frac{n}{n^2 + j^2} .$$

Solution. The sum

$$\sum_{j=1}^{n^2} \frac{n}{n^2 + j^2} = \sum_{j=1}^{n^2} \left(\frac{1/n}{1 + (j/n)^2} \right)$$

can be regarded as a Riemann sum for the function $f(x) = 1/(1 + x^2)$ over the interval $[0, n]$ (see Figure 7.4). Unfortunately, it is not really a Riemann sum, because the interval over which it is taken is not a fixed interval; thus, as $n \to \infty$, we don't get a definite integral. We can say, however, that for each n

$$\sum_{j=1}^{n^2} \left(\frac{n}{n^2 + j^2} \right) \leqslant \int_0^{n^2} \frac{dx}{1 + x^2} = \arctan n^2.$$

To get a lower bound for the sum under consideration, let k be a fixed positive integer, and fix the interval $[0, k]$. Then, for any n greater than k,

$$\sum_{j=1}^{kn} \frac{n}{n^2 + j^2} = \sum_{j=1}^{kn} \frac{1/n}{1 + (j/n)^2}$$

is a Riemann sum for $f(x) = 1/(1 + x^2)$ over the interval $[0, k]$. Also,

$$\sum_{j=1}^{kn} \frac{n}{n^2 + j^2} < \sum_{j=1}^{n^2} \frac{n}{n^2 + j^2} .$$

Putting all of this together, we have

$$\sum_{j=1}^{kn} \frac{n}{n^2 + j^2} < \sum_{j=1}^{n^2} \frac{n}{n^2 + j^2} < \arctan n^2,$$

so, by the squeeze principle,

$$\lim_{n\to\infty} \sum_{j=1}^{kn} \frac{1/n}{1+(j/n)^2} \leqslant \lim_{n\to\infty} \sum_{j=1}^{n^2} \frac{n}{n^2+j^2} \leqslant \lim_{n\to\infty} \arctan n^2,$$

$$\int_0^k \frac{dx}{1+x^2} \leqslant \lim_{n\to\infty} \sum_{j=1}^{n^2} \frac{n}{n^2+j^2} \leqslant \tfrac{1}{2}\pi.$$

But, since k was an arbitrary positive integer, we must have

$$\tfrac{1}{2}\pi = \lim_{k\to\infty} \arctan k \leqslant \lim_{n\to\infty} \sum_{j=1}^{n^2} \frac{n}{n^2+j^2} \leqslant \tfrac{1}{2}\pi.$$

It follows that the desired limit equals $\tfrac{1}{2}\pi$.

Another important application of inequalities to the evaluation of limits is based on the following important fact.

Monotonic, bounded sequences converge.

That is to say, if $\{a_n\}$ is a sequence of real numbers such that $a_{n+1} \geqslant a_n$ for all sufficiently large n (or $a_{n+1} \leqslant a_n$ for all sufficiently large n), and if for some constant K, $a_n \leqslant K$ for all n (or $a_n \geqslant K$ respectively), then the sequence $\{a_n\}$ converges.

For example, to prove that the sequence $(1 + 1/n)^n$ converges, it is sufficient to prove it is monotonic (increasing in this case) and bounded above (by 3; see 7.1.5).

7.6.4. If $\{a_n\}$ is a sequence such that for $n \geqslant 1$

$$(2 - a_n)a_{n+1} = 1,$$

prove that $\lim_{n\to\infty} a_n$ exists and is equal to 1.

Solution. First we will prove that if the sequence converges, it must converge to 1. The argument is standard when a sequence is defined recursively as it is here. Let $\lim_{n\to\infty} a_n = L$. Then, taking the limit of each side of the recurrence relation $(2 - a_n)a_{n+1} = 1$, we see that $(2 - L)L = 1$, or equivalently, $(L - 1)^2 = 0$, from which we conclude that $L = 1$.

Now, to prove that the sequence converges, we will prove that it is bounded, and "eventually" becomes monotonic. (For another solution of this problem, see 1.1.11.)

Suppose that for some a_n, $0 < a_n < 1$. Then

$$a_{n+1} - a_n = \frac{1}{2 - a_n} - a_n = \frac{1 - (2 - a_n)a_n}{2 - a_n}$$

$$= \frac{(1 - a_n)^2}{2 - a_n} > 0$$

and $a_{n+1} = 1/(2 - a_n) < 1$. Therefore, $a_n < a_{n+1} < a_{n+2} < \cdots < 1$, so the sequence is monotonic and bounded and therefore converges. Thus, it suffices to prove that for some n, $0 < a_n < 1$. There are several cases.

If $a_1 < 0$, then $0 < a_2 < 1/(2 - a_1) < 1$, so we're done by the preceding argument.

If $a_1 > 2$, then $a_2 = 1/(2 - a_1) < 0$, so again we're done.

If $a_1 = 1$, then $a_n = 1$ for all n.

It remains to check the case $1 < a_1 \leqslant 2$. Some playing around with special cases in this interval leads to the following (each of which can be proved by induction).

The sequence is not defined if a_1 has the form $(n + 1)/n$. For if $a_1 = (n + 1)/n$, then (one can show) $a_n = 2$ and consequently a_{n+1} is not defined. If a_1 belongs to the interval

$$\left(\frac{n+1}{n}, \frac{n}{n-1} \right) \quad \text{for } n > 1,$$

then (one can show that) a_{n+1} lies in the interval $(0, 1)$ and the proof is complete by previous reasoning.

Thus, in all cases (for which the sequence is defined) the sequence converges.

7.6.5. Let $f(x)$ be a function such that $f(1) = 1$ and for $x \geqslant 1$

$$f'(x) = \frac{1}{x^2 + f^2(x)}.$$

Prove that $\lim_{x \to \infty} f(x)$ exists and is less than $1 + \frac{1}{4}\pi$.

Solution. By the fundamental theorem of calculus

$$f(x) - f(1) = \int_1^x f'(x) \, dx.$$

Observe that $f(x)$ is increasing; moreover, $f(x) \geqslant 1$ for all $x \geqslant 1$, since $f(1) = 1$ and $f'(x) > 0$. Therefore

$$f(x) - f(1) = \int_1^x \frac{dx}{x^2 + f^2(x)} \leqslant \int_1^x \frac{dx}{1 + x^2}$$

$$= \arctan x \Big]_1^x$$

$$= \arctan x - \arctan 1$$

$$< \tfrac{1}{2}\pi - \tfrac{1}{4}\pi = \tfrac{1}{4}\pi.$$

Thus, $f(x)$ is increasing and bounded above by $1 + \frac{1}{4}\pi$, and consequently, $\lim_{x \to \infty} f(x)$ exists and is less than $1 + \frac{1}{4}\pi$.

7.6.6. Consider all the natural numbers which represented in the decimal system have no 9 among their digits. Prove that the series formed by the reciprocals of these numbers converges.

Solution. Let S_m denote the mth partial sum of the series under consideration. The sequence $\{S_m\}$ is monotone increasing, so to prove convergence, we need only prove that the sequence is bounded.

For a given partial sum S_m, let n denote the number of digits in the integer m. The number of integers of exactly n digits which have no 9 in their decimal representation is $8 \times 9^{n-1}$ (the first digit cannot be zero). Therefore, the sum of their reciprocals is less than $8 \times 9^{n-1}/10^{n-1}$. Thus

$$S_m < 8 + 8 \times \tfrac{9}{10} + 8 \times \left(\tfrac{9}{10}\right)^2 + \cdots + 8 \times \left(\tfrac{9}{10}\right)^{n-1}$$

$$< 8\left[1 + \tfrac{9}{10} + \left(\tfrac{9}{10}\right)^2 + \cdots \right] = 80,$$

and the proof is complete.

Problems

7.6.7. Prove the inequalities which follow and apply the squeeze principle to evaluate a limit:

(a) $\dfrac{n}{\sqrt{n^2 + n}} < \displaystyle\sum_{i=1}^{n} \dfrac{1}{\sqrt{n^2 + i}} < \dfrac{n}{\sqrt{n^2 + 1}}$.

(b) $b < (a^n + b^n)^{1/n} < b(2)^{1/n}, \, 0 < a < b$.

(c) $e^{1-1/(2n)} < (1 + 1/n)^n < e^{1-1/(2n)+1/(3n^2)}$.

7.6.8. Prove that each of the following sequences converges, and find its limit:

(a) $\sqrt{1}, \sqrt{1 + \sqrt{1}}, \sqrt{1 + \sqrt{1 + \sqrt{1}}}, \sqrt{1 + \sqrt{1 + \sqrt{1 + \sqrt{1}}}}, \ldots$.

(b) $\sqrt{2}, \sqrt{2 + \sqrt{2}}, \sqrt{2 + \sqrt{2 + \sqrt{2}}}, \sqrt{2 + \sqrt{2 + \sqrt{2 + \sqrt{2}}}}, \ldots$.

7.6.9. Prove that the sequence $\{a_n\}$ defined by

$$a_n = 1 + \frac{1}{2} + \cdots + \frac{1}{n} - \log n$$

converges.

7.6.10. Prove that the sequence $\{a_n\}$ defined by

$$a_{n+1} = \frac{6(1 + a_n)}{7 + a_n}$$

converges, and find its limit.

7.6.11. Let a_1 and b_1 be any two positive numbers, and define $\{a_n\}$ and $\{b_n\}$ by

$$a_n = \frac{2a_{n-1}b_{n-1}}{a_{n-1} + b_{n-1}}, \qquad b_n = \sqrt{a_{n-1}b_{n-1}} \ .$$

Prove that the sequences $\{a_n\}$ and $\{b_n\}$ converge and have the same limit.

7.6.12. $S_1 = \log a$, and $S_n = \sum_{i=1}^{n-1} \log(a - S_i)$, $n > 1$. Show that

$$\lim_{n \to \infty} S_n = a - 1.$$

(Hint: Note that $S_{n+1} = S_n + \log(a - S_n)$.)

7.6.13. The sequence $Q_n(x)$ of polynomials is defined by

$$Q_1(x) = 1 + x, \qquad Q_2(x) = 1 + 2x,$$

and for $m \geqslant 1$,

$$Q_{2m+1}(x) = Q_{2m}(x) + (m + 1)xQ_{2m-1}(x),$$

$$Q_{2m+2}(x) = Q_{2m+1}(x) + (m + 1)xQ_{2m}(x).$$

Let x_n be the largest real solution of $Q_n(x) = 0$. Prove that $\{x_n\}$ is an increasing sequence and that $\lim_{n \to \infty} x_n = 0$.

7.6.14. Prove that if $\sum_{n=1}^{\infty} a_n^2$ converges, so does $\sum_{n=1}^{\infty}(a_n/n)$.

7.6.15. Prove that

$$\lim_{n \to \infty} \frac{2^2 \times 4^2 \times 6^2 \times \cdots \times (2n)^2}{(1 \times 3)(3 \times 5) \cdots ((2n - 1)(2n + 1))}$$

$$= \lim_{n \to \infty} \frac{2^2 \times 4^2 \times 6^2 \times \cdots \times (2n)^2}{1^2 \times 3^2 \times 5^2 \times \cdots \times (2n - 1)^2} \left(\frac{1}{2n + 1} \right) = \frac{1}{2} \pi.$$

(Hint: For $0 < \theta < \frac{1}{2}\pi$, $\int_0^{\pi/2}\sin^{2n+1}\theta \, d\theta < \int_0^{\pi/2}\sin^{2n}\theta \, d\theta < \int_0^{\pi/2}\sin^{2n-1}\theta \, d\theta$. Apply the result of 2.5.14 together with the squeeze principle.)

Additional Examples

6.1.5, 6.3.7, 6.4.4, 6.6.2, Section 6.8, 6.94. Also, see examples of "repeated bisection" in Section 6.1.

Chapter 8. Geometry

In this chapter we will look at some of the most common techniques for solving problems in Euclidean geometry. In addition to the classical synthetic methods of Euclid, we will see how algebra, trigonometry, analysis, vector algebra, and complex numbers can be useful tools in the study of geometry.

8.1. Classical Plane Geometry

In this section we will review the ideas and methods characteristic of classical plane geometry: namely, the study of those properties of triangles, quadrilaterals, and circles that remain invariant under motion (e.g., translation, rotation, reflection). We will be concerned with synthetic geometry, which builds on an understanding of the basic notations of congruency, similarity, proportion, concurrency, arcs and chords of circles, inscribed angles, etc. In addition, we wish to draw attention to the importance of algebraic and trigonometric techniques for proving results in traditional Euclidean plane geometry.

8.1.1. Find the area of a convex octagon that is inscribed in a circle and has four consecutive sides of length 3 units and the remaining four sides of length 2 units. Give the answer in the form $r + s\sqrt{t}$, with r, s, and t positive integers.

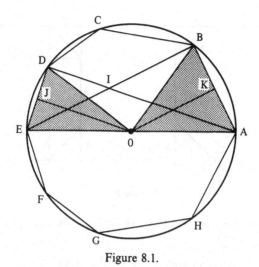

Figure 8.1.

We will give several solutions to this problem to illustrate the variety of methods that are at ones disposal in this subject.

Solution 1. Let the vertices be labeled $ABCDEFGH$ as shown in Figure 8.1 where $AB = BC = GH = HA = 3$ and $CD = DE = EF = FG = 2$. Let O denote the center of the circle.

We first find the area of $\triangle OAB$ and $\triangle ODE$. For this, it suffices to find the altitudes OK and OJ.

Notice that $OK = \frac{1}{2}EB$; this follows because O is the midpoint of EA and K is the midpoint of AB. Similarly, $OJ = \frac{1}{2}AD$, and therefore, it suffices to find DI, IA, EI, and IB, where I is the intersection of AD and EB.

By angle–side–angle, $\triangle DBC \cong \triangle DBI$, and therefore $DI = 2$ and $IB = 3$. Furthermore, since $\triangle ADE$ and $\triangle ABE$ are each inscribed in a semicircle, $\angle ADE$ and $\angle ABE$ are right angles. Therefore, $\triangle IBA$ and $\triangle EDI$ are isosceles right triangles, and it follows that $IA = 3\sqrt{2}$ and $EI = 2\sqrt{2}$.

We can now find the area of the octagon:

$$\text{Area} = 4\left[\frac{1}{2} \times 3\left(\frac{3 + 2\sqrt{2}}{2}\right)\right] + 4\left[\frac{1}{2} \times 2\left(\frac{2 + 3\sqrt{2}}{2}\right)\right] = 13 + 12\sqrt{2}.$$

Solution 2. Perhaps the easiest solution is based on recognizing that the area of the octagon is the same as either of those shown in Figure 8.2, having alternating sides of lengths 2 and 3. The area can be computed by subtracting four triangular regions from a square, or by adding the areas of

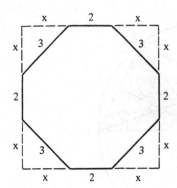 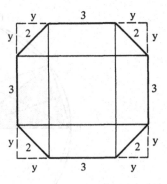

Figure 8.2.

a square, four rectangles, and four triangles. Thus, for the diagram on the left, we have $(x = \frac{3}{2}\sqrt{2})$

$$\text{octagon area} = (2x + 2)^2 - 4(\tfrac{1}{2}x^2)$$

$$= 2x^2 + 8x + 4$$

$$= 2(\tfrac{3}{2}\sqrt{2})^2 + 8 \times \tfrac{3}{2}\sqrt{2} + 4$$

$$= 13 + 12\sqrt{2}.$$

Or, for working from the inside on the figure on the right $(y = \sqrt{2})$,

$$\text{octagon area} = 9 + 4(3y) + 4(\tfrac{1}{2}y^2)$$

$$= 9 + 12\sqrt{2} + 2 \times 2$$

$$= 13 + 12\sqrt{2}.$$

Solution 3. Let R denote the radius of the circle. The area of the octagon is equal to four times the area in quadrilateral $OABC$ (see Figure 8.3). Clearly

$$\text{Area } OABC = \text{Area } \triangle OAC + \text{Area } \triangle ABC,$$

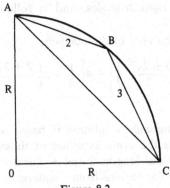

Figure 8.3.

By Heron's formula for the area of a triangle,

$$\text{Area } \triangle ABC = \sqrt{s(s - 2)(s - 3)(s - \sqrt{2}\,R)}$$

where $s = \frac{1}{2}(2 + 3 + \sqrt{2}\,R) = \frac{5}{2} + \frac{1}{2}\sqrt{2}\,R$. This leads to

Area $OABC$

$$= \tfrac{1}{2}R^2 + \sqrt{\left(\tfrac{5}{2} + \tfrac{1}{2}\sqrt{2}\,R\right)\left(\tfrac{1}{2} + \tfrac{1}{2}\sqrt{2}\,R\right)\left(-\tfrac{1}{2} + \tfrac{1}{2}\sqrt{2}\,R\right)\left(\tfrac{5}{2} - \tfrac{1}{2}\sqrt{2}\,R\right)}$$

$$= \tfrac{1}{2}R^2 + \sqrt{\left(\tfrac{25}{4} - \tfrac{1}{2}R^2\right)\left(-\tfrac{1}{4} + \tfrac{1}{2}R^2\right)}\ .$$

By the law of cosines (using $\angle B$ in $\triangle ABC$) we get

$$2R^2 = 4 + 9 - 2 \times 2 \times 3 \cos 135°$$

$$= 13 + 12 \times \tfrac{1}{2}\sqrt{2}\ ,$$

and therefore,

$$R^2 = \tfrac{13}{2} + 3\sqrt{2}\ .$$

The final result then follows after substituting this value for R^2 into the preceding equation for Area $OABC$.

Solution 4. In Figure 8.4, D and E are the feet of perpendiculars drawn from B to OA and OC respectively. Let $x = OE$ and $y = OD$, and let R be the radius of the circle. Then

$$\text{area of octagon} = 4(\text{area of quadrilateral } OABC)$$

$$= 4(\text{Area } \triangle OAB + \text{Area } \triangle OCB)$$

$$= 4\left[\tfrac{1}{2}Rx + \tfrac{1}{2}Ry\right]$$

$$= 2R(x + y).$$

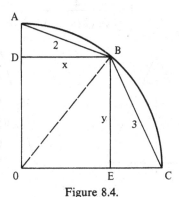

Figure 8.4.

Our plan is to express $x + y$ in terms of R, and then use the fact that $R^2 = \frac{13}{2} + 3\sqrt{2}$ (see the last solution).

The Pythagorean theorem applied to $\triangle ABD$ yields $x^2 + (R - y)^2 = 4$, or equivalently, $2R(R - y) = 4$ (note: $x^2 + y^2 = R^2$). Similarly, from $\triangle EBC$ we have $y^2 = 9 - (R - x)^2$, or equivalently, $2R(R - x) = 9$. Adding $R - y = 4/(2R)$ and $R - x = 9/(2R)$ yields $2R - (x + y) = 13/(2R)$, or equivalently $x + y = (4R^2 - 13)/(2R)$. Substituting, we find

$$\text{area of octagon} = 2R\left[\frac{4R^2 - 13}{2R}\right] = 4R^2 - 13$$

$$= 4\left(\frac{13}{2} + 3\sqrt{2}\right) - 13 = 13 + 12\sqrt{2}\,.$$

Solution 5. The octagaon can be cut like a pie into eight triangular pieces with equal sides of length R (equal to the radius of the circumscribed circle) and with bases 3 and 2. Let H and h denote the altitudes of these triangles as shown in Figure 8.5. Then

$$\text{area of octagon} = 4\left(\frac{1}{2}3 \cdot H\right) + 4\left(\frac{1}{2}2 \cdot h\right)$$

$$= 6 \cdot H + 4 \cdot h.$$

With α and β as shown in Figure 8.5, we have the following relationships: $\alpha + \beta = \pi/4$; $\sin\alpha = 3/(2R)$; $\cos\alpha = H/R$; $\sin\beta = 1/R$; $\cos\beta = h/R$. From these, we find

$$R = \frac{1}{\sin\beta} = \frac{1}{\sin(\frac{1}{4}\pi - \alpha)}$$

$$= \frac{1}{\frac{1}{2}\sqrt{2}\cos\alpha - \frac{1}{2}\sqrt{2}\sin\alpha} = \frac{2}{\sqrt{2}}\left(\frac{1}{\cos\alpha - \sin\alpha}\right)$$

$$= \frac{2}{\sqrt{2}}\left(\frac{1}{H/R - 3/2R}\right) = \frac{2}{\sqrt{2}}\left(\frac{2R}{2H - 3}\right).$$

It follows that

$$1 = \frac{4}{\sqrt{2}\,(2H - 3)}\,,$$

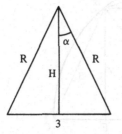 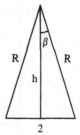

Figure 8.5.

or equivalently,

$$H = \tfrac{3}{2} + \sqrt{2} \ .$$

Using this,

$$h = R \cos \beta = R\left[\cos(\tfrac{1}{4}\pi - \alpha)\right] = R\left[\tfrac{1}{2}\sqrt{2}\cos\alpha + \tfrac{1}{2}\sqrt{2}\sin\alpha\right]$$

$$= \tfrac{1}{2}\sqrt{2}\, R\left[\frac{H}{R} + \frac{3}{2R}\right] = \tfrac{1}{4}\sqrt{2}\,[2H + 3]$$

$$= \tfrac{1}{4}\sqrt{2}\left[2(\tfrac{3}{2} + \sqrt{2}) + 3\right]$$

$$= 1 + \tfrac{3}{2}\sqrt{2} \ .$$

Substituting,

$$\text{area of octagon} = 6(\tfrac{3}{2} + \sqrt{2}) + 4(1 + \tfrac{3}{2}\sqrt{2}) = 13 + 12\sqrt{2}.$$

8.1.2. If A and B are fixed points on a given circle and XY is a variable diameter of the same circle, determine the locus of the points of intersection of lines AX and BY. (You may assume that AB is not a diameter.)

Solution. Consider Figure 8.6, where A and B are fixed points on the circumference of a given circle. Let B' be the point on the circle diametrically opposite of B. Let P and P' denote respectively the intersection of AX and BY when this intersection lies inside or outside the circle (depending upon which side of the line BB' the point X falls; see figure).

In the first case, $\angle APB = 90° + \tfrac{1}{2}(\text{Arc}\,AB)$, and this is a constant value for all diameters which result in an "inside" intersection point P. This implies that P lies on the circle formed by those points making a constant angle (namely, $90° + \tfrac{1}{2}(\text{Arc}\,AB)$) with the constant base AB.

In the second case, $\angle AP'B = 90° - \tfrac{1}{2}(\text{Arc}\,AB)$, and this is a constant value for all diameters which result in an "outside" intersection point P'.

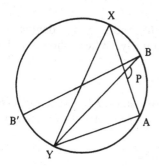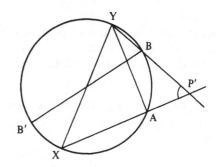

Figure 8.6.

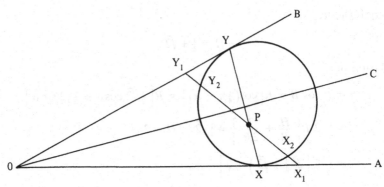

Figure 8.7.

Thus, P' lies on a circle which passes through A and B. Furthermore, $\angle APB$ and $\angle AP'B$ are supplementary angles ($\angle APB + \angle AP'B = 90° + \frac{1}{2}(\text{Arc } AB) + 90° - \frac{1}{2}(\text{Arc } AB) = 180°$), and therefore $APBP'$ is a cyclic quadrilateral; that is to say, P and P' lie on the same circle through A and B.

8.1.3. P is an interior point of the angle whose sides are the rays OA and OB. Locate X on OA and Y on OB so that the line segment XY contains P and so that the product of distances $(PX)(PY)$ is a minimum.

Solution. This problem was solved in 6.4.2 by using methods of analysis. Here we will solve it geometrically.

Let OC be the line bisecting $\angle AOB$, and let L denote the line through P which is perpendicular to OC. Let X and Y denote the intersections of L with OA and OB respectively (see Figure 8.7).

Now, $OX = OY$, so there is a circle tangent to OA at X and OB at Y. Let $X_1 Y_1$ be any other segment containing P with X_1 on OA and Y_1 on OB. Let X_2 and Y_2 be the intersections of $X_1 Y_1$ with the circle. Then $(PX)(PY) = (PX_2)(PY_2) < (PX_1)(PY_1)$, so $(PX)(PY)$ is the minimum.

8.1.4. Let P be an interior point of triangle ABC, and let x, y, z denote the distances from P to BC, AC, and AB respectively. Where should P be located to maximize the product xyz?

Solution. Let a, b, c denote the lengths of the sides BC, AC, and AB respectively (Figure 8.8). By the arithmetic-mean–geometric-mean inequality,

$$\sqrt[3]{(ax)(by)(cz)} \leqslant \frac{ax + by + cz}{3}.$$

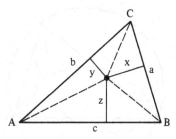

Figure 8.8.

But we know that $ax + by + cz = 2A$, where A is the area of the triangle. Thus, the maximum value of xyz is $8A^3/(27abc)$, and this occurs if and only if $ax = by = cz$.

We will show that $ax = by = cz$ if and only if P is located at the centroid of $\triangle ABC$. For this, suppose that CP intersects AB at D. Let α, β, γ, δ be the angles as shown in Figure 8.9. It is known that

$$\frac{b \sin \beta}{a \sin \alpha} = \frac{AD}{DB} .$$

(This relationship is useful in many problems. To see that it is true, apply the law of sines to $\triangle ADC$ and to $\triangle CDB$ to get

$$\frac{AD}{\sin \beta} = \frac{b}{\sin \gamma} \quad \text{and} \quad \frac{DB}{\sin \alpha} = \frac{a}{\sin \delta} .$$

Using these equations it follows that

$$\frac{b \sin \beta}{a \sin \alpha} = \frac{AD \sin \gamma}{DB \sin \delta} = \frac{AD}{DB} ,$$

since γ and δ are obviously supplementary.)

Using the above equation, we have

$$\frac{AD}{DB} = \frac{b \sin \beta}{a \sin \alpha} = \frac{by/(CP)}{ax/(CP)} = \frac{by}{ax} ,$$

and it follows that $AD = DB$ if and only if $by = ax$. Thus, $ax = by$ if and only if P is on the median line from C.

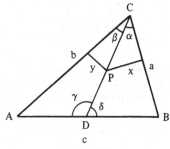

Figure 8.9.

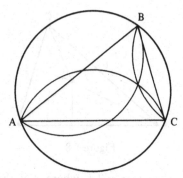

Figure 8.10.

In a similar manner, $ax = cz$ if and only if P is on the median line from B. It follows that $ax = by = cz$ if and only if P is the centroid of $\triangle ABC$.

Problems

8.1.5. Show that a triangle must be equilateral if any pair of the following centers coincide: incenter, circumcenter, centroid, orthocenter.

8.1.6. An acute triangle is inscribed in a circle. The resulting three minor arcs of the circle are reflected about the corresponding sides of the triangle (i.e., arc AB is reflected about side AB, etc.; see Figure 8.10). Are the reflected arcs concurrent?

8.1.7. Let C_1 and C_2 be circles of radius 1, tangent to each other and to the x-axis, with the center of C_1 on the y-axis. Now construct a sequence of circles C_n such that C_{n+1} is tangent to C_{n-1}, C_n, and the x-axis.

(a) Find the radius r_n of C_n.
(b) Show that the length of the common tangent included between its contacts with two consecutive circles C_n and C_{n+1} is $\binom{n}{2}^{-1}$ for $n \geqslant 2$.
(c) From part (b) and the geometry of the problem, show that

$$\sum_{n=2}^{\infty} \binom{n}{2}^{-1} = 2.$$

8.1.8. If a, b, c are the sides of a triangle ABC, t_a, t_b, t_c are the angle bisectors, and T_a, T_b, T_c are the angle bisectors extended until they are chords of the circle circumscribing the triangle ABC, prove that

$$abc = \sqrt{T_a T_b T_c t_a t_b t_c} \, .$$

(Hint: Prove that $T_a t_a = bc$, etc.)

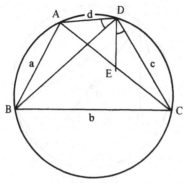

Figure 8.11.

8.1.9.

(a) Let P be a point inside angle XOY. Let AB be a segment through P with A on OX and B on OY, and such that $AP = PB$. Let MN be another segment through P with M on OX and N on OY. Prove that the area of triangle MON is greater than or equal to the area of triangle AOB.

(b) Let the tangents from a point A outside a given circle touch the circle at points D and E. Take P to be any point on the minor arc DE, and let the tangent at P intersect AD at B and AE at C. Show that the perimeter of $\triangle ABC$ is independent of the position of P (on the minor arc DE).

(c) In the setup of part (b), let MN be any other line through P which intersects AD and AE in M and N respectively. Prove that the perimeter of $\triangle ABC$ is smaller than the perimeter of $\triangle AMN$.

8.1.10. A quadrilateral $ABCD$ is inscribed in a circle (see Figure 8.11). Let $x = BD$, $y = AC$, and a, b, c, d be the lengths of the sides as indicated. Construct $\angle CDE$ equal to $\angle ADB$.

(a) Prove that $\triangle CDE \sim \triangle ADB$ and hence that $EC \cdot x = ac$.

(b) Prove that $\triangle ADE \sim \triangle BCD$ and hence $AE \cdot x = bd$.

(c) From parts (a) and (b), prove Ptolemy's theorem (an important fact about cyclic quadrilaterals): In a cyclic quadrilateral the product of the diagonals is equal to the sum of products of the opposite sides.

8.1.11.

(a) A line from vertex A of an equilateral triangle ABC meets the opposite side BC in a point P and the circumcircle in Q. Prove that

$$\frac{1}{PQ} = \frac{1}{BQ} + \frac{1}{CQ}.$$

(b) Using the notation of part (a), prove that $AQ^4 + BQ^4 + CQ^4$ is consant for all positions of Q on the minor arc BC. (Hint: For a trigonometric approach, let $x = AQ$, $y = BQ$, $z = CQ$, and $\theta = \angle BAQ$.

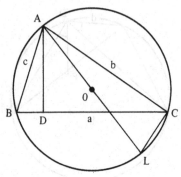

Figure 8.12.

Show that $x = (2/\sqrt{3})\sin\theta$, $z = (1/\sqrt{3})[\cos\theta - \sin\theta]$, $y = x + z$. Also, see 8.4.6.)

8.1.12. In Figure 8.12 we are given an inscribed triangle ABC. Let R denote the circumradius; let h_a denote the altitude AD.

(a) Show that triangles ABD and ALC are similar, and hence that $h_a \cdot 2R = bc$.

(b) Show that the area of $\triangle ABC$ is $abc/4R$.

8.1.13. The radius of the inscribed circle of a triangle is 4, and the segments into which one side is divided by the point of contact are 6 and 8. Determine the other two sides.

8.1.14. Triangles ABC and DEF are inscribed in the same circle. Prove that

$$\sin A + \sin B + \sin C = \sin D + \sin E + \sin F$$

if and only if the perimeters of the given triangles are equal.

8.1.15. In the following figure, CD is a half chord perpendicular to the diameter AB of the semicircle with center O. A circle with center P is inscribed as shown in Figure 8.13, touching AB at E and arc BD at F.

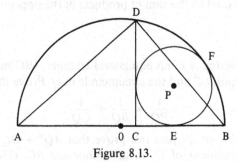

Figure 8.13.

Prove that $\triangle EDA$ is isosceles. (Hint: Label the figure and make good use of the Pythagorean theorem.)

8.1.16. Find the length of a side of an equilateral triangle in which the distances from its vertices to an interior point are 5, 7, and 8.

Additional Examples

1.2.1, 1.3.14, 1.4.2, 1.6.1, 1.6.10, 1.8.3, 1.8.7.

8.2. Analytic Geometry

The introduction of a coordinate system makes it possible to attack many geometry problems by way of algebra and analysis.

8.2.1. Let P be a point on an ellipse with foci F_1 and F_2, and let d be the distance from the center of the ellipse to the line tangent to the ellipse at P (Figure 8.14). Prove that $(PF_1)(PF_2)d^2$ is constant as P moves on the ellipse.

Solution. Place coordinates on the plane in such a way that the ellipse has the equation

$$\frac{x^2}{a^2} + \frac{y^2}{b^2} = 1, \qquad 0 < b \leqslant a.$$

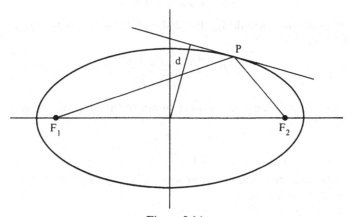

Figure 8.14.

The technique is straightforward: compute PF_1, PF_2, and d (as functions of the x-coordinate of P), and check to see if the required product is a constant.

Let the coordinate of P be (α, β). The focal points F_1 and F_2 have coordinates $(\pm c, 0)$, where $c^2 = a^2 - b^2$. Therefore we have

$$PF_1 = \sqrt{\beta^2 + (\alpha + c)^2}\,,$$

$$PF_2 = \sqrt{\beta^2 + (\alpha - c)^2}\,.$$

To find d^2, it is necessary to write the equation of the tangent to the ellipse at $P(\alpha, \beta)$. To find the slope of the tangent at P, we compute the derivative:

$$\frac{2x}{a^2} + \frac{2y}{b^2}\, y' = 0,$$

so that

$$y' = \frac{-2x/a^2}{2y/b^2} = -\frac{b^2x}{a^2y}\,.$$

It follows that the equation of the tangent at $P(\alpha, \beta)$ is

$$y - \beta = -\frac{b^2\alpha}{a^2\beta}(x - \alpha),$$

or equivalently,

$$a^2\beta y + b^2\alpha x = b^2\alpha^2 + a^2\beta^2.$$

But $\alpha^2/a^2 + \beta^2/b^2 = 1$, since $P(\alpha, \beta)$ is a point on the ellipse, and therefore $\alpha^2 b^2 + a^2\beta^2 = a^2 b^2$. Hence the equation of the tangent at $P(\alpha, \beta)$ is

$$\alpha b^2 x + \beta a^2 y - a^2 b^2 = 0.$$

Now recall the formula for the distance D from a point $Q(c, d)$ to the line $Ax + By + C = 0$:

$$D = \frac{|Ac + Bd + C|}{\sqrt{A^2 + B^2}}\,.$$

In our case, the distance d from the origin to the tangent line is

$$d = \frac{a^2 b^2}{\sqrt{\alpha^2 b^4 + \beta^2 a^4}}\,.$$

We now need to examine the product $d^2(PF_1)(PF_2)$. We can eliminate β in each of these factors, since

$$\beta^2 = \frac{a^2 b^2 - \alpha^2 b^2}{a^2}\,.$$

We have

$$d^2 = \frac{a^4 b^4}{\alpha^2 b^4 + ((a^2 b^2 - \alpha^2 b^2)/a^2)a^4}$$

$$= \frac{a^4 b^4}{\alpha^2 b^4 + a^4 b^2 - \alpha^2 a^2 b^2}$$

$$= \frac{a^4 b^4}{b^2 [\alpha^2 b^2 - \alpha^2 a^2] + a^4 b^2}$$

$$= \frac{a^4 b^4}{b^2(-c^2 \alpha^2) + a^4 b^2}$$

$$= \frac{a^4 b^2}{a^4 - c^2 \alpha^2},$$

and

$$PF_1^2 = \beta^2 + (\alpha + c)^2$$

$$= \frac{a^2 b^2 - \alpha^2 b^2}{a^2} + \alpha^2 + 2\alpha c + c^2$$

$$= \frac{a^2 b^2 - \alpha^2 b^2 + a^2 \alpha^2 + 2a^2 \alpha c + a^2 c^2}{a^2}$$

$$= \frac{a^2(b^2 + c^2) + \alpha^2(a^2 - b^2) + 2a^2 c}{a^2}$$

$$= \frac{a^4 + 2a^2 c\alpha + c^2 \alpha^2}{a^2}$$

$$= \frac{(a^2 + c\alpha)^2}{a^2}.$$

Similarly,

$$PF_2^2 = \frac{(a^2 - c\alpha)^2}{a^2}.$$

Thus,

$$d^2 (PF_1)(PF_2) = \left(\frac{a^4 b^2}{a^4 - c^2 \alpha^2} \right) \left(\frac{a^2 + c\alpha}{a} \right) \left(\frac{a^2 - c\alpha}{a} \right)$$

$$= a^2 b^2.$$

This completes the proof.

8.2.2. Suppose that (x_1, y_1), (x_2, y_2), (x_3, y_3) are three points on the parabola $y^2 = ax$ which have the property that their normal lines intersect in a common point. Prove that $y_1 + y_2 + y_3 = 0$.

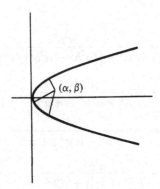

Figure 8.15.

Solution. The solution falls out as a by-product of the following analysis. Let (α, β) be the coordinates of the intersection of the three normal lines (Figure 8.15), and let (x, y) be an arbitrary point on the parabola. The slope of the line through (x, y) and (α, β) is $(y - \beta)/(x - \alpha)$. The slope of the tangent to the parabola at (x, y) is $y' = a/(2y)$, and therefore, the slope of the normal line at (x, y) is $-2y/a$. It follows that (x_1, y_1), (x_2, y_2), (x_3, y_3) satisfy the equation

$$\frac{y - \beta}{x - \alpha} = -\left(\frac{2y}{a}\right).$$

Replacing x by y^2/a, this equation is

$$y - \beta = -\left(\frac{2y}{a}\right)\left(\frac{y^2}{a} - \alpha\right),$$

$$a^2(y - \beta) = -2y^3 + 2a\alpha y.$$

Thus y_1, y_2, y_3 are the three roots of the cubic equation

$$2y^3 + a(a - 2)y - a^2 = 0.$$

Now, remembering how the coefficients of a cubic equation are related to the roots (see Section 4.3), we see that $y_1 + y_2 + y_3 = 0$ (the coefficient of y^2 is zero).

8.2.3. A straight line cuts the asymptotes of a hyperbola in points A and B and the curve in points P and Q. Prove that $AP = BQ$.

Solution. We may assume the hyperbola and the straight line have the equations

$$xy = 1 \tag{1}$$

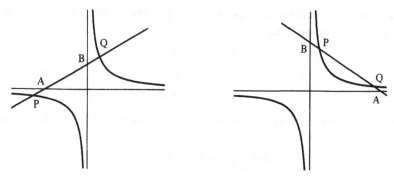

Figure 8.16.

and

$$\frac{x}{a} + \frac{y}{b} = 1 \qquad (2)$$

respectively. (The hyperbola can be taken to have this form by appropriate scaling followed by a rotation, each of which takes straight lines into straight lines and preserves ratios of line segments.)

The asymptotes of the hyperbola are the x and y axes (Figure 8.16); so let A be the x-intercept of the line, and let B be the y-intercept. Let (x_1, y_1) and (x_2, y_2) be the coordinates of P and Q. Substituting $y = 1/x$ into (2) yields

$$x^2 - ax + a/b = 0,$$

and since x_1 and x_2 are roots of this equation, we know that

$$x_1 + x_2 = a.$$

Similarly, substituting $x = 1/y$ into (2) yields

$$y^2 - by + b/a = 0,$$

and this implies that

$$y_1 + y_2 = b.$$

It follows that

$$
\begin{aligned}
AP^2 &= (x_1 - a)^2 + y_1^2 \\
&= (a - x_2 - a)^2 + (b - y_2)^2 \\
&= x_2^2 + (b - y_2)^2 \\
&= BQ^2,
\end{aligned}
$$

and the result follows.

8.2.4. Determine all the straight lines lying in the surface $z = xy$.

Solution. The parametric equation for the line through (a_1, a_2, a_3) with direction (d_1, d_2, d_3) is given by

$$x = a_1 + d_1 t,$$

$$y = a_2 + d_2 t,$$

$$z = a_3 + d_3 t.$$

For such a line to lie in the surface $z = xy$ it is necessary and sufficient that for all t,

$$a_3 + d_3 t = (a_1 + d_1 t)(a_2 + d_2 t)$$

$$= a_1 a_2 + (a_2 d_1 + a_1 d_2)t + d_1 d_2 t^2.$$

It follows that $d_1 d_2 = 0$, and d_1 and d_2 cannot both be zero, since this would imply that $d_1 = d_2 = d_3 = 0$, a contradiction.

If $d_2 = 0$, then

$$a_3 + d_3 t = a_2(a_1 + d_1 t),$$

or

$$z = a_2 x.$$

If $d_1 = 0$, then

$$a_3 + d_3 t = a_1(a_2 + d_2 t),$$

or

$$z = a_1 y.$$

Thus, the only straight lines in the surface $z = xy$ are of the form $z = ax$, $y = a$ or of the form $z = ay$, $x = a$, where a is an arbitrary constant.

8.2.5. An equilateral triangle ABC is projected orthogonally from a given plane P to another plane P'. Show that the sum of the squares of the sides of the resulting triangle $A'B'C'$ (Figure 8.17) is independent of the orientation of the triangle ABC in P.

Solution. First, some observations about how lengths are transformed under this projection. Suppose that AB is a line segment in P of length one, and that it makes an angle ϕ with the line L of intersection of P and P'. Let θ denote the angle between the planes. Locate C so that $\triangle ABC$ is a right triangle, and AC is parallel to L (see figure). Triangle ABC will project into a right triangle $A'B'C'$. Furthermore, AC and $A'C'$ have the same length, and $B'C' = BC \cos\theta$. Since $AC = \cos\phi$ and $BC = \sin\phi$, it follows that

$$A'B' = \sqrt{(\cos\phi)^2 + (\sin\phi\cos\theta)^2} \ .$$

Now, let ABC denote an arbitrary equilateral triangle in P. We may suppose that the length of the side is one. Suppose that AB makes an angle

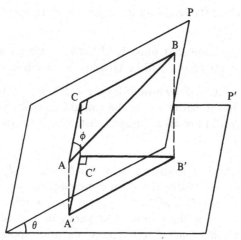

Figure 8.17.

ϕ with L. Then BC and CA will make angles of $\phi + \frac{1}{3}\pi$ and $\phi + \frac{2}{3}\pi$ with L. Applying the result obtained above, we find that the sum of the squares of the sides of triangle $A'B'C'$ is

$$\left[(\cos\phi)^2 + (\sin\phi\cos\theta)^2\right] + \left[(\cos(\phi + \tfrac{1}{3}\pi))^2 + (\sin(\phi + \tfrac{1}{3}\pi)\cos\theta)^2\right]$$
$$+ \left[(\cos(\phi + \tfrac{2}{3}\pi))^2 + (\sin(\phi + \tfrac{2}{3}\pi)\cos\theta)^2\right],$$

which reduces to

$$3\cos^2\theta + \tfrac{3}{2}\sin^2\theta,$$

which is independent of ϕ.

Problems

8.2.6. Let the triangle ABC be inscribed in a circle, let P denote the centroid of the triangle, and let O denote the circumcenter. Suppose that A, B, C have coordinates $(0,0)$, $(a,0)$, and (b,c) respectively.

(a) Express the coordinates of P and O in terms of a, b, c.
(b) Extend line segments AP, BP, and CP to meet the circle in points D, E, and F respectively. Show that

$$\frac{AP}{PD} + \frac{BP}{PE} + \frac{CP}{PF} = 3.$$

(Hint: One way to proceed is the following: Let x denote OP, and let R denote the radius of the circumcircle. Then

$$\frac{AP}{PD} + \frac{BP}{PE} + \frac{CP}{PF} = \frac{AP^2 + BP^2 + CP^2}{R^2 - x^2}.$$

Now express each of the terms on the right side in terms of a, b, c [using the results of part (a)].)

8.2.7. Find the relation that must hold between the parameters a, b, c so that the line $x/a + y/b = 1$ will be tangent to the circle $x^2 + y^2 = c^2$.

8.2.8. Equilateral triangles whose sides are $1, 3, 5, 7, \ldots$ are placed so that the bases lie corner to corner along the straight line. Show that the vertices lie on a parabola and are all at integral distances from its focus.

8.2.9.

(a) Tangents are drawn from two points (a, b) and (c, d) on the parabola $y = x^2$. Find the coordinates of their intersection.
(b) Two tangent lines, L_1 and L_2, are drawn from a point T to a parabola; let P and Q denote the points of tangency of L_1 and L_2 respectively. Let L be any other tangent to the parabola, and suppose L intersects L_1 and L_2 at R and S respectively. Prove that

$$\frac{TR}{TP} + \frac{TS}{TQ} = 1.$$

8.2.10. A parabola with equation $y^2 = ax$ is cut in four points by the circle $(x - h)^2 + (y - k)^2 = r^2$. Determine the product of the distances of the four points of intersection from the axis of the parabola.

8.2.11. Let b and c be fixed real numbers, and let the ten points (j, y_j), $j = 1, 2, \ldots, 10$, lie on the parabola $y = x^2 + bx + c$. For $j = 1, 2, \ldots, 9$, let I_j be the point of intersection of the tangents to the given parabola at (j, y_j) and $(j + 1, y_{j+1})$. Determine the polynomial function $y = g(x)$ of least degree whose graph passes through all nine points I_j.

8.2.12. Prove or disprove: there is at least one straight line normal to the graph of $y = \cosh x$ at a point $(a, \cosh a)$ and also normal to the graph of $y = \sinh x$ at a point $(c, \sinh c)$.

8.2.13.

(a) Show that the tangent lines to ellipse $x^2/a^2 + y^2/b^2 = 1$ have the form

$$y = \alpha x \pm (a^2\alpha^2 + b^2)^{1/2},$$

and vary in position with different values of α. (Because of the great utility of this form, particularly in problems of tangency which do not involve the consideration of the point of contact, this is called the *magical equation of the tangent*.)
(b) Find the equation of the tangents to the ellipse $3x^2 + y^2 = 3$ which have slope of one.
(c) Find the area of the triangle formed by a tangent to the ellipse (say of slope m) and the two coordinate axes.

8.2.14.

(a) Let D be the disk $x^2 + y^2 < 1$. Let the point A have coordinates $(r, 0)$, where $0 < r < 1$. Describe the set of points P in D such that the open disk whose center is the midpoint of AP and whose radius is $AP/2$ is a subset of D.

(b) Let D be the disk $x^2 + y^2 < 1$. Let points A and B be selected at random in D. Find the probability that the open disk whose center is the midpoint of AB and whose radius is $AB/2$ is a subset of D.

8.2.15. Given an ellipse $x^2/a^2 + y^2/b^2 = 1$, $a \neq b$, find the equation of the set of all points from which there are two tangents to the ellipse whose slopes are reciprocals.

8.2.16. If two chords of a conic are mutually bisecting, prove that the conic cannot be a parabola.

8.2.17. Prove that the graph of a cubic equation is symmetric about its point of inflection. (Note: If the cubic equation is $f(x) = ax^3 + bx^2 + cx + d$, the x-coordinate of the inflection point is $-b/3a$.)

Additional Examples

1.3.11, 1.5.3, 1.5.8, 1.6.4, 3.1.4, 4.3.6, 4.3.7.

8.3. Vector Geometry

In this section we will think of vectors as quantities which have both magnitude and direction. Examples of vector quantities include force, velocity, and acceleration. We shall see that vectors can also be used advantageously in geometry problems.

We will represent vectors by arrows (i.e., directed line segments) in the Euclidean plane. The direction of the arrow indicates the direction of the vector, and the length of the arrow indicates the magnitude of the vector.

Two vectors are *equal* if they have the same length and the same direction. It is important to realize that two vectors may be equal without being collinear.

If P and Q are two points, the vector from P to Q will be denoted by $\overrightarrow{PQ}$. The length, or magnitude, of $\overrightarrow{PQ}$ will be denoted by $|\overrightarrow{PQ}|$.

The sum, $\vec{A} + \vec{B}$, of vectors $\vec{A}$ and $\vec{B}$ is given by the parallelogram law (see Figure 8.18), or equivalently, by completing the triangle as in Figure 8.19. The difference, $\vec{A} - \vec{B}$, of $\vec{B}$ from $\vec{A}$, is shown geometrically in Figure 8.20.

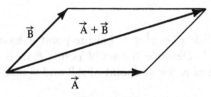

Figure 8.18.

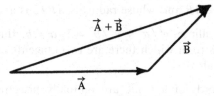

Figure 8.19.

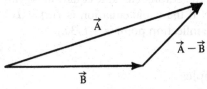

Figure 8.20.

Place coordinates on the plane and denote the origin by 0. Each point P in the plane determines a unique vector, $\overrightarrow{OP}$, called the position vector of P; we will often denote this vector simply by $\vec{P}$ (instead of $\overrightarrow{OP}$).

Suppose that $\vec{P}$ and $\vec{Q}$ are the position vectors of two points P and Q (Figure 8.21). Let R be a point on the directed line segment PQ which divides PQ in the ratio $m : n$. (Figure 8.22). Then the position vector of R is

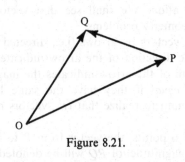

Figure 8.21.

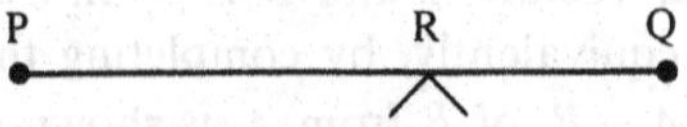

Figure 8.22.

given by

$$\vec{R} = \vec{P} + \frac{m}{m+n}(\vec{Q} - \vec{P})$$

$$= \frac{(m+n)\vec{P} + m(\vec{Q} - \vec{P})}{m+n}$$

$$= \left(\frac{n}{m+n}\right)\vec{P} + \left(\frac{m}{m+n}\right)\vec{Q}.$$

It is instructive to think of $\vec{R}$ in physical terms in the following way. Imagine a weightless bar PQ with a mass of $n/(m+n)$ at P and a mass of $m/(m+n)$ at Q. The center of mass of the resulting system will be at a point X on PQ where "the seesaw balances," that is, at the point X where

$$\left(\frac{n}{m+n}\right)PX = \left(\frac{m}{m+n}\right)XQ.$$

But this is the same as

$$\frac{PX}{XQ} = \frac{m}{n}.$$

Thus, X divides PQ into the ratio $m:n$; in other words, $\vec{X} = \vec{R} = (n/(m+n))\vec{P} + (m/(m+n))\vec{Q}$. The coefficients $n/(m+n)$ and $m/(m+n)$ can be thought of as "weighting factors." Increasing the proportion of "weight" at P moves the point R toward P and decreases the ratio $m:n$, etc.

8.3.1. In a triangle ABC the points D, E, F trisect the sides so that $BC = 3BD$, $CA = 3CE$, and $AB = 3AF$ (Figure 8.23). Show that triangles ABC and DEF have the same centroid.

Solution. We will first show that the position vector of the centroid of an arbitrary triangle PQR is given by $\frac{1}{3}\vec{P} + \frac{1}{3}\vec{Q} + \frac{1}{3}\vec{R}$. To see this, remember that the centroid of triangle PQR is located at a point $\frac{2}{3}$ of the way from P to the midpoint of QR. From the discussion preceding the problem, we know that the position vector for the midpoint of QR is $\frac{1}{2}\vec{Q} + \frac{1}{2}\vec{R}$, and

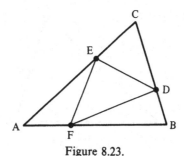

Figure 8.23.

therefore, the position vector of the centroid of PQR is $\frac{1}{3}\vec{P} + \frac{2}{3}(\frac{1}{2}\vec{Q} + \frac{1}{2}\vec{R})$, and this is equal to $\frac{1}{3}\vec{P} + \frac{1}{3}\vec{Q} + \frac{1}{3}\vec{R}$, as claimed.

Because of the way D, E, and F are defined, we have

$$\vec{D} = \tfrac{2}{3}\vec{B} + \tfrac{1}{3}\vec{C},$$
$$\vec{E} = \tfrac{2}{3}\vec{C} + \tfrac{1}{3}\vec{A},$$
$$\vec{F} = \tfrac{2}{3}\vec{A} + \tfrac{1}{3}\vec{B},$$

and therefore

$$\text{centroid of } \triangle DEF = \tfrac{1}{3}\vec{D} + \tfrac{1}{3}\vec{E} + \tfrac{1}{3}\vec{F}$$
$$= \tfrac{1}{3}\left[\tfrac{2}{3}\vec{B} + \tfrac{1}{3}\vec{C}\right] + \tfrac{1}{3}\left[\tfrac{2}{3}\vec{C} + \tfrac{1}{3}\vec{A}\right] + \tfrac{1}{3}\left[\tfrac{2}{3}\vec{A} + \tfrac{1}{3}\vec{B}\right]$$
$$= \tfrac{1}{3}\vec{A} + \tfrac{1}{3}\vec{B} + \tfrac{1}{3}\vec{C} = \text{centroid of } \triangle ABC.$$

8.3.2. Prove that it is possible to construct a triangle with sides equal and parallel to the medians of a given triangle.

Solution. Consider a triangle ABC, and let D, E, F be the midpoints of sides BC, AC, and AB respectively (see Figure 8.24). Then

$$\vec{AD} = \vec{AB} + \tfrac{1}{2}\vec{BC},$$
$$\vec{BE} = \vec{BC} + \tfrac{1}{2}\vec{CA},$$
$$\vec{CF} = \vec{CA} + \tfrac{1}{2}\vec{AB}.$$

Adding these, we find that $\vec{AD} + \vec{BE} + \vec{CF} = (\vec{AB} + \vec{BC} + \vec{CA}) + \frac{1}{2}(\vec{BC} + \vec{CA} + \vec{AB}) = 0 + (\frac{1}{2})\cdot 0 = 0$. This implies that the vectors $\vec{AD}$, $\vec{BE}$, and $\vec{CF}$ form a triangle. But $\vec{AD}$, $\vec{BE}$, and $\vec{CF}$ are equal in magnitude and direction to the medians of triangle ABC.

Before considering the next examples we will develop the following basic principle. Suppose that P, Q, and R are points which are not collinear

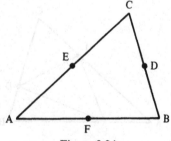

Figure 8.24.

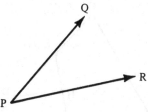

Figure 8.25.

(Figure 8.25). If $a\overrightarrow{PQ} + b\overrightarrow{PR} = c\overrightarrow{PQ} + d\overrightarrow{PR}$, then $a = c$ and $b = d$. For if the condition holds and if $a \neq c$, then

$$\overrightarrow{PQ} = \left(\frac{d-b}{a-c}\right)\overrightarrow{PR},$$

and this implies that P, Q, R are collinear (vectors $\overrightarrow{PQ}$ and $\overrightarrow{PR}$ have the point P in common), which is a contradiction. Therefore $a = c$. In a similar manner $b = d$.

8.3.3. Prove that the line joining one vertex of a parallelogram to the midpoint of an opposite side trisects a diagonal of the parallelogram.

Solution. Label the parallelogram by A, B, C, D as shown in Figure 8.26, let F be the midpoint of DC, and let E be the intersection of AF and BD. Note that $\overrightarrow{AB} = \overrightarrow{DC}$ and $\overrightarrow{AD} = \overrightarrow{BC}$, because as vectors they have the same magnitude and the same direction.

The point E is at the intersection of two lines. We can express this algebraically by saying that there exist constants a and b such that

$$\overrightarrow{AE} = a\overrightarrow{AF},$$
$$\overrightarrow{AE} = \overrightarrow{AB} + b\overrightarrow{BD}.$$

Therefore,

$$\overrightarrow{AB} + b\overrightarrow{BD} = a\overrightarrow{AF}.$$

The idea is to express each of the vectors in this last equation in terms of $\overrightarrow{AB}$ and $\overrightarrow{AD}$, and then we will make use of the principle discussed prior to

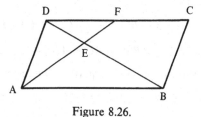

Figure 8.26.

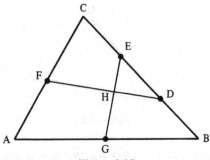

Figure 8.27.

the statement of the problem. Thus we have

$$\overrightarrow{AB} + b(\overrightarrow{AD} - \overrightarrow{AB}) = a(\overrightarrow{AD} + \tfrac{1}{2}\overrightarrow{AB}),$$

$$(1 - b)\overrightarrow{AB} + b\overrightarrow{AD} = \tfrac{1}{2}a\overrightarrow{AB} + a\overrightarrow{AD}.$$

It follows that

$$1 - b = \tfrac{1}{2}a,$$

$$b = a.$$

These equations imply that $a = b = \tfrac{2}{3}$, and the result follows.

8.3.4. In triangle ABC (Figure 8.27), let D and E be the trisection points of sides BC with D between B and E, let F be the midpoint of side AC, and let G be the midpoint of side AB. Let H be the intersection of segments EG and DF. Find the ratio $EH : HG$.

Solution. The plan is exactly as in the preceding problem. There are constants a and b such that

$$\overrightarrow{AG} + a\overrightarrow{GE} = \overrightarrow{AF} + b\overrightarrow{FD}.$$

Now express each of the vectors in terms of $\overrightarrow{AB}$ and $\overrightarrow{AC}$:

$$\overrightarrow{AG} = \tfrac{1}{2}\overrightarrow{AB},$$

$$\overrightarrow{GE} = \overrightarrow{GB} + \overrightarrow{BE} = \tfrac{1}{2}\overrightarrow{AB} + \tfrac{2}{3}\overrightarrow{BC}$$

$$= \tfrac{1}{2}\overrightarrow{AB} + \tfrac{2}{3}(\overrightarrow{AC} - \overrightarrow{AB}) = -\tfrac{1}{6}\overrightarrow{AB} + \tfrac{2}{3}\overrightarrow{AC},$$

$$\overrightarrow{AF} = \tfrac{1}{2}\overrightarrow{AC},$$

$$\overrightarrow{FD} = \overrightarrow{FA} + \overrightarrow{AB} + \overrightarrow{BD} = -\tfrac{1}{2}\overrightarrow{AC} + \overrightarrow{AB} + \tfrac{1}{3}\overrightarrow{BC}$$

$$= -\tfrac{1}{2}\overrightarrow{AC} + \overrightarrow{AB} + \tfrac{1}{3}(\overrightarrow{AC} - \overrightarrow{AB}) = -\tfrac{1}{6}\overrightarrow{AC} + \tfrac{2}{3}\overrightarrow{AB}.$$

Substituting these into the previous equation, we have

$$\tfrac{1}{2}\overrightarrow{AB} + a\left[-\tfrac{1}{6}\overrightarrow{AB} + \tfrac{2}{3}\overrightarrow{AC}\right] = \tfrac{1}{2}\overrightarrow{AC} + b\left[-\tfrac{1}{6}\overrightarrow{AC} + \tfrac{2}{3}\overrightarrow{AB}\right],$$

$$(\tfrac{1}{2} - \tfrac{1}{6}a)\overrightarrow{AB} + \tfrac{2}{3}a\overrightarrow{AC} = \tfrac{2}{3}b\overrightarrow{AB} + (\tfrac{1}{2} - \tfrac{1}{6}b)\overrightarrow{AC}.$$

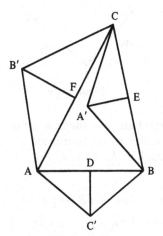

Figure 8.28.

It follows that

$$\tfrac{1}{2} - \tfrac{1}{6}a = \tfrac{2}{3}b,$$

$$\tfrac{2}{3}a = \tfrac{1}{2} - \tfrac{1}{6}b,$$

or equivalently,

$$a + 4b = 3,$$

$$4a + b = 3.$$

The solution to this system is $a = b = \tfrac{3}{5}$, and it follows that $EH:HG = 2:3$.

8.3.5. Given a triangle ABC, construct similar isosceles triangles ABC' and ACB' outwards on the respective bases AB and AC, and BCA' inwards on the base BC (Figure 8.28). Show that $AB'A'C'$ is a parallelogram.

Solution. In vector language, our problem is to show that $\overrightarrow{AB'} + \overrightarrow{AC'} = \overrightarrow{AA'}$.

Let D, E, F be the midpoints of sides AB, BC, AC respectively. Then

$$\overrightarrow{AB'} = \overrightarrow{AF} + \overrightarrow{FB'} = \tfrac{1}{2}\overrightarrow{AC} + \overrightarrow{FB'}$$

$$\overrightarrow{AC'} = \overrightarrow{AD} + \overrightarrow{DC'} = \tfrac{1}{2}\overrightarrow{AB} + \overrightarrow{DC'}$$

$$\overrightarrow{AA'} = \overrightarrow{AB} + \overrightarrow{BE} + \overrightarrow{EA'}$$

$$= \overrightarrow{AB} + \tfrac{1}{2}(\overrightarrow{AC} - \overrightarrow{AB}) + \overrightarrow{EA'} = \tfrac{1}{2}\overrightarrow{AB} + \tfrac{1}{2}\overrightarrow{AC} + \overrightarrow{EA'}.$$

To put $\overrightarrow{FB'}$, $\overrightarrow{DC'}$, and $\overrightarrow{EA'}$ into terms of $\overrightarrow{AB}$ and $\overrightarrow{AC}$, we introduce the following notation (useful in other problems as well). Given points P and Q, let $|\overrightarrow{PQ}$ denote the vector obtained by rotating $\overrightarrow{PQ}$, with unchanged

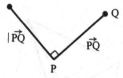

Figure 8.29.

magnitude, through a right angle in the positive direction, as in Figure 8.29. Now, suppose the isosceles triangles erected on the sides of ABC have height-to-base ratio equal to k; that is, $FB'/AC = DC'/AB = EA'/BC = k$. Then,

$$\overrightarrow{AB'} = \tfrac{1}{2}\overrightarrow{AC} + \overrightarrow{FB'} = \tfrac{1}{2}\overrightarrow{AC} + k\,|\overrightarrow{AC},$$

$$\overrightarrow{AC'} = \tfrac{1}{2}\overrightarrow{AB} + \overrightarrow{DC'} = \tfrac{1}{2}\overrightarrow{AB} - k\,|\overrightarrow{AB},$$

$$\overrightarrow{AA'} = \tfrac{1}{2}\overrightarrow{AB} + \tfrac{1}{2}\overrightarrow{AC} + \overrightarrow{EA}$$

$$= \tfrac{1}{2}\overrightarrow{AB} + \tfrac{1}{2}\overrightarrow{AC} + k\,|\overrightarrow{BC}$$

$$= \tfrac{1}{2}\overrightarrow{AB} + \tfrac{1}{2}\overrightarrow{AC} + k\,|(\overrightarrow{AC} - \overrightarrow{AB})$$

$$= \tfrac{1}{2}\overrightarrow{AB} + \tfrac{1}{2}\overrightarrow{AC} + k\,|\overrightarrow{AC} - k\,|\overrightarrow{AB}.$$

(Note that $|(\vec{P} + \vec{Q}) = |\vec{P} + |\vec{Q}$, and $|a\vec{P} = a|\vec{P}$ for an arbitrary constant a.) These expressions for $\overrightarrow{AB'}$, $\overrightarrow{AC'}$, and $\overrightarrow{AA'}$ show that $\overrightarrow{AB'} + \overrightarrow{AC'} = \overrightarrow{AA'}$, and thus the solution is complete.

Given vectors $\overrightarrow{PQ}$ and $\overrightarrow{RS}$, the *dot product* $\overrightarrow{PQ}\cdot\overrightarrow{RS}$ is defined by the formula

$$\overrightarrow{PQ}\cdot\overrightarrow{RS} = |\overrightarrow{PQ}||\overrightarrow{RS}|\cos\theta,$$

where θ is the angle between the vectors, $0 \leqslant \theta \leqslant 180°$.

It can be shown that for arbitrary vectors $\vec{A}$, $\vec{B}$, $\vec{C}$,

$$\vec{A}\cdot\vec{B} = \vec{B}\cdot\vec{A}$$

and

$$\vec{A}\cdot(\vec{B} + \vec{C}) = \vec{A}\cdot\vec{B} + \vec{A}\cdot\vec{C}.$$

Notice that if $\vec{A}$ and $\vec{B}$ are perpendicular, then $\vec{A}\cdot\vec{B} = 0$. Conversely, if $\vec{A}\cdot\vec{B} = 0$, then either $\vec{A} = 0$ or $\vec{B} = 0$, or $\vec{A}$ and $\vec{B}$ are perpendicular. Also, notice that $\vec{A}\cdot\vec{A} = |\vec{A}|^2$.

8.3.6. In triangle ABC (Figure 8.30), $AB = AC$, D is the midpoint of BC, E is the foot of the perpendicular drawn D to AC, and F is the midpoint of DE. Prove that AF is perpendicular to BE.

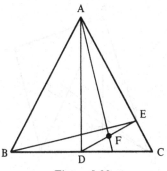

Figure 8.30.

Solution. This is the same problem as 1.5.3, but here we will give a proof using vector notation. We have

$$\overrightarrow{AF} \cdot \overrightarrow{BE} = (\overrightarrow{AE} + \overrightarrow{EF}) \cdot (\overrightarrow{BD} + \overrightarrow{DE})$$

$$= \overrightarrow{AE} \cdot \overrightarrow{BD} + \overrightarrow{EF} \cdot \overrightarrow{BD} + \overrightarrow{EF} \cdot \overrightarrow{DE}$$

$$= (\overrightarrow{AD} + \overrightarrow{DE}) \cdot \overrightarrow{BD} + \overrightarrow{EF} \cdot \overrightarrow{BD} + \overrightarrow{EF} \cdot \overrightarrow{DE}$$

$$= \overrightarrow{DE} \cdot \overrightarrow{BD} + \overrightarrow{EF} \cdot \overrightarrow{BD} + \overrightarrow{EF} \cdot \overrightarrow{DE}$$

$$= \overrightarrow{DE} \cdot \overrightarrow{DC} - \frac{\overrightarrow{DE} \cdot \overrightarrow{DC}}{2} - \frac{\overrightarrow{DE} \cdot \overrightarrow{DE}}{2}$$

$$= \frac{\overrightarrow{DE} \cdot \overrightarrow{DC}}{2} - \frac{\overrightarrow{DE} \cdot \overrightarrow{DE}}{2}$$

$$= \overrightarrow{DE} \cdot \left(\frac{\overrightarrow{DC} - \overrightarrow{DE}}{2} \right)$$

$$= \frac{\overrightarrow{DE} \cdot \overrightarrow{EC}}{2}$$

$$= 0.$$

The concept of vector makes sense in Euclidean 3-space just as it does in the Euclidean plane. Just as in the case of the plane, vectors have length and direction and are represented as arrows or directed line segments (but now in 3-space). They are added by the parallelogram law, and can be manipulated to prove results in solid geometry.

8.3.7. If two altitudes of a tetrahedron are coplanar, the edge joining the two vertices from which these altitudes issue is orthogonal to the opposite edge of the tetrahedron.

Solution. Suppose that AP and BZ are altitudes from A and B respectively, and suppose they intersect in a point H (Figure 8.31).

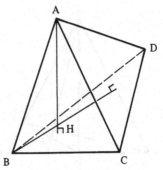

Figure 8.31.

$\overrightarrow{AH}$ is orthogonal to each of $\overrightarrow{BC}$, $\overrightarrow{CD}$, and $\overrightarrow{BD}$, and $\overrightarrow{BH}$ is orthogonal to $\overrightarrow{CD}$, $\overrightarrow{AD}$, $\overrightarrow{AC}$. We wish to show that $\overrightarrow{AB}$ is orthogonal to $\overrightarrow{CD}$. For this, we compute the dot product:

$$\overrightarrow{AB} \cdot \overrightarrow{CD} = (\overrightarrow{HB} - \overrightarrow{HA}) \cdot \overrightarrow{CD} = \overrightarrow{HB} \cdot \overrightarrow{CD} - \overrightarrow{HA} \cdot \overrightarrow{CD} = 0 - 0 = 0.$$

This completes the proof.

8.3.8. Prove that if the opposite sides of a skew (nonplanar) quadrilateral have the same lengths, then the line joining the midpoints of the two diagonals is perpendicular to these diagonals.

Solution. Let A, B, C, D denote the vertices of the quadrilateral, and let P and Q be the midpoints of AC and BD respectively (Figure 8.32). We are given that $|\overrightarrow{AD}| = |\overrightarrow{BC}|$ and $|\overrightarrow{AB}| = |\overrightarrow{CD}|$. Squaring, and translating into dot-product language, we have

$$\overrightarrow{AD} \cdot \overrightarrow{AD} = \overrightarrow{BC} \cdot \overrightarrow{BC},$$
$$\overrightarrow{AB} \cdot \overrightarrow{AB} = \overrightarrow{CD} \cdot \overrightarrow{CD},$$

or equivalently,

$$(\vec{D} - \vec{A}) \cdot (\vec{D} - \vec{A}) = (\vec{C} - \vec{B}) \cdot (\vec{C} - \vec{B}), \tag{1}$$
$$(\vec{B} - \vec{A}) \cdot (\vec{B} - \vec{A}) = (\vec{D} - \vec{C}) \cdot (\vec{D} - \vec{C}).$$

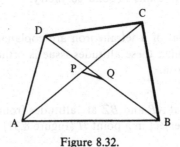

Figure 8.32.

We wish to prove that $\overrightarrow{PQ}$ is perpendicular to $\overrightarrow{AC}$ and $\overrightarrow{BD}$; in vector language we wish to show that

$$\overrightarrow{PQ} \cdot \overrightarrow{AC} = 0,$$
$$\overrightarrow{PQ} \cdot \overrightarrow{BD} = 0,$$

or equivalently,

$$(\vec{Q} - \vec{P}) \cdot (\vec{C} - \vec{A}) = 0,$$
$$(\vec{Q} - \vec{P}) \cdot (\vec{D} - \vec{B}) = 0.$$

Substituting $\vec{P} = \frac{1}{2}(\vec{A} + \vec{C})$ and $\vec{Q} = \frac{1}{2}(\vec{B} + \vec{D})$, these equations yield

$$(\vec{B} + \vec{D} - \vec{A} - \vec{C}) \cdot (\vec{C} - \vec{A}) = 0,$$
$$(\vec{B} + \vec{D} - \vec{A} - \vec{C}) \cdot (\vec{D} - \vec{B}) = 0. \tag{2}$$

Our problem then is equivalent to showing that the equations (1) imply the equations (2).

Expanding (1), we have

$$\vec{D} \cdot \vec{D} - 2\vec{A} \cdot \vec{D} + \vec{A} \cdot \vec{A} = \vec{C} \cdot \vec{C} - 2\vec{B} \cdot \vec{C} + \vec{B} \cdot \vec{B},$$
$$\vec{B} \cdot \vec{B} - 2\vec{A} \cdot \vec{B} + \vec{A} \cdot \vec{A} = \vec{D} \cdot \vec{D} - 2\vec{C} \cdot \vec{D} + \vec{C} \cdot \vec{C}.$$

Adding these, we get

$$-2\vec{A} \cdot \vec{D} - 2\vec{A} \cdot \vec{B} + 2\vec{A} \cdot \vec{A} = 2\vec{C} \cdot \vec{C} - 2\vec{B} \cdot \vec{C} - 2\vec{C} \cdot \vec{D},$$
$$-(\vec{B} + \vec{D}) \cdot \vec{A} + \vec{A} \cdot \vec{A} = \vec{C} \cdot \vec{C} - \vec{C} \cdot (\vec{B} + \vec{D}),$$
$$(\vec{B} + \vec{D}) \cdot (\vec{C} - \vec{A}) - (\vec{C} \cdot \vec{C} - \vec{A} \cdot \vec{A}) = 0,$$
$$(\vec{B} + \vec{D}) \cdot (\vec{C} - \vec{A}) - (\vec{C} + \vec{A}) \cdot (\vec{C} - \vec{A}) = 0,$$
$$(\vec{B} + \vec{D} - \vec{C} - \vec{A}) \cdot (\vec{C} - \vec{A}) = 0.$$

This is the first of the two equations in (2). To get the second of the equations in (2), take the difference of the equations (1). The details are just as in the previous computation.

In a similar way, adding and subtracting the equations in (2) yield the equations in (1), which means that the converse theorem is also true: namely, if the line joining the midpoints of the two diagonals of a skew quadrilateral is perpendicular to these diagonals, then the opposite sides of the quadrilateral are of equal length.

Problems

8.3.9. In a triangle ABC the points D, E, and F trisect the sides so that $BC = 3BD$, $CA = 3CE$, and $AB = 3AF$. Similarly, the points G, H, and I trisect the sides of triangle DEF so that $EF = 3EG$, $FD = 3FH$, and $DE = 3DI$. Prove that the sides of $\triangle GHI$ are parallel to the sides of

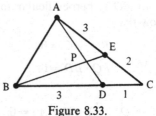

Figure 8.33.

$\triangle ABC$ and that each side of the smaller triangle is $\frac{1}{3}$ as long as its parallel side in the larger triangle.

8.3.10. The sides of AD, AB, CB, CD of the quadrilateral $ABCD$ are divided by the points E, F, G, H so that $AE : ED = AF : FB = CG : GB = CH : HD$. Prove that $EFGH$ is a parallelogram.

8.3.11.

(a) In triangle ABC (Figure 8.33), points D and E divide sides BC and AC in such a way that $BD/DC = 3$ and $AE/EC = \frac{3}{2}$. Let P denote the intersection of AD and BE. Find the ratio $BP : PE$.

(b) In triangle ABC (Figure 8.34), points E and F divide sides AC and AB respectively so that $AE/EC = 4$ and $AF/FB = 1$. Suppose D is a point on side BC, let G be the intersection of EF and AD, and suppose D is situated so that $AG/GD = \frac{3}{2}$. Find the ratio BD/DC.

8.3.12. On the sides of an arbitrary parallelogram $ABCD$, squares are constructed lying exterior to it. Prove that their centers M_1, M_2, M_3, M_4 are themselves the vertices of a square.

8.3.13. On the sides of an arbitrary convex quadrilateral $ABCD$, equilateral triangles ABM_1, BCM_2, CDM_3, and DAM_4 are constructed so that the first and third of them are exterior to the quadrilateral, while the second and fourth are on the same side of sides BC and DA as in the quadrilateral itself. Prove that the quadrilateral $M_1M_2M_3M_4$ is a parallelogram.

8.3.14. On the sides of an arbitrary convex quadrilateral $ABCD$, squares are constructed, all lying external to the quadrilateral, with centers M_1, M_2,

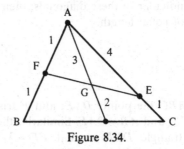

Figure 8.34.

M_3, M_4. Show that $M_1M_3 = M_2M_4$ and that M_1M_3 is perpendicular to M_2M_4.

8.3.15. Similar isosceles triangles BCX, CAY, and ABZ are constructed externally on the sides of a triangle ABC. Show that the centroids of $\triangle ABC$ and $\triangle XYZ$ coincide.

8.3.16. The altitudes of a triangle ABC are extended externally to points A', B' and C' respectively, where $AA' = k/h_a$, $BB' = k/h_b$, and $CC' = k/h_c$. Here, k is a constant and h_a denotes the length of the altitude of ABC from vertex A, etc. Prove that the centroid of the triangle $A'B'C'$ coincides with the centroid of ABC.

8.3.17. Let ABC be an acute angled triangle. Construct squares externally on the three sides. Extend the altitudes from the three vertices until they meet the far sides of the squares on the opposite sides. Then the squares are cut into two rectangles. Prove that "adjacent rectangles" from different squares are equal in area. That is, prove that area i = area i' for $i = 1, 2, 3$ (see Figure 8.35). (Use the dot product to give a one-line proof.) What happens as ABC becomes a right triangle?

8.3.18. In a tetrahedron, two pairs of opposite edges are orthogonal. Prove that the third pair of opposite edges must also be orthogonal.

8.3.19. Let O be a given point, let $P_1, P_2, \ldots, P_n$ be vertices of a regular n-gon, $n \geqslant 7$, and let $Q_1, Q_2, \ldots, Q_n$ be given by

$$\overrightarrow{OQ_i} = \overrightarrow{OP_i} + \overrightarrow{P_{i+1}P_{i+2}}, \qquad i = 1, 2, \ldots, n$$

$(P_{n+1} = P_1, P_{n+2} = P_2)$. Prove that $Q_1, Q_2, \ldots, Q_n$ are vertices of a regular n-gon.

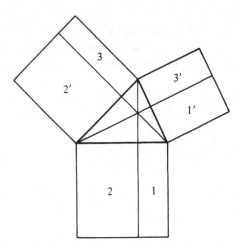

Figure 8.35.

8.4. Complex Numbers in Geometry

In this section we will build on the geometry of complex numbers introduced in Section 3.5.

8.4.1. $A_1, A_2, \ldots, A_n$ are vertices of a regular polygon inscribed in a circle of radius r and center O. P is a point on OA_1 extended beyond A_1. Show that

$$\prod_{k=1}^{n} PA_k = OP^n - r^n.$$

Solution. Consider Figure 8.36 as representing the complex plane with the center of the circle at the origin, and with the vertices A_i at the n roots of $z^n - r^n = 0$. Specifically, we set the affix of A_k to be $z_k = re^{2\pi(k-1)i/n}$. (The *affix* of a point Q in the plane is the complex number which corresponds to Q.) With these coordinates, P corresponds to a real number, which we will denote by z. Then

$$\prod_{k=1}^{n} PA_k = \prod_{k=1}^{n} |z - z_k|$$

$$= \left| \prod_{k=1}^{n} (z - z_k) \right|$$

$$= |z^n - r^n|$$

$$= z^n - r^n \qquad (z \text{ and } r \text{ are real, and } z > r)$$

$$= OP^n - r^n.$$

8.4.2. Given a point P on the circumference of a unit circle and the vertices $A_1, A_2, \ldots, A_n$ of an inscribed regular polygon of n sides, prove that $PA_1^2 + PA_2^2 + \cdots + PA_n^2$ is a constant.

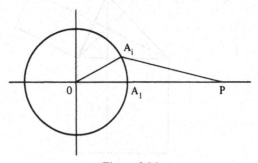

Figure 8.36.

Solution. Again, let $A_1, A_2, \ldots, A_n$ correspond to the n roots of unity; specifically, let the affix of A_k be $z_k = e^{2\pi ki/n}$, $k = 1, 2, \ldots, n$. Let the affix of P be z. Then, since for complex numbers w, $|w|^2 = w\bar{w}$, we have

$$\sum_{k=1}^{n} PA_k^2 = \sum_{k=1}^{n} |z - z_k|^2$$

$$= \sum_{k=1}^{n} (z - z_k)(\bar{z} - \bar{z}_k)$$

$$= \sum_{k=1}^{n} (z\bar{z} - z_k\bar{z} - z\bar{z}_k + z_k\bar{z}_k)$$

$$= \sum_{k=1}^{n} z\bar{z} - \left(\sum_{k=1}^{n} z_k\right)\bar{z} - z\left(\sum_{k=1}^{n} \bar{z}_k\right) + \sum_{k=1}^{n} z_k\bar{z}_k.$$

But $\sum_{k=1}^{n} z_k = 0$, since the z_k's are the roots of $z^n - 1 = 0$ and the coefficient of z^{n-1} is zero. Therefore,

$$\sum_{k=1}^{n} PA_k^2 = \sum_{k=1}^{n} z\bar{z} + \sum_{k=1}^{n} z_k\bar{z}_k$$

$$= \sum_{k=1}^{n} |z|^2 + \sum_{k=1}^{n} |z_k|^2$$

$$= n + n \qquad (|z| = 1 \text{ and } |z_k| = 1)$$

$$= 2n.$$

8.4.3. Prove that if the points in the complex plane corresponding to two distinct complex numbers z_1 and z_2 are two vertices of an equilateral triangle, then the third vertex corresponds to $-\omega z_1 - \omega^2 z_2$, where ω is an imaginary cube root of unity.

Solution. Points z_1, z_2, z_3 form an equilateral triangle if and only if $z_3 - z_1 = (z_2 - z_1)e^{\pm \pi i/3}$. Thus, given z_1 and z_2, z_3 must have the form

$$z_3 = (1 - e^{\pm \pi i/3})z_1 + e^{\pm \pi i/3}z_2$$

$$= -[-1 + e^{\pm \pi i/3}]z_1 - [-e^{\pm \pi i/3}]z_2.$$

We can see from the geometrical interpretation of these quantities (Figure 8.37) that $-1 + e^{\pm \pi i/3}$ and $-e^{\pm \pi i/3}$ are the imaginary cube roots of unity. Alternatively, we can verify this algebraically:

$$-1 + e^{\pm \pi i/3} = -1 + \cos(\pm \tfrac{1}{3}\pi) + i\sin(\pm \tfrac{1}{3}\pi)$$

$$= -1 + \tfrac{1}{2} \pm \tfrac{1}{2}\sqrt{3}\,i$$

$$= -\tfrac{1}{2} \pm \tfrac{1}{2}\sqrt{3}\,i,$$

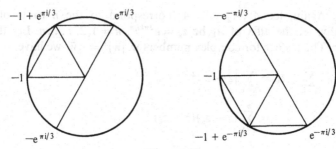

Figure 8.37.

and

$$-e^{\pm \pi i/3} = -\left[\cos(\pm \tfrac{1}{3}\pi) + i\sin(\pm \tfrac{1}{3}\pi)\right]$$
$$= -\tfrac{1}{2} \mp \tfrac{1}{2}\sqrt{3}\, i.$$

Conversely, suppose $z_3 = -\omega z_1 - \omega^2 z_2$, where ω is an imaginary cube root of unity. Then

$$\omega = -1 + e^{\pi i/3} \quad \text{and} \quad \omega^2 = -e^{\pi i/3}$$

or

$$\omega = -1 + e^{-\pi i/3} \quad \text{and} \quad \omega^2 = -e^{-\pi i/3},$$

and the previous arguments show that z_1, z_2, z_3 form an equilateral triangle.

8.4.4. Equilateral triangles are erected externally on the sides of an arbitrary triangle ABC. Prove that the centers (centroids) of these three equilateral triangles form an equilateral triangle.

Solution. Let a, b, c be the affixes of A, B, C respectively (in the complex plane), with x, y, z the affixes of the centers of the equilateral triangles as shown in Figure 8.38. Let $\omega = e^{2\pi i/3}$. Then $\omega^2 + \omega + 1 = 0$ (ω is a cube roots of unity, so $0 = \omega^3 - 1 = (\omega - 1)(\omega^2 + \omega + 1)$). Also, note that $e^{\pi i/3} = -\omega^2$ and $e^{-\pi i/3} = -\omega$.

The centroid of $\triangle ABC$ has affix $\frac{1}{3}(a + b + c)$. In a similar way, $x, y,$ and z are given by

$$x = \tfrac{1}{3}\left[a + c + \left[a - \omega^2(c - a)\right]\right] = \tfrac{1}{3}\left[(2 + \omega^2)a + (1 - \omega^2)c\right],$$
$$y = \tfrac{1}{3}\left[a + b + \left[a - \omega(b - a)\right]\right] = \tfrac{1}{3}\left[(2 + \omega)a + (1 - \omega)b\right],$$
$$z = \tfrac{1}{3}\left[b + c + \left[b - \omega(c - b)\right]\right] = \tfrac{1}{3}\left[(2 + \omega)b + (1 - \omega)c\right].$$

To show that x, y, z forms an equilateral triangle, it suffices to show that

$$z - x = -\omega^2(y - x),$$

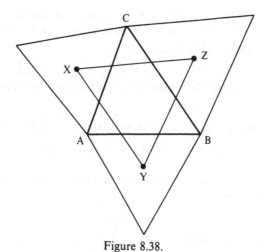

Figure 8.38.

We have

$$3(z - x) = -(2 + \omega^2)a + (2 + \omega)b + (-\omega + \omega^2)c$$

$$-3\omega^2(y - x) = 3\omega^2(x - y) = (\omega^4 - \omega^3)a - (\omega^2 - \omega^3)b + (\omega^2 - \omega^4)c.$$

But

$$\omega^4 - \omega^3 = \omega - 1 = (-1 - \omega^2) - 1 = -(2 + \omega^2),$$

$$-(\omega^2 - \omega^3) = -\omega^2 + 1 = (1 + \omega) + 1 = 2 + \omega,$$

$$\omega^2 - \omega^4 = \omega^2 - \omega,$$

and therefore, the coefficients of a, b, c in the above expressions for $z - x$ and $-\omega^2(y - x)$ are equal. It follows that x, y, z forms an equilateral triangle.

Problems

8.4.5. Let A_0, A_1, A_2, A_3, A_4 divide a unit circle (circle of radius 1) into five equal parts. Prove that the chords A_0A_1, A_0A_2 satisfy

$$(A_0A_1 \cdot A_0A_2)^2 = 5.$$

8.4.6. Given a point P on the circumference of a unit circle and the vertices $A_1, A_2, \ldots, A_n$ of an inscribed regular polygon of n sides, prove that $PA_1^4 + PA_2^4 + \cdots + PA_n^4$ is a constant (i.e., independent of the position of P on the circumference).

8.4.7. Let G denote the centroid of triangle ABC. Prove that

$$3(GA^2 + GB^2 + GC^2) = AB^2 + BC^2 + CA^2.$$

8.4.8. Let $ABCDEF$ be a hexagon in a circle of radius r. Show that if $AB = CD = EF = r$, then the midpoints of BC, DE, and FA are the vertices of an equilateral triangle.

8.4.9. If z_1, z_2, z_3 are such that $|z_1| = |z_2| = |z_3| = 1$ and $z_1 + z_2 + z_3 = 0$, show that z_1, z_2, z_3 are the vertices of an equilateral triangle inscribed in a unit circle.

8.4.10. Show that z_1, z_2, z_3 form an equilateral triangle if and only if

$$z_1^2 + z_2^2 + z_3^2 = z_1 z_2 + z_2 z_3 + z_3 z_1.$$

8.4.11. The three points in the complex plane which correspond to the roots of the equation

$$z^3 - 3pz^2 + 3qz - r = 0$$

are the vertices of a triangle.

(a) Prove that the centroid of the triangle is the point corresponding to p.
(b) Prove that the triangle is equilateral if and only if $p^2 = q$.

Glossary of Symbols and Definitions

Centroid (of a triangle)	The point where the medians of a triangle intersect. (A median of a triangle is a line joining a vertex to the midpoint of the opposite side.)
Circumcenter (of a triangle)	The center of the circumscribed circle (the circle passing through the three vertices of the triangle). The point where the perpendicular bisectors of the sides of the triangle intersect.
Convex hull	The smallest convex set which contains all the points of the set.
Convex set	A set that contains the line segment joining any two of its points.
Fibonacci sequence	The sequence of numbers defined as $F_1 = 1$, $F_2 = 1$, and $F_n = F_{n-1} + F_{n-2}$ for $n > 2$. The sequence begins 1, 1, 2, 3, 5, 8, 13, 21, 34, 55, ...
Function	
even	A function f with the property that $f(-x) = f(x)$ for all x.
odd	A function f with the property that $f(-x) = -f(x)$ for all x.
convex	A real-valued function defined in the interval (a, b) such that for each x, y, z with $a < x < y < z < b$, $f(x) \leqslant L(x)$, where $L(x)$ is the linear function coinciding with $f(x)$ at x and z.

concave	A function which is the negative of a convex function.
$[\![\ ]\!]$	The greatest integer function; for each real number x, $[\![x]\!]$ is the largest integer less than or equal to x.
Incenter (of a triangle)	The center of the incircle of a triangle. The incircle, or inscribed circle, is the circle tangent to the sides of the triangle. The incenter is the point where the bisectors of the angles of the triangle intersect.
Lattice point	A point in the Euclidean plane (or R^n) whose coordinates are integers.
Orthocenter (of a triangle)	The point of intersection of the three altitudes of a triangle.
Pascal's triangle	A triangular array of numbers whose nth row ($n = 0, 1, 2, \ldots$) is composed of the coefficients of the expansion of $(a + b)^n$.
Pythagorean triple	A set of three positive integers which satisfy the equation $x^2 + y^2 = z^2$.
Set	
$\quad S\text{-}T$	The subset consisting of those elements in the set S that are not in the set T.
$\quad k$-subset	A subset of k elements.
Triangular numbers	The numbers in the sequence $1, 3, 6, 10, \ldots$ whose nth term is $n(n + 1)/2$.

Sources

1.1.2. 1974 Putnam Exam
1.1.3. 1979 Putnam Exam
1.1.4. 1976 International Olympiad
1.1.5. 1971 Putnam Exam
1.1.6. *The Mathematics Student*, Vol. 26, No. 2, November 1978
1.1.7. B. G. Eke, *Mathematical Spectrum*, Vol. 9, No. 3, 1976–1977, p. 97.
1.1.8. 1968 Putnam Exam
1.1.9. D. H. Browne, *American Mathematical Monthly*, Vol. 53, No. 2, February 1946, p. 97
1.1.10. See 1.5.10
1.1.11. See 7.6.4
1.1.12. 1972 Putnam Exam

1.2.1. J. E. Trevor, *American Mathematical Monthly*, Vol. 42, No. 8, October 1935, p. 508
1.2.2. 1972 Putnam Exam
1.2.4. Richard A. Howland, A Generalization of the Handshake Problem, *Crux Mathematicorum*, Vol. 6, No. 8, October 1980, pp. 237–238
1.2.7. W. R. Ransom, *American Mathematical Monthly*, Vol. 60, No. 9, November 1953, p. 627; Also, see W. A. Wickelgren, *How to Solve Problems*, W. H. Freeman, San Francisco, 1974, pp. 163–166
1.2.9. 1961 Putnam Exam
1.2.10. See 6.6.4

1.3.3. 1981 International Olympiad

1.3.5. 1955 Putnam Exam
1.3.6. Romae Cormier and Roger Eggleton, Counting by Correspondence, *Mathematics Magazine*, Vol. 49, No. 4, September 1976, pp.
181–186
1.3.7. *Mathematical Spectrum*, Vol. 2, No. 2, 1969–1970, p. 70
1.3.11. USSR Olympiad
1.3.12. 1956 Putnam Exam
1.3.13. 1957 Putnam Exam

1.4.1. William R. Klinger, *Two-Year College Mathematics Journal*, Vol.
12, No. 2, March 1981, p. 154
1.4.2. Fred A. Miller, *School Science and Mathematics*, Vol. 81, No. 2,
February 1981, p. 165
1.4.3. USSR Olympiad

1.5.1. J. A. Renner, *American Mathematical Monthly*, Vol. 44, No. 10,
December 1937, p. 666
1.5.3. M. T. Salhab, *American Mathematical Monthly*, Vol. 68, No. 7,
September 1961, p. 667. Solution by D. C. Stevens
1.5.4. Michael Golomb, *American Mathematical Monthly*, Vol. 87, No. 6,
June–July 1980, p. 489
1.5.5. John Clement, Jack Lockhead, and George Monk, Translation
Difficulties in Learning Mathematics, *American Mathematical
Monthly*, Vol. 88, No. 4, April 1981, pp. 286–290
1.5.8. 1963 Putnam Exam
1.5.10. 1974 Putnam Exam

1.6.1. 1977 International Olympiad
1.6.3. 1980 Putnam Exam
1.6.4. G. P. Henderson, *Crux Mathematicorum*, Vol. 5, No. 6, June–July,
1979, p. 171
1.6.5. Zalman Usiskin, *Two-Year College Mathematics Journal*, Vol. 12,
No. 2, March 1981, p. 155
1.6.6b. See 4.3.9
1.6.7. Alvin J. Paullay and Sidney Penner, *Two-Year College Mathematics Journal*, Vol. 11, No. 5, November 1980, p. 336

1.7.3. See 2.3.1
1.7.8. Murray Klamkin, *Crux Mathematicorum*, Vol. 5, No. 9, November
1979, p. 259

1.8.4. 1965 Putnam Exam
1.8.6a. Léo Sauvé, *Eureka*, Vol. 1, No. 1, November 1975, p. 88.
1.8.6b. Victor Linis, *Eureka*, Vol. 4, No. 4, June 1975, p. 28.
1.8.8. Edwin A. Maxwell, *Fallacies in Mathematics*, Cambridge Press,
London, 1961, p. 42

1.9. This proof that the harmonic series diverges is due to Leonard Gillman
1.9.1. 1907 Hungarian Olympiad
1.9.2. 1982 U.S.A. Olympiad
1.9.4. 1981 U.S.A. Olympiad
1.9.5. 1962 Putnam Exam

1.10.1. 1971 Putnam Exam
1.10.4. 1954 Putnam Exam
1.10.5. 1973 Putnam Exam
1.10.8. 1906 Hungarian Olympiad
1.10.9. Thomas E. Moore, *Two-Year College Mathematics Journal*, Vol. 12, No. 1, January 1981, p. 63
1.10.10. 1954 Putnam Exam

1.11.1. Paul Erdös
1.11.2. 1979 Putnam Exam
1.11.3. 1965 Putnam Exam
1.11.5. *Mathematical Spectrum*, Vol. 1, No. 2, 1968–1969, p. 60

1.12.6c. 1982 Putnam Exam
1.12.7. Mau-Keung Siu, Inventor's Paradox, *Two-Year College Journal of Mathematics*, Vol. 12, No. 4, September 1981, p. 267

2.1.2. Leo Moser, *American Mathematical Monthly*, Vol. 69, No. 8, October 1962, p. 809
2.1.6b. C. S. Venkataraman, *American Mathematical Monthly*, Vol. 59, No. 6, June 1952, p. 410
2.1.7. 1962 Putnam Exam
2.1.8. Leonard Cohen, *American Mathematical Monthly*, Vol. 68, No. 1, January 1961, p. 62

2.2.2. 1978 Putnam Exam
2.2.6. Murray Klamkin, *Crux Mathematicorum*, Vol. 5, No. 1, January 1979, p. 13
2.2.7. See 2.6.1
2.2.8. S. W. Golomb and A. W. Hales, *American Mathematical Monthly*, Vol. 69, No. 8, October 1962, p. 809

2.3.1. Solution due to A. Liu, *Crux Mathematicorum*, Vol. 4, No. 9, November 1978, pp. 272–274
2.3.2. J. L. Brown, *American Mathematical Monthly*, Vol. 68, No. 10, December 1961, p. 1005

2.4.2. Douglas Hensley, *American Mathematical Monthly*, Vol. 87, No. 7, September 1980, p. 577

2.4.4. 1967 Putnam Exam
2.4.5. Murray Klamkin, *American Mathematical Monthly*, Vol. 61, No. 6, June 1954, p. 423
2.4.6. George Polya, *Induction and Analogy in Mathematics*, Princeton University Press, Princeton, N.J., 1954, pp. 118–119

2.5.3. David Wheeler, *Crux Mathematicorum*, Vol. 4, No. 3, March 1978, p. 74. Solution given by Bob Prielipp.
2.5.4. 1954 Putnam Exam
2.5.6b. 1969 Putnam Exam
2.5.6d. E. M. Scheuer, *American Mathematical Monthly*, Vol. 66, No. 9, November 1959, p. 813
2.5.9. 1980 Canadian Olympiad
2.5.13. 1982 Canadian Olympiad

2.6.1. 1958 Putnam Exam
2.6.2. 1954 Putnam Exam
2.6.3. 1976 U.S.A. Olympiad
2.6.4. 1980 Putnam Exam
2.6.6. 1978 Putnam Exam
2.6.7b. Michael Brozinsky, *School Science and Mathematics*, Vol. 81, No. 6, October 1981, p. 532
2.6.9. 1975 Canadian Olympiad
2.6.10. 1928 Hungarian Olympiad
2.6.11a. C. W. Bostwick, *American Mathematical Monthly*, Vol. 65, No. 6, June–July 1958, p. 446
2.6.12. Leo Moser, *American Mathematical Monthly*, Vol. 60, No. 10, December 1953, p. 713

3.1.1. Steve Galovich, *American Mathematical Monthly*, Vol. 84, No. 6, June–July 1977, p. 487
3.1.5. 1959 International Olympiad
3.1.6. 1981 U.S.A. Olympiad
3.1.10b. 1956 Putnam Exam
3.1.14. William J. LeVeque, *Elementary Theory of Numbers*, Addison-Wesley, Reading, Mass., 1962, p. 34

3.2.1. See 3.2.11
3.2.5. Andy Vince, *American Mathematical Monthly*, Vol. 72, No. 3, March 1965, p. 316
3.2.6. R. S. Luthar, *American Mathematical Monthly*, Vol. 83, No. 7, August–September 1976, p. 566
3.2.7. 1894 Hungarian Olympiad
3.2.9. 1955 Putnam Exam
3.2.10. Albert A. Mullin, *American Mathematical Monthly*, Vol. 84, No. 5, May 1977, p. 386

3.2.11. *The Mathematics Student*, Vol. 26, No. 3, December 1978

3.2.12. Larry Lass, *American Mathematical Monthly*, Vol. 71, No. 3, March 1964, p. 317

3.2.13b. Hugh L. Montgomery, *American Mathematical Monthly*, Vol. 82, No. 9, November 1975, p. 936

3.2.14e. 1899 Hungarian Olympiad

3.2.15f. 1976 U.S.A. Olympiad

3.2.17. Michael Brozinsky, *School Science and Mathematics*, Vol. 81, No. 4, p. 352

3.2.18. 1954 Putnam Exam

3.2.22. 1900 Hungarian Olympiad

3.2.24. Hal Forsey, *Mathematics Magazine*, Vol. 53, No. 4, September 1980, p. 244

3.2.25. N. S. Mendelsohn, *American Mathematical Monthly*, Vol. 66, No. 10, December 1959, p. 915

3.3.5. W. C. Rufus, *American Mathematical Monthly*, Vol. 51, No. 6, June–July 1944, p. 348

3.3.6. 1981 Hungarian Olympiad

3.3.7. 1960 Putnam Exam

3.3.8. 1980 Putnam Exam

3.3.9. 1972 U.S.A. Olympiad

3.3.13. Murray Klamkin, *Mathematics Magazine*, Vol. 27, No. 1, January 1953, p. 56

3.3.14. 1947 Putnam Exam

3.3.17. *The Mathematics Student*, Vol. 27, No. 1, October 1979

3.3.18. 1967 Putnam Exam

3.3.19c. 1956 Putnam Exam

3.3.22d. Harvey Berry, *American Mathematical Monthly*, Vol. 59, No. 3, March 1952, p. 180

3.3.24. H. J. Godwin, *Mathematical Spectrum*, Vol. 11, No. 1, 1978–1979, p. 28

3.3.25. Norman Schaumberger, *Two-Year College Mathematics Journal*, Vol. 12, No. 1, January 1981, p. 185

3.4.1. *Mathematical Spectrum*, Vol. 1, No. 2, 1968–1969, p. 59

3.4.2. 1981 Canadian Olympiad

3.4.3. 1975 International Olympiad

3.4.4. 1977 Putnam Exam

3.4.7. USSR Olympiad

3.4.8. 1962 International Olympiad

3.4.9a. H. G. Dworschak, *Eureka*, Vol. 1, No. 9, November 1975, p. 86

3.4.9d. Leo Moser, *American Mathematical Monthly*, Vol. 58, No. 10, December 1951, p. 700

3.4.11. USSR Olympiad

3.4.12. C. H. Braunholtz, *American Mathematical Monthly*, Vol. 70, No. 6, June–July 1963, p. 675

3.5.1. H. G. Dworschak, *Eureka*, Vol. 2, No. 3, March 1976, p. 50
3.5.2. L. Mirsky, *Mathematical Spectrum*, Vol. 13, No. 2, 1980–1981, p. 58
3.5.5. 1980 U.S.A. Olympiad

4.1.1. F. G. B. Maskell, *Crux Mathematicorum*, Vol. 4, No. 6, June–July 1978, p. 164. Solution by Bob Prielipp
4.1.2. 1977 Putnam Exam
4.1.3. 1976 Putnam Exam
4.1.4. 1979 British Olympiad
4.1.6. 1969 International Olympiad
4.1.7. 1975 Putnam Exam
4.1.10. Murray Klamkin, *Crux Mathematicorum*, Vol. 5, No. 4, April 1979, p. 105

4.2.2. USSR Olympiad
4.2.3. 1970 Canadian Olympiad
4.2.4. 1940 Putnam Exam
4.2.5. Azriel Rosenfeld, *American Mathematical Monthly*, Vol. 69, No. 8, October 1962, p. 809. Solution by Murray Klamkin
4.2.10. Murray Klamkin, *Crux Mathematicorum*, Vol. 5, No. 10, 1979, p. 290
4.2.16a. 1952 Putnam Exam
4.2.16b. 1963 Putnam Exam
4.2.18. 1977 U.S.A. Olympiad
4.2.21. *The Mathematics Student*, Vol. 28, No. 5, February 1981

4.3.1. 1971 Putnam Exam
4.3.5. 1956 Putnam Exam
4.3.6. 1977 Putnam Exam
4.3.7. Solution by G. P. Henderson, *Crux Mathematicorum*, Vol. 5, No. 6, June–July 1979, p. 171
4.3.8. 1899 Hungarian Olympiad
4.3.9. Hayo Ahlberg, *Crux Mathematicorum*, Vol. 7, No. 5, May 1981, p. 639
4.3.10. Murray Klamkin, *Crux Mathematicorum*, Vol. 5, No. 9, November 1979, p. 259
4.3.12a. *Mathematical Spectrum*, Vol. 3, No. 1, 1970–1971, p. 28
4.3.12b. *Mathematical Spectrum*, Vol. 9, No. 1, 1976–1977, p. 32
4.3.15. 1962 Putnam Exam
4.3.17c. 1977 Putnam Exam
4.3.17d. J. M. Gandhi, *American Mathematical Monthly*, Vol. 66, No. 1, January 1959, p. 61

4.4.3.   1972 Putnam Exam
4.4.4.   I. N. Herstein, *Topics in Algebra*, Xerox College Publishing, 1964, p. 41
4.4.6.   Guy Torchinelli, *American Mathematical Monthly*, Vol. 71, No. 3, March 1964, p. 317. Solution by Francis P. Callahan
4.4.7.   1972 Putnam Exam
4.4.8.   R. L. Graham and F. D. Parker, *American Mathematical Monthly*, Vol. 70, No. 2, February 1963, p. 210. Solution by J. A. Schatz
4.4.9.   Solomon W. Golomb, *American Mathematical Monthly*, Vol. 85, No. 7, August–September 1978, p. 593
4.4.10. F. S. Carter, *Mathematics Magazine*, Vol. 49, No. 4, September 1976, p. 211
4.4.12. F. M. Sioson, *American Mathematical Monthly*, Vol. 70, No. 8, October 1963, p. 891
4.4.13. 1968 Putnam Exam
4.4.14. 1977 Putnam Exam
4.4.20. Leo Moser, *American Mathematical Monthly*, Vol. 67, No. 3, March 1960, p. 290
4.4.21. T. J. Kearns, *American Mathematical Monthly*, Vol. 69, No. 1, January 1962, p. 57
4.4.22. Murray Klamkin, *Crux Mathematicorum*, Vol. 6, No. 3, March 1980, p. 73
4.4.23. J. Linkovskii-Condé, *American Mathematical Monthly*, Vol. 87, No. 2, February 1980, p. 137
4.4.24. 1982 U.S.A. Olympiad
4.4.25. Seth Warner, *Classical Modern Algebra*, Prentice-Hall, Englewood Cliffs, N.J., 1971, p. 134
4.4.28. See 4.4.25
4.4.29. 1968 Putnam Exam
4.4.30c. 1957 Putnam Exam
4.4.31. 1979 Putnam Exam

5.1.5.   Peter Orno, *Mathematics Magazine*, Vol. 54, No. 4, September 1981, p. 213. Solution by Harry Sedinger
5.1.7a. W. C. Waterhouse, *American Mathematical Monthly*, Vol. 70, No. 10, December 1963, p. 1099
5.1.7b. Roger B. Eggleton, *American Mathematical Monthly*, Vol. 71, No. 8, October 1964, p. 913
5.1.10a. Murray Klamkin, *Crux Mathematicorum*, Vol. 5, No. 5, May 1979, p. 129
5.1.14. Andy Liu, *Crux Mathematicorum*, Vol. 4, No. 7, August–September 1978, p. 192
5.1.15. Donald Knuth, Take-home problem, Stanford University, Fall 1974. Also, see C. F. Pinska, *American Mathematical Monthly*, Vol. 65, No. 4, April 1958, p. 284

5.2.2.   1975 Putnam Exam

5.2.5.   A. D. Sands, *American Mathematical Monthly*, Vol. 87, No. 1, January 1980, p. 60

5.2.6.   W. L. Nicholson, *American Mathematical Monthly*, Vol. 70, No. 8, October 1963, p. 893

5.2.8.   L. L. Garner, *American Mathematical Monthly*, Vol. 67, No. 8, October 1960, p. 807

5.2.9.   1981 Putnam Exam

5.2.13. 1982 Canadian Olympiad

5.2.16. 1977 Putnam Exam

5.3.3.   1978 Putnam Exam

5.3.4.   Gabriel Klambauer, *American Mathematical Monthly*, Vol. 87, No. 2, February 1980, pp. 128–130

5.3.7b. 1977 Putnam Exam

5.3.8.   British Scholarship Problem

5.3.10. Gabriel Klambauer, *American Mathematical Monthly*, Vol. 87, No. 2, February 1980, pp. 128–130

5.3.11c. 1981 Putnam Exam

5.3.12. Leo Moser, *American Mathematical Monthly*, Vol. 55, No. 7, September 1948, p. 427

5.3.13. Michael Aissen, *American Mathematical Monthly*, Vol. 76, No. 9, November 1969, p. 1063

5.4.2.   1972 Putnam Exam

5.4.5.   1951 Putnam Exam

5.4.6.   V. N. Murty, *Two-Year College Mathematics Journal*, Vol. 11, No. 4, September 1980, p. 276

5.4.7.   1939 Putnam Exam

5.4.12. A. J. Douglas, *Mathematical Spectrum*, Vol. 5, No. 2, 1972/73, p. 67

5.4.15. 1975 Putnam Exam

5.4.20. 1981 Putnam Exam

5.4.21. 1980 Putnam Exam

5.4.23. V. N. Murty, *Two-Year College Mathematics Journal*, Vol. 11, No. 4, September 1980, p. 276

5.4.27. 1970 Putnam Exam

6.1.2.   Ko-Wei-Lih, *American Mathematical Monthly*, Vol. 88, No. 6, June–July 1981, p. 444

6.1.4.   1947 Putnam Exam

6.1.6.   Albert Wilansky, *American Mathematical Monthly*, Vol. 65, No. 9, November 1958, p. 708

6.2.3.   1979 Putnam Exam
6.2.7.   Joseph Silverman, *Mathematics Magazine*, Vol. 51, No. 2, March 1978, p. 127
6.2.8.   1979 Putnam Exam
6.2.11. 1970 Putnam Exam
6.2.13. 1959 Putnam Exam

6.3.2.   1967 Putnam Exam

6.4.2.   1976 Putnam Exam
6.4.6.   1981 U.S.A. Olympiad
6.4.7.   1981 Canadian Olympiad

6.5.4.   1981 Putnam Exam
6.5.5b. 1958 Putnam Exam
6.5.9.   1973 Putnam Exam

6.6.1.   1946 Putnam Exam
6.6.4.   Sidney Penner, *Mathematics Magazine*, Vol. 49, No. 3, May 1976, p. 150
6.6.5.   1976 Putnam Exam
6.6.6.   Peter Orno, *Mathematics Magazine*, Vol. 51, No. 4, September 1978, p. 245
6.6.9.   G. Z. Chang, *Mathematics Magazine*, Vol. 54, No. 3, May 1981, p. 140

6.7.1.   1956 Putnam Exam
6.7.2.   1955 Putnam Exam
6.7.4e. 1946 Putnam Exam
6.7.5.   1979 Putnam Exam
6.7.6.   Bernard Vanbrugghe

6.8.1.   1976 Putnam Exam
6.8.2.   1970 Putnam Exam
6.8.6.   Victor Linis, *Crux Mathematicorum*, Vol. 2, No. 9, November 1976, p. 203
6.8.9.   1964 Putnam Exam
6.8.10. 1977 Putnam Exam

6.9.2.   H. G. Dworschak, *Eureka*, Vol. 1, No. 8, October 1975, p. 77
6.9.3.   1946 Putnam Exam
6.9.4.   Marius Solomon, *American Mathematical Monthly*, Vol. 77, No. 6, June–July 1977, p. 487

6.9.5.   1938 Putnam Exam

6.9.11.   D. H. Browne, *American Mathematical Monthly*, Vol. 53, No. 1, January 1946, p. 36

7.1.3.   1980 U.S.A. Olympiad

7.1.4.   Solution by Angus Rodgers, *Mathematical Spectrum*, Vol. 5, No. 1, 1972–1973, p. 31

7.1.15.   Victor Linis, *Eureka*, Vol. 2, No. 2, February 1976, p. 29

7.2.2.   1975 Putnam Exam

7.2.3.   Murray Klamkin, *Crux Mathematicorum*, Vol. 5, No. 2, February 1979, p. 45

7.2.9a.   Freddy Storey, *American Mathematical Monthly*, Vol. 68, No. 10, December 1961, p. 1009

7.3.1.   J. L. Brenner, *Two-Year College Mathematics Journal*, Vol. 12, No. 1, January 1981, p. 64

7.3.2.   Mark Kleiman, *Mathematics Magazine*, Vol. 50, No. 1, January 1977, p. 49

7.3.3.   1978 U.S.A. Olympiad

7.3.4.   1977 Putnam Exam

7.3.8.   Michael Ecker, *Crux Mathematicorum*, Vol. 7, No. 7, August–September 1981, p. 208

7.3.9.   Dan Sokolowsky, *Crux Mathematicorum*, Vol. 6, No. 8, October 1980, p. 259

7.3.10.   1981 International Olympiad

7.4.1.   T. B. Cruddis, *Mathematical Spectrum*, Vol. 10, No. 1, 1977–1978, p. 31

7.4.2.   T. S. Bolis, *American Mathematical Monthly*, Vol. 82, No. 7, August–September 1975, p. 756

7.4.4.   1973 Putnam Exam

7.4.6.   Gideon Schwarz, *American Mathematical Monthly*, Vol. 88, No. 2, February 1981, p. 148

7.4.8.   USSR Olympiad

7.4.9.   1978 Putnam Exam

7.4.12.   H. G. Dworschak, *Eureka*, Vol. 2, No. 5, May 1976, p. 98

7.4.13.   See 7.4.12

7.4.19.   See 7.3.1

7.5.1.   1980 Putnam Exam

7.5.4.   Murray Klamkin, *Crux Mathematicorum*, Vol. 6, No. 10, December 1980, p. 312

7.5.7.   See 7.4.12

7.5.8.   Ralph Boas, *American Mathematical Monthly*, Vol. 85, No. 6, June–July 1978, p. 495

7.6.1. Marius Solomon, *Mathematics Magazine*, Vol. 49, No. 2, March 1976, p. 95. Solution by Jordan Levy

7.6.2. 1954 Putnam Exam

7.6.3. 1961 Putnam Exam

7.6.4. 1947 Putnam Exam

7.6.5. 1947 Putnam Exam

7.6.12. 1957 Putnam Exam

7.6.13. 1978 Putnam Exam

8.1.1. 1978 Putnam Exam

8.1.2. 1976 U.S.A. Olympiad

8.1.3. 1976 Putnam Exam

8.1.7. Leon Bankoff, *Crux Mathematicorum*, Vol. 6, No. 3, March 1980, p. 90

8.1.8. Jack Garfunkel, *Mathematics Magazine*, Vol. 50, No. 3, May 1977, p. 164

8.1.9. John A. Tierney, *Eureka*, Vol. 2, No. 5, May 1976, p. 103

8.1.15. Zelda Katz, *Pi Mu Epsilon Journal*, Vol. 7, No. 4, Spring 1981, p. 265

8.1.16. Norman Schaumberger, *Two-Year College Mathematics Journal*, Vol. 12, No. 2, March 1981, p. 155

8.2.1. 1976 Putnam Exam

8.2.4. 1938 Putnam Exam

8.2.5. Murray Klamkin, *Mathematics Magazine*, Vol. 49, No. 4, September 1976, p. 211

8.2.6. K. R. S. Sastry, *Mathematics Magazine*, Vol. 54, No. 2, March 1981, p. 84

8.2.8. Norman Anning, *American Mathematical Monthly*, Vol. 27, No. 10, December 1920, p. 482

8.2.11. 1980 Putnam Exam

8.2.12. 1979 Putnam Exam

8.2.14a. Roger L. Creech, *Mathematics Magazine*, Vol. 53, No. 1, January 1980, p. 49

8.2.14b. Roger L. Creech, *Mathematics Magazine*, Vol. 54, No. 1, January 1981, p. 35

8.2.16. Murray Klamkin, *Crux Mathematicorum*, Vol. 7, No. 2, February 1981, p. 65

8.3.1. 1978 U.S.A. Olympiad

8.3.5. M. Slater, *American Mathematical Monthly*, Vol. 88, No. 1, January 1981, pp. 66–67. Solution by Jordi Dou.

8.3.6. See 1.5.3

8.3.8. 1977 U.S.A. Olympiad

8.3.18. H. G. Dworschak, *Eureka*, Vol. 2, No. 3, March 1976, p. 46

Index

Problem Books in Mathematics *(continued)*

Unsolved Problems in Number Theory (2nd ed.)
by *Richard K. Guy*

An Outline of Set Theory
by *James M. Henle*

Demography Through Problems
by *Nathan Keyfitz and John A. Beekman*

Theorems and Problems in Functional Analysis
by *A.A. Kirillov and A.D. Gvishiani*

Exercises in Classical Ring Theory
by *T.Y. Lam*

Problem-Solving Through Problems
by *Loren C. Larson*

Winning Solutions
by *Edward Lozansky and Cecil Rosseau*

A Problem Seminar
by *Donald J. Newman*

Exercises in Number Theory
by *D.P. Parent*

Contests in Higher Mathematics:
Miklós Schweitzer Competitions 1962-1991
by *Gábor J. Székely (editor)*

9 780387 961712